T0074500

Next-Generation Sequencing Data Analysis

Next-generation DNA and RNA sequencing has revolutionized biology and medicine. With sequencing costs continuously dropping and our ability to generate large datasets rising, data analysis becomes more important than ever. *Next-Generation Sequencing Data Analysis* walks readers through next-generation sequencing (NGS) data analysis step by step for a wide range of NGS applications.

For each NGS application, this book covers topics from experimental design, sample processing, sequencing strategy formulation, to sequencing read quality control, data preprocessing, read mapping or assembly, and more advanced stages that are specific to each application. Major applications include:

- RNA-seq: Both bulk and single cell (separate chapters)
- Genotyping and variant discovery through whole genome/exome sequencing
- Clinical sequencing and detection of actionable variants
- *De novo* genome assembly
- ChIP-seq to map protein-DNA interactions
- Epigenomics through DNA methylation sequencing
- Metagenome sequencing for microbiome analysis

Before detailing the analytic steps for each of these applications, the book presents introductory cellular and molecular biology as a refresher mostly for data scientists, the ins and outs of widely used NGS platforms, and an overview of computing needs for NGS data management and analysis. The book concludes with a chapter on the changing landscape of NGS technologies and data analytics.

The second edition of this book builds on the well-received first edition by providing updates to each chapter. Two brand new chapters have been added to meet rising data analysis demands on single-cell RNA-seq and clinical sequencing. The increasing use of long-reads sequencing has also been reflected in all NGS applications. This book discusses concepts and principles that underlie each analytic step, along with software tools for implementation. It highlights key features of the tools while omitting tedious details to provide an easy-to-follow guide for practitioners in life sciences, bioinformatics, biostatistics, and data science. Tools introduced in this book are open source and freely available.

Next-Generation Sequencing Data Analysis

Second Edition

Xinkun Wang

CRC Press
Taylor & Francis Group
Boca Raton London New York

CRC Press is an imprint of the
Taylor & Francis Group, an **informa** business
A CHAPMAN & HALL BOOK

Second edition published 2024
by CRC Press
6000 Broken Sound Parkway NW, Suite 300, Boca Raton, FL 33487-2742

and by CRC Press
4 Park Square, Milton Park, Abingdon, Oxon, OX14 4RN

CRC Press is an imprint of Taylor & Francis Group, LLC

First edition published by CRC Press 2016

ISBN: 9780367349899 (hbk)
ISBN: 9781032505701 (pbk)
ISBN: 9780429329180 (ebk)

DOI: 10.1201/9780429329180

Typeset in Palatino
by Newgen Publishing UK

Contents

Preface to the Second Edition

When I started working on the second edition of *Next-Generation Sequencing Data Analysis*, my primary goal was to add new chapters and contents on clinical sequencing, single-cell sequencing, and third-generation sequencing (i.e., long reads) data analyses. These contents were either absent, or only briefly discussed, in the first edition. For example, data processing for clinical applications, where NGS has a direct impact on public health, was absent, and a new chapter that covers clinical sequencing data QA/QC, standard analysis pipeline, and clinical interpretation is beneficial to the community. The dramatic growth in single-cell sequencing also warrants a new chapter, because extracting rich biological information at the single-cell resolution requires a new set of tools different from what is used to analyze "bulk" sequencing data. Although long-read sequencing was covered in the first edition of this book, technologies have since then made significant improvements and achieved wide usage. Such developments require extensive updates to nearly all of the applications, from RNA-seq to metagenomics.

After meeting my primary goal, I set out to update the rest of the book. This took much longer than I had initially planned. The challenges were two-fold. The first was updating the list of tools for each of the NGS applications. Thanks to the productivity of the bioinformatics community, most NGS applications have seen an abundance of tool development, and as a result many new tools have emerged. Updating this large number of new tools took quite some time. The second challenge was selectively introducing new and existing tools, instead of overwhelming readers with a long list of tools that have ever existed. While the tools presented in this edition are by no means the most representative among all tools available, I made every effort to select most of the effective open source tools in existence as of late 2022, drawing information from benchmarking studies, citations, and recent updates.

In writing this edition, I have developed a renewed appreciation of the intensity, excitement, and multiplicity of expertise in the NGS field. For instance, there is an increasing convergence of expertise from artificial intelligence, computer science, and high-performance computing. At the same time, because of the highly dynamic nature of the field, it becomes increasingly challenging to keep abreast of the latest developments. This new edition represents an effort from a practitioner in the field towards the goal of informing readers on recent NGS data analysis tools. I would like to express my gratitude to the many researchers and clinicians I have interacted with in my role as the director of the Northwestern University NGS Facility. It is their need for the latest NGS technologies that has kept me up to date with the NGS field.

Author

Xinkun Wang is Research Professor and the Director of the Next-Generation Sequencing Facility at Northwestern University in Chicago. Dr. Wang's first foray into the genomics field was during his doctoral training, performing microarray-based gene expression analysis. From 2005 to 2015, he was the founding director of the University of Kansas Genomics Facility, prior to moving to Northwestern to head the Northwestern University Sequencing Facility (NUSeq) in late 2015. Dr. Wang is a renowned expert on genomics technologies and data mining and their applications to the biomedical field. Besides his monographic publications, he has published extensively in neuroscience, with a focus on brain aging and neurodegenerative diseases (mostly Alzheimer's disease). Dr. Wang has served as principal investigator on dozens of grants. Dr. Wang's other professional activities include serving on journal editorial boards, and as reviewers for journals, publishers, and funding agencies.

Dr. Wang is a member of American Society of Human Genetics, Association of Biomolecular Resource Facilities, the Honor Society of Phi Kappa Phi, and Society for Neuroscience. Dr. Wang was born in Shandong province, China, and is a first-generation college graduate. His off-work hobbies include cycling and Alpine skiing.

Part I

Introduction to Cellular and Molecular Biology

1

The Cellular System and the Code of Life

1.1 The Cellular Challenge

A cell, although minuscule with a diameter of less than 50 μm, works wonders if you compare it to any human-made system. Moreover, it perpetuates itself using the information coded in its DNA. In case you ever had the thought of designing an artificial system that shows this type of sophistication, you would know the many insurmountable challenges such a system needs to overcome. A cell has a complicated internal system, containing many types of molecules and parts. To sustain the system, a cell needs to perform a wide variety of tasks, the most fundamental of which are to maintain its internal order, prevent its system from malfunctioning or breaking down, and reproduce or even improve the system, in an environment that is constantly changing.

Energy is needed to maintain the internal order of the cellular system. Without constant energy input, the entropy of the system will gradually increase, as dictated by the second law of thermodynamics, and ultimately lead to the destruction of the system. Besides energy, raw "building" material is also constantly needed to renew its internal parts or build new ones if needed, as the internal structure of a cell is dynamic and responds to constant changes in environmental conditions. Therefore, to maintain the equilibrium inside and with the environment, it requires a constant influx of energy and raw material, and excretion of its waste. Guiding the capture of the requisite energy and raw material for its survival and the perpetuation of the system is the information encoded in its DNA sequence.

During the course of evolution a great number of organisms no longer function as a single cell. The human body, for example, contains trillions of cells. In a multicellular system, each cell becomes specialized to perform a specific function, e.g., β-cells in our pancreas synthesize and release insulin, and cortical neurons in the brain perform neurobiological functions that underlie learning and memory. Despite this "division of labor," the challenges a single-cell organism faces still hold true for each one of these cells. Instead of dealing with the external environment directly, they interact with and respond to changes in their microenvironment.

DOI: 10.1201/9780429329180-2

1.2 How Cells Meet the Challenge

Many cells, like algae and plant cells, directly capture energy from the sun or other energy sources. Other cells (or organisms) obtain energy from the environment as heterotrophs. For raw material, cells can either fix carbon dioxide in the air using the energy captured into simple organic compounds, which are then converted to other requisite molecules, or directly obtain organic molecules from the environment and convert them to requisite materials. In the meantime, existing cellular components can also be broken down when not needed for the re-use of their building material. This process of energy capture and utilization, and synthesis, interconversion, and breaking down for re-use of molecular material, constitutes the cellular metabolism. Metabolism, the most fundamental characteristic of a cell, involves numerous biochemical reactions.

Reception and transduction of various signals in the environment are crucial for cellular survival. Reception of signals relies on specific receptors situated on the cell surface, and for some signals, those inside the cell. Transduction of incoming signals usually involves cascades of events in the cell, through which the original signals are amplified and modulated. In response, cellular metabolic profile is altered. The cellular signal reception and transduction network is composed of circuits that are organized into various pathways. Malfunctioning of these pathways can have a detrimental effect on cellular response to the environment and eventually its survival.

Perpetuation and evolution of the cellular system rely on DNA replication and cell division. The replication of DNA (to be detailed in Chapter 2) is a high-fidelity, but not error-free, process. While maintaining the stability of the system, this process also provides the mechanism for the diversification and evolution of the cellular system. The cell division process is also tightly regulated, for the most part to ensure equal transfer of the replicated DNA into daughter cells. For the majority of multicellular organisms that reproduce sexually, in the process of germ cell formation the DNA is replicated once but cell division occurs twice, leading to the reduction of DNA material by half in the gametes. The recombination of DNA from female and male gametes leads to further diversification in the offspring.

1.3 Molecules in Cells

Different types of molecules are needed to carry out the various cellular processes. In a typical cell, water is the most abundant representing 70% of the total cell weight. Besides water, there are a large variety of small and large molecules. The major categories of small molecules include inorganic ions

(Na^+, K^+, Ca^{2+}, Cl^-, Mg^{2+}, etc.), monosaccharides, fatty acids, amino acids, and nucleotides. Major varieties of large molecules are polysaccharides, lipids, proteins, and nucleic acids (DNA and RNA). Among these components, the inorganic ions are important for signaling (e.g., waves of Ca^{2+} represent important intracellular signal), cell energy storage (e.g., in the form of Na^+ /K^+ cross-membrane gradient), or protein structure/function (e.g., Mg^{2+} is an essential cofactor for many metalloproteins). Carbohydrates (including monosaccharides and polysaccharides), fatty acids, and lipids are major energy-providing molecules in the cell. Lipids are also the major component of cell membrane. Proteins, which are assembled from 20 types of amino acids in different order and length, underlie almost all cellular activities, including metabolism, signal transduction, DNA replication, and cell division. They are also the building blocks of many subcellular structures, such as cytoskeleton (see next section). Nucleic acids carry the code of life in their nearly endless nucleotide permutations, which not only provides instructions on the assembly of all proteins in cells but also exerts control on how such assembly is carried out based on environmental conditions.

1.4 Intracellular Structures or Spaces

Cells maintain a well-organized internal structure (Figure 1.1). Based on the complexity of their internal structure, cells are divided into two major categories: prokaryotic and eukaryotic cells. The fundamental difference between them is whether a nucleus is present. Prokaryotic cells, being more primordial of the two, do not have a nucleus, and as a result their DNA is located in a nucleus-like but non-enclosed area. Prokaryotic cells also lack organelles, which are specialized and compartmentalized intracellular structures that carry out different cellular functions (detailed next). Eukaryotic cells, on the other hand, contain a distinct nucleus dedicated for DNA storage, maintenance, and expression. Furthermore, they contain various organelles including endoplasmic reticulum (ER), Golgi apparatus, cytoskeleton, mitochondrion, and chloroplast (plant cells). The following is an introduction to the various intracellular structures and spaces, including the nucleus, the organelles, and other subcellular structures and spaces such as the cell membrane and cytoplasm.

1.4.1 Nucleus

Since DNA stores the code of life, it must be protected and properly maintained to avoid possible damage and ensure accuracy and stability. As proper execution of the genetic information embedded in the DNA is critical to the normal functioning of a cell, gene expression must also be tightly regulated under

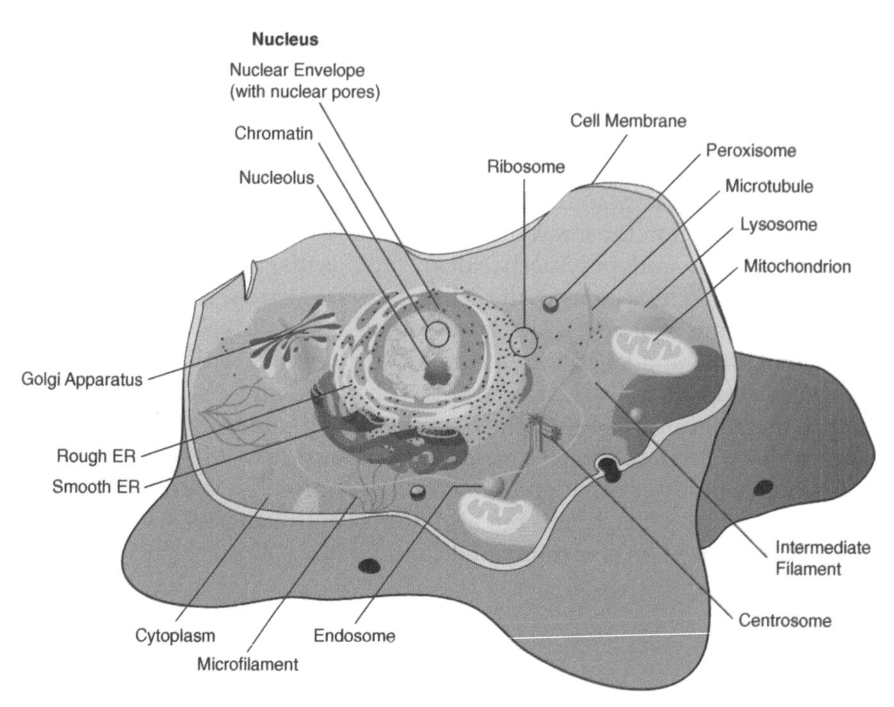

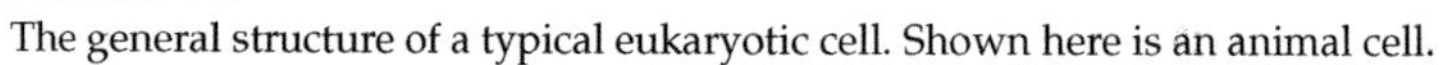

FIGURE 1.1
The general structure of a typical eukaryotic cell. Shown here is an animal cell.

all conditions. The nucleus, located in the center of most cells in eukaryotes, offers a well-protected environment for DNA storage, maintenance, and gene expression. The nuclear space is enclosed by nuclear envelope consisting of two concentric membranes. To allow movement of proteins and RNAs across the nuclear envelope, which is essential for gene expression, there are pores on the nuclear envelope that span the inner and outer membrane. The mechanical support of the nucleus is provided by the nucleoskeleton, a network of structural proteins including lamins and actin among others. Inside the nucleus, long strings of DNA molecules, through binding to certain proteins called histones, are heavily packed to fit into the limited nuclear space. In prokaryotic cells, a nucleus-like irregularly shaped region that does not have a membrane enclosure called the nucleoid provides a similar but not as well-protected space for DNA.

1.4.2 Cell Membrane

The cell membrane serves as a barrier to protect the internal structure of a cell from the outside environment. Biochemically, the cell membrane, as well as all other intracellular membranes such as the nuclear envelope, assumes a lipid bilayer structure. While offering protection to their internal structure, the cell membrane is also where cells exchange materials, and concurrently energy, with the outside environment. Since the membrane is made of lipids, most water-soluble substances, including ions, carbohydrates, amino acids, and nucleotides, cannot directly cross it. To overcome this barrier, there are channels, transporters, and pumps, all of which are specialized proteins, on the cell membrane. Channels and transporters facilitate passive movement, that is, in the direction from high to low concentration, without consumption of cellular energy. Pumps, on the other hand, provide active transportation of the molecules, since they transport the molecules against the concentration gradient and therefore consume energy.

The cell membrane is also where a cell receives most incoming signals from the environment. After signal molecules bind to their specific receptors on the cell membrane, the signal is relayed to the inside, usually eliciting a series of intracellular reactions. The ultimate cellular response that the signal induces is dependent on the nature of the signal, as well as the type and condition of the cell. For example, upon detecting insulin in the blood via the insulin receptor in their membrane, cells in the liver respond by taking up glucose from the blood for storage.

1.4.3 Cytoplasm

Inside the cell membrane, cytoplasm is the thick solution that contains the majority of cellular substances, including all organelles in eukaryotic cells but excluding the nucleus in eukaryotic cells and the DNA in prokaryotic cells. The general fluid component of the cytoplasm that excludes the organelles is called the cytosol. The cytosol makes up more than half of the cellular volume and is where many cellular activities take place, including a large number of metabolic steps such as glycolysis and interconversion of molecules, and most signal transduction steps. In prokaryotic cells, due to the lack of the nucleus and other specialized organelles, the cytosol is almost the entire intracellular space and where most cellular activities take place.

Besides water, the cytosol contains large amounts of small and large molecules. Small molecules, such as inorganic ions, provide an overall biochemical environment for cellular activities. In addition, ions such as Na^+, K^+, and Ca^{2+} also have substantial concentration differences between the cytosol and the extracellular space. Cells spend a lot of energy maintaining these concentration differences, and use them for signaling and metabolic purposes. For example, the concentration of Ca^{2+} in the cytosol is normally kept very low at $\sim 10^{-7}$ M whereas in the extracellular space it is $\sim 10^{-3}$ M. The rushing in of Ca^{2+} under certain conditions through ligand- or voltage-gated channels serves as an important messenger, inducing responses in a number of signaling pathways, some of which lead to altered gene expression. Besides small molecules, the cytosol also contains large numbers of macromolecules. Far from being simply randomly diffusing in the cytosol, these large molecules form molecular machines that collectively function as a "bustling metropolitan city" [1]. These supra-macromolecular machines are usually assembled out of multiple proteins, or proteins and RNA. Their emergence and disappearance are dynamic and regulated by external and internal conditions.

1.4.4 Endosome, Lysosome, and Peroxisome

Endocytosis is the process that cells bring in macromolecules, or other particulate substances such as bacteria or cell debris, into the cytoplasm from the surroundings. Endosome and lysosome are two organelles that are involved in this process. To initiate endocytosis, part of the cell membrane forms a pit, engulfs the external substances, and then an endocytotic vesicle pinches off from the cell membrane into the cytosol. Endosome, normally in the size range of 300–400 nm in diameter, forms from the fusion of these endocytotic vesicles. The internalized materials contained in the endosome are sent to other organelles such as lysosome for further digestion.

The lysosome is the principal site for intracellular digestion of internalized materials as well as obsolete components inside the cell. Like the condition in our stomach, the inside of the lysosome is acidic (pH at 4.5–5.0), providing an ideal condition for the many digestive enzymes within. These enzymes can break down proteins, DNA, RNA, lipids, and carbohydrates. Normally the lysosome membrane keeps these digestive enzymes from leaking into the cytosol. Even in the event of these enzymes leaking out of the lysosome, they can do little harm to the cell, since their digestive activities are heavily dependent on the acidic environment inside the lysosome whereas the pH of the cytosol is slightly alkaline (around 7.2).

Peroxisome is morphologically similar to the lysosome, but it contains a different set of proteins, mostly oxidative enzymes that use molecular oxygen to extract hydrogen from organic compounds to form hydrogen peroxide. The hydrogen peroxide can then be used to oxidize other substrates, such as phenols or alcohols, via peroxidation reaction. As an example, liver and kidney cells use these reactions to detoxify various toxic substances that enter the body. Another function of the peroxisome is to break down long-chain fatty acids into smaller molecules by oxidation. Despite its important functions, the origin of the peroxisome is still not entirely clear. One theory proposes that this organelle has an endosymbiotic origin [2]. If this theory holds true, all genes in the genome of the original endosymbiotic organism must have been transferred to the nuclear genome. A more recent hypothesis, however, is that they had an endogenous origin from the endomembrane system, similar to the lysosome and the Golgi apparatus (see next section) [3].

1.4.5 Ribosome

Ribosome is the protein assembly factory in cells, translating genetic information carried in messenger RNAs (mRNAs) into proteins. There are vast numbers of ribosomes, usually from thousands to millions, in a typical cell. While both prokaryotic and eukaryotic ribosomes are composed of two components (or subunits), eukaryotic ribosomes are larger than their prokaryotic counterparts. In eukaryotic cells, the two ribosomal subunits are first assembled inside the nucleus in a region called the nucleolus and then shipped out to the cytoplasm. In the cytoplasm, ribosomes can be either free, or get attached to another organelle (the ER). Biochemically, ribosomes contain more than 50 proteins and several ribosomal RNA (rRNA) species. Because ribosomes are highly abundant in cells, rRNAs are the most abundant in total RNA extracts, accounting for 85% to 90% of all RNA species. For profiling cellular RNA populations using next-gen sequencing (NGS), rRNAs are usually not of interest despite their abundance and therefore need to be depleted to avoid generation of overwhelming amounts of sequencing reads from them.

1.4.6 Endoplasmic Reticulum

As indicated by the name, ER is a network of membrane-enclosed spaces throughout the cytosol. These spaces interconnect and form a single internal environment called the ER lumen. There are two types of ERs in cells: rough ER and smooth ER. The rough ER is where all cell membrane proteins, such as ion channels, transporters, pumps, and signal molecule receptors, as well as secretory proteins, such as insulin, are produced and sorted. The characteristic surface roughness of this type of ER comes from the ribosomes that bind to them on the outside. Proteins destined for cell membrane or secretion, once emerging from these ribosomes, are threaded into the ER lumen. This ER-targeting process is mediated by a signal sequence, or "address tag," located at the beginning part of these proteins. This signal sequence is subsequently cleaved off inside ER before the protein synthesis process is complete. Functionally different from the rough ER, the smooth ER plays an important role in lipid synthesis for the replenishment of cellular membranes. Besides membrane and secretory protein preparation and lipid synthesis, one other important function of ER is to sequester Ca^{2+} from the cytosol. In Ca^{2+}-mediated cell signaling, shortly after entry of the calcium wave into the cytosol, most of the incoming Ca^{2+} needs to be pumped out of the cell and/or sequestered into specific organelles such as ER and mitochondria.

1.4.7 Golgi Apparatus

Besides ER, the Golgi apparatus also plays an indispensable role in sorting as well as dispatching proteins to the cell membrane, extracellular space, or other subcellular destinations. Many proteins synthesized in the ER are sent to the Golgi apparatus via small vesicles for further processing before being sent to their final destinations. Therefore the Golgi apparatus is sometimes metaphorically described as the "post office" of the cell. The processing carried out in this organelle includes chemical modification of some of the proteins, such as adding oligosaccharide side chains, which serves as "address labels." Other important functions of the Golgi apparatus include synthesizing carbohydrates and extracellular matrix materials, such as the polysaccharide for the building of the plant cell wall.

1.4.8 Cytoskeleton

Cellular processes like the trafficking of proteins in vesicles from ER to the Golgi apparatus, or the movement of a mitochondrion from one intracellular location to another, are not simply based on diffusion. Rather, they follow certain protein-made skeletal structure inside the cytosol, that is, the cytoskeleton, as tracks. Besides providing tracks for intracellular transport, the cytoskeleton, like the skeleton in the human body, plays an equally important role in maintaining cell shape, and protecting the cell framework from physical stresses as the lipid bilayer cell membrane is fragile and vulnerable to such stresses. In eukaryotic cells, there are three major types of cytoskeletal structures: microfilament, microtubule, and intermediate filament. Each type is made of distinct proteins and has their own unique characteristics and functions. For example, microfilament and microtubule are assembled from actins and tubulins, respectively, and have different thickness (the diameter is around 6 nm for microfilament and 23 nm for microtubule). While biochemically and structurally different, both the microfilament and the microtubule have been known to provide tracks for mRNA transport in the form of large ribonucleoprotein complexes to specific intracellular sites, such as the distal end of a neuronal dendrite, for targeted protein translation [4]. Besides its role in intracellular transportation, the microtubule also plays a key role in cell division through attaching to the duplicated chromosomes and moving them equally into two daughter cells. In this process, all microtubules involved are organized around a small organelle called a centrosome. Previously thought to be only present in eukaryotic cells, cytoskeletal structure has also been discovered in prokaryotic cells [5].

1.4.9 Mitochondrion

The mitochondrion is the "powerhouse" in eukaryotic cells. While some energy is produced from the glycolytic pathway in the cytosol, most energy is generated from the Krebs cycle and the oxidative phosphorylation process that take place in the many mitochondria contained in a cell. The number of mitochondria in a cell is ultimately dependent on its energy demand. The more energy a cell needs, the more mitochondria it has. Structurally, the mitochondrion is an organelle enclosed by two membranes. The outer membrane is highly permeable to most cytosolic molecules, and as a result the intermembrane space between the outer and inner membranes is similar to the cytosol. Most of the energy releasing process occurs in the inner membrane and in the matrix, that is, the space enclosed by the inner membrane. For the energy release, high-energy electron carriers generated from the Krebs cycle in the matrix are fed into an electron transport chain embedded in the inner membrane. The energy released from the transfer of high-energy electrons through the chain to molecular oxygen (O_2), the final electron acceptor, creates a proton gradient across the inner membrane. This proton gradient serves as the energy source for the synthesis of ATP, the universal energy currency in cells. In prokaryotic cells, since they do not have this organelle, ATP synthesis takes place on their cytoplasmic membrane instead.

The origin of the mitochondrion, based on the widely accepted endosymbiotic theory, is an ancient α-Proteobacterium. So not surprisingly, the mitochondrion carries its own DNA, but the genetic information contained in the mitochondrial DNA (mtDNA) is extremely limited compared to the nuclear DNA. The human mitochondrial DNA, for example, is 16,569 bp in size coding for 37 genes, including 22 for transfer RNAs (tRNAs), 2 for rRNAs, and 13 for mitochondrial proteins. While it is much smaller compared to the nuclear genome, there are multiple copies of mtDNA molecules in each mitochondrion. Since cells usually contain hundreds to thousands of mitochondria, there are a large number of mtDNA molecules in each cell. In comparison, most cells only contain two copies of the nuclear DNA. As a result, when sequencing cellular DNA samples, sequences derived from mitochondrial DNA usually comprise a notable, sometimes substantial, percentage of total generated reads. Although small, the mitochondrial genomic system is fully functional and has the entire set of protein factors for mtDNA transcription, translation, and replication. As a result of its activity, when cellular RNA molecules are sequenced, those transcribed from the mitochondrial genome also generate significant amounts of reads in the sequence output.

The many copies of mtDNA molecules in a cell may not all have the same sequence due to mutations in individual molecules. Heteroplasmy occurs when cells contain a heterogeneous set of mtDNA molecules. In general, mitochondrial DNA has a higher mutation rate than its nuclear counterpart. This is because the transfer of high-energy electrons along the electron transport chain can produce reactive oxygen species as byproducts, which can oxidize and cause mutations in mtDNA. To make this situation even worse, the DNA repair capability in mitochondria is rather limited. Increased heteroplasmy has been associated with higher risk of developing aging-related diseases, including Alzheimer's disease, heart disease, and Parkinson's disease [6]. Furthermore, mitochondrial DNA mutations have been known to underlie aging and cancer development [7]. Certain hereditary mtDNA mutations also underlie maternally inherited diseases that mostly affect the nervous system and muscle, both of which are characterized by high energy demand.

1.4.10 Chloroplast

In animal cells, the mitochondrion is the only organelle that contains an extranuclear genome. Plant and algae cells have another extranuclear genome besides the mitochondrion, the plastid genome. Plastid is an organelle that can differentiate into various forms, the most prominent of which is the chloroplast. The chloroplast carries out photosynthesis through capturing the energy in sunlight and fixing it into carbohydrates using carbon dioxide as substrate, and releasing oxygen in the same process. For energy capturing, the green pigment called chlorophyll first absorbs energy from sunlight, which is then transferred through an electron transport chain to build up a proton gradient to drive the synthesis of ATP. Despite the energy source, the buildup of proton gradient for ATP synthesis in the chloroplast is very similar to that for ATP synthesis in the mitochondrion. The chloroplast ATP derived from the captured light energy is then spent on CO_2 fixation. Similar to the mitochondrion, the chloroplast also has two membranes: a highly permeable outer membrane and a much less permeable inner membrane. The photosynthetic electron transport chain, however, is not located in the inner membrane, but in the membrane of a series of sac-like structures called thylakoids located in the chloroplast stroma (analogous to the mitochondrial matrix).

Plastid is believed to be evolved from an endosymbiotic cyanobaterium, which has gradually lost the majority of its genes in its genome over millions of years. The current size of most plastid genomes is 120–200 kb, coding for rRNAs, tRNAs, and proteins. In higher plants there are around 100 genes coding for various proteins of the photosynthetic system [8]. The transmission of plastid DNA (ptDNA) from parent to offspring is more complicated than the maternal transmission of mtDNA usually observed in animals. Based on the transmission pattern, it can be classified into three types: 1) maternal, inheritance only through the female parent; 2) paternal, inheritance only through the male parent; or 3) bioparental, inheritance through both parents [9]. Similar to the situation in mitochondrion, there exist multiple copies of ptDNA in each plastid, and as a result there are large numbers of ptDNA molecules in each cell with potential heteroplasmy. Transcription from these ptDNA also generates copious amounts of RNAs in the organelle. Therefore, sequence reads from ptDNA or RNA comprise part of the data when sequencing plant and algae DNA or RNA samples, along with those from mtDNA or RNA.

1.5 The Cell as a System

1.5.1 The Cellular System

From the above description of a typical cell, it is obvious that the cell is a self-organizing system, containing many different molecules and structures that work together coherently. Unlike other non-biological systems, including natural and artificial systems such as a car or a computer, the cell system is unique as it continuously renews and perpetuates itself without violating the laws of the physical world. It achieves this by obtaining energy from and exchanging materials with its environment. The cellular system is also characterized by its autonomy, that is, all of its activities are self-regulated. This autonomy is conferred by the genetic instructions coded in the cell's DNA. Besides such characteristics, the cell system is highly robust, as its homeostasis is not easily disturbed by changes in its surroundings. This robustness is a result of billions of years of evolution, which has led to the building of tremendous complexity into the system. To study this complexity, biologists have been mostly taking a reductionist approach to studying the different cellular molecules and structures piece by piece. This approach has been highly successful and much knowledge has been gathered on most parts of the system. For a cell to function as a single entity, however, these different parts do not work alone. To study how it operates as a whole, the different parts need to be studied in the context of the entire system and therefore a holistic approach is also needed. It has become more and more clear to researchers in the life science community that the interactions between the different cellular parts are equally, if not more, important as any part alone.

1.5.2 Systems Biology of the Cell

Systems biology is an emerging field that studies the complicated interactions among the different parts of biological systems. It is an application of the systems theory to the biological field. Introduced by the biologist Ludwig von Bertalanffy in the 1940s, this theory aims to investigate the principles common to all complex systems, and to describe these principles using mathematical models. This theory is applicable to many disciplines including physics, sociology, and biology, and one goal of this theory is to unify the principles of systems as uncovered from the different disciplines. It is expected, therefore, that principles uncovered from other systems may be applicable to biological systems and provide guidance to better understanding of their working.

In the traditional reductionist approach, a single gene or protein is the basic functioning unit. In systems biology, however, the basic unit is a genetic circuit. Genetic circuit can be defined as a group of genes (or the proteins they code) that work together to perform a certain task. There are a multitude of tasks in a cell that need to be carried out by genetic circuits, from the transduction of extracellular signal to the inside, the step-by-step breakdown of energy molecules (such as glucose) to release energy, to the replication of DNA prior to cell division. It is these genetic circuits that underlie cellular behavior and physiology. If the information or material flux in a genetic circuit is blocked or goes awry, the whole system will be influenced, which might lead to the malfunctioning of the system and likely a diseased state.

Based on the hierarchical organization principle of systems, gene circuits interact with each other and form a complicated genetic network. Mapping out a genetic network is a higher goal of systems biology. Genetic network has been shown to share some common characteristics with non-biological networks such as the human society or the Internet [10]. One of such characteristics is modularity, referring to the fact that genes (or proteins) that work together to achieve a common goal often form a module and the module is used as a single functional unit when needed. Another common characteristic is the existence of hub or anchor nodes in the network, as a small number of highly connected genes (or proteins) in a genetic network serve as hubs or anchors through which other genes (or proteins) are connected to each other.

1.5.3 How to Study the Cellular System

Research into the systems biology of the cell is largely enabled by technological advancements in genomics, proteomics, and metabolomics. High-throughput genomics technologies, for example, allow simultaneous analysis of tens of thousands of genes in an organism's genome. Genome refers to the whole set of genetic material in an organism's DNA, including both protein-coding and non-coding sequences. Similarly, proteome and metabolome are defined as the complement of proteins and metabolites (small molecules), respectively, in a cell or population of cells. Proteomics, through simultaneous separation and identification of proteins in a proteome, provides answers to the questions of how many proteins are present in the target cell(s) and at what abundance levels. Metabolomics, on the other hand, through analyzing a large number of metabolites simultaneously, monitors the metabolic status of target cells.

The development of modern genomics technologies was mostly initiated when the human genome was sequenced by the Human Genome Project. The completion of the sequencing of this genome and the genomes of other organisms, and the concurrent development of genomics technologies, have for the first time offered an opportunity to study the systems properties of the cell. The first big wave of genomics technologies was mostly centered on microarray, which enables analysis of the transcriptome and subsequently study of genome-wide sequence polymorphism and the epigenome. By studying all RNAs transcribed in a cell or population of cells, transcriptomic analysis investigates what genes are active and how active. Determination of genome-wide sequence variations among individuals in a population enables examination of the relationship between certain genomic polymorphisms and cellular dysfunctions, phenotypic traits, or diseases. Epigenomic studies provide answers to the question how the genomic information encoded in the DNA sequence is regulated by the code conferred by chemical modifications of DNA bases. More recently, the development of NGS technologies provides more power, coverage, and resolution to the study of the genome, the transcriptome, and the epigenome (for details on the development of NGS technologies see Chapter 4). These NGS technologies, along with recent technological developments in proteomics and metabolomics, further empower the study of the cellular system.

References

1. Vale RD. The molecular motor toolbox for intracellular transport. *Cell* 2003, 112(4):467–480.
2. de Duve C. Peroxisomes and related particles in historical perspective. *Ann N Y Acad Sci* 1982, 386:1–4.
3. Gabaldon T. Evolution of the peroxisomal proteome. *Subcell Biochem* 2018, 89:221–233.
4. Das S, Vera M, Gandin V, Singer RH, Tutucci E. Intracellular mRNA transport and localized translation. *Nat Rev Mol Cell Biol* 2021, 22(7):483–504.
5. Mayer F. Cytoskeletons in prokaryotes. *Cell Biol Int* 2003, 27(5):429–438.
6. Chocron ES, Munkacsy E, Pickering AM. Cause or casualty: the role of mitochondrial DNA in aging and age-associated disease. *Biochim Biophys Acta Mol Basis Dis* 2019, 1865(2):285–297.
7. Smith ALM, Whitehall JC, Greaves LC. Mitochondrial DNA mutations in ageing and cancer. *Mol Oncol* 2022, 16(18):3276–3294.
8. de Vries J, Archibald JM. Plastid genomes. *Curr Biol* 2018, 28(8):R336–R337.
9. Harris SA, Ingram R. Chloroplast DNA and biosystematics: the effects of intraspecific diversity and plastid transmission. *Taxon* 1991:393–412.
10. Roy U, Grewal RK, Roy S. Complex Networks and Systems Biology. In: *Systems and Synthetic Biology*. Springer; 2015: 129–150.

2

DNA Sequence: The Genome Base

2.1 The DNA Double Helix and Base Sequence

Among the different types of molecules in cells, DNA has a structure that makes it ideal to code the blueprint of life. The building blocks of DNA are nucleotides, which are made up of three chemical groups: a five-carbon sugar (deoxyribose), phosphate, and one of four nucleobases. The spatial structure of DNA is a double helix comprised of two strands. The backbone of each strand is made of the sugar moiety and phosphate, which are invariably connected in an alternating fashion and therefore do not carry genetic information. The "rungs" that connect the two strands are composed of nucleobases, which are where the information is stored. Since the discovery of this structure in 1953 by Watson and Crick, the elegance and simplicity of this structure has fascinated generations of biologists, chemists, and scientists from other fields.

There are four different types of nucleobases (or simply bases) in DNA: two purines (adenine, usually abbreviated as A; and guanine, G), and two pyrimidines (cytosine, C; and thymine, T). Nucleobases in the two DNA strands that form the rung structure interact via hydrogen bonding in a fixed manner: A always pairing with T, and C with G. This complementary base-pairing pattern enables the DNA molecule to assume the most thermodynamically favorable structure. The fixed pairing pattern between the bases makes it easy to provide coding for life and to replicate for perpetuation.

The almost endless arrangements of the base pairs in DNA provide the basis for its role as the genetic information carrier. The information embedded in the DNA base sequence dictates what, when, and how many proteins are made in a cell at a certain point of time. At a deeper level, the information codes for the entire operating logic of the cellular system. It contains all instructions needed to form a new life and for it to grow, develop, and reproduce. From the medical point of view, alterations or polymorphisms in the DNA base sequence can predispose us to certain diseases as well as underlie our responses to medications.

DOI: 10.1201/9780429329180-3

2.2 How DNA Molecules Replicate and Maintain Fidelity

The DNA's double helix structure and complementary base-pairing make it robust to copy the bio-information it carries through its replication. To replicate, the two strands of the parent DNA molecule are first unwound by an enzyme called helicase. The two unwound strands then serve as templates for the synthesis of new complementary strands, giving rise to two offspring DNA molecules. The enzyme that carries out the new strand synthesis is called DNA polymerase, which assembles nucleotides into a new strand by adding one nucleotide at a time to a pre-existing primer sequence based on complementary base-pairing with the template strand (Figure 2.1). Biochemically, the enzyme catalyzes the formation of a covalent phosphodiester bond between the 5′-phosphate group of the incoming complementary nucleotide and the 3′-hydroxyl group on the elongating strand end. Besides elongating the new DNA strand, most DNA polymerases also have proofreading capability. If a nucleotide that is not complementary to the template is attached to the end of the elongating strand (i.e., mis-pairing), the enzyme will turn around and cleave the wrong nucleotide off. This proofreading activity is important to maintain the high fidelity of the DNA replication process. Mutations, or sudden changes of nucleotide sequence in DNA, would occur much more frequently without this activity.

Many sequencing technologies are based on the process of DNA replication. These technologies, often referred to as sequencing-by-synthesis, use this process to read nucleotide sequence off one strand of the sequencing DNA target. Corresponding to the components required in the DNA replication process, these sequencing systems require the following basic components: (1) sequencing DNA target, which provides the template; (2) nucleotides; (3) a primer; and (4) a DNA polymerase. Since the DNA polymerase extends the new strand by attaching one nucleotide at a time, detecting the attached nucleotide after each extension cycle generates a readout of the nucleotide sequence on the template DNA strand. To facilitate the detection, the nucleotides used in sequencing reactions are usually chemically modified, including labeling with fluorescent tags. Chapter 4 focuses on the evolution of sequencing technologies.

Besides the high fidelity of DNA polymerases, an efficient DNA repair system is also crucial to maintain genome stability and keep mutation rate low. Even under normal conditions, DNA nucleotide sequence can be accidently altered by many physical and chemical factors in the environment, including intracellularly generated reactive oxygen and nitrogen species, radiation in the environment (such as UV, X-ray, or γ-ray), and other chemical mutagens. If left uncorrected, these changes will accumulate and cause disturbances to normal cell function or even cause cell death, leading to diseases. To maintain the fidelity of DNA molecules, cells invest heavily on DNA repair enzymes. These enzymes constantly scan genomic DNA and make repairs if damage is

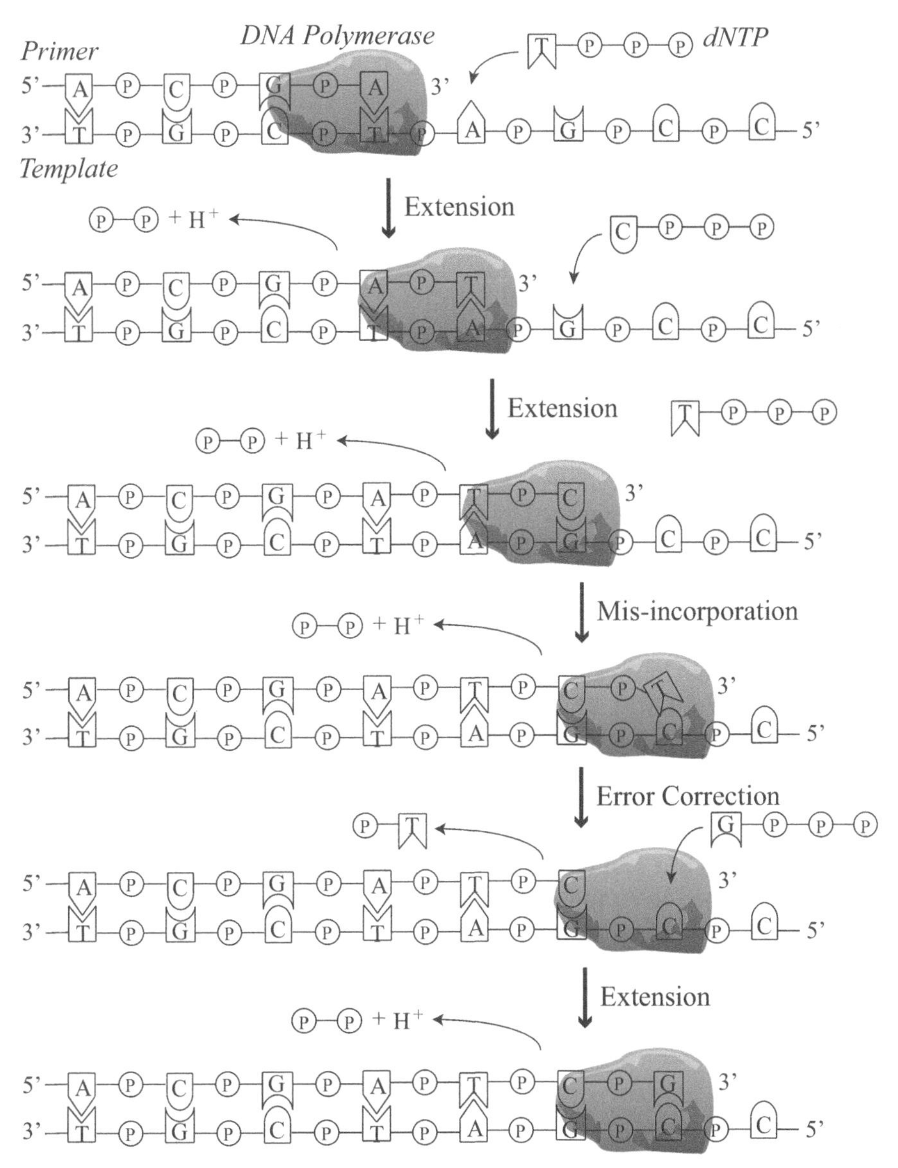

FIGURE 2.1
The DNA replication process. To initiate the process, a primer, which is a short DNA sequence complementary to the start region of the DNA template strand, is needed for DNA polymerase to attach nucleotides and extend the new strand. The attachment of nucleotides is based on complementary base-pairing with the template. If an error occurs due to mis-pairing, the DNA polymerase removes the mis-paired nucleotide using its proofreading function. Due to the biochemical structure of the DNA molecule, the direction of the new strand elongation is from its 5′ end to 3′ end (the template strand is in the opposite direction; the naming of the two ends of each DNA strand as 5′ and 3′ is from the numbering of carbon atoms in the nucleotide sugar ring).

detected. The serious consequences of a weakened DNA repair system can be exemplified by mutations in *BRCA2*, a gene coding for a DNA repair enzyme, which lead to breast and ovarian cancers.

2.3 How the Genetic Information Stored in DNA Is Transferred to Protein

While the logic of the cellular system is written in the nucleotide sequence of its genomic DNA, almost all cellular activities are executed by the wide array of proteins in the cell's proteome. The bio-information flow from DNA to protein, known as the central dogma (Figure 2.2), provides a fundamental framework for modern molecular biology and genetic engineering. Based on this framework, a gene's DNA sequence is first transcribed to make mRNA, and then the nucleotide sequence in mRNA is used to guide the assembly of amino acids into a protein. The translation of the mRNA nucleotide sequence to the protein amino acid sequence is based on the triplet genetic code. A continuous segment of DNA that contains the full set of triplet codon for protein translation, from start to stop, is often called an open reading frame (or ORF). The synthesis of one type of bio-polymer molecule based on information stored in another bio-polymer is one of the greatest "inventions" of nature.

Since its initial introduction, the central dogma has been gradually modified with increased sophistication. In its original form, one gene is translated into one protein via one mRNA. This one gene–one protein paradigm was later found to be too simplistic, as one gene can generate multiple forms of

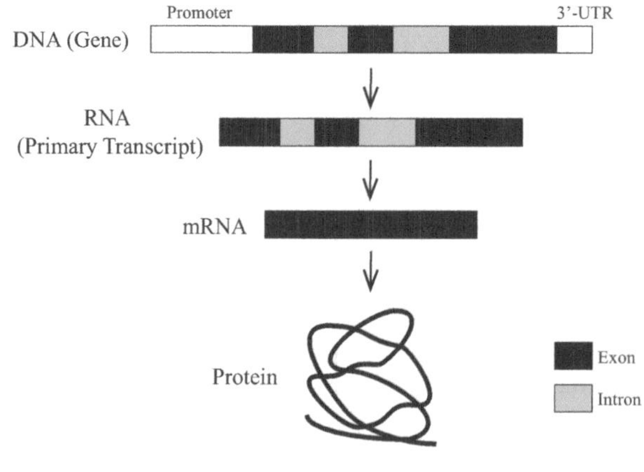

FIGURE 2.2
The central dogma.

proteins through alternative splicing (next chapter). In addition, the information flow between DNA and RNA is not simply one-way from DNA to RNA, but RNA can also be reverse transcribed to DNA in some organisms. On the additional role of RNAs in this information flow, some non-protein-coding RNAs can silence gene expression through mechanisms such as inhibiting gene transcription or translation, or protect genome through mechanisms like preventing the movement of transposable elements (or transposons, mobile DNA elements that copy themselves to different genomic loci) (also see next chapter). Furthermore, chemical modifications of DNA and some DNA-interacting proteins constitute the epigenome, which also regulates the flow of genetic information.

2.4 The Genomic Landscape

2.4.1 The Minimal Genome

After understanding the flow of bio-information from DNA to protein, the next question is what is the minimum amount of genetic information needed to make the cellular system tick, that is, what constitutes the minimal genome. Attempts to define the minimal genome started in the late 1950s, shortly after the discovery of the double helix structure of DNA. The answer to this important question is not straightforward, however, as the amount of genetic information needed for a minimal life form is dependent on the specific environment it lives in. Considering the basic functions that a cell has to perform, the minimal genome needs to contain genes at least for DNA replication, RNA synthesis and processing, protein translation, energy and molecular metabolism. A small bacterium, *Mycoplasma genitalium*, often used as a model of a naturally existing minimal genome for a free-living organism, has a genome of 580 kilobase pairs (kbp) and 504 genes [1]. An artificially designed and completely synthetic minimal genome reported in 2016 to support a viable cell contains 531 kbp sequence and 473 genes [2].

2.4.2 Genome Sizes

For the least sophisticated organisms, such as *Mycoplasma genitalium*, a minimal genome is sufficient. For increased organismal complexity, more genetic information and, therefore, a larger genome is needed. As a result, there is a positive correlation between organismal complexity and genome size, especially in prokaryotes. In eukaryotes, however, this correlation becomes much weaker, largely due to the existence of non-coding DNA elements in varying amounts in different eukaryotic genomes (for details on non-coding

TABLE 2.1

Genome Sizes and Total Gene Numbers in Major Model Organisms* (Ordered by Genome Size)

Organism	Genome Size (bp)	Number of Coding Genes
Mycoplasma genitalium (Bacterium)	580,076	504
Haemophilus influenzae (Bacterium)	1,846,259	1,715
Escherichia coli (Bacterium)	4,641,652	4,288
Saccharomyces cerevisiae (Yeast)	12,157,105	6,016
Caenorhabditis elegans (Nematode)	100,286,401	19,984
Arabidopsis thaliana (Thale cress)	119,668,634	27,562
Drosophila melanogaster (Fruit fly)	143,726,002	13,968
Oryza sativa (Rice)	374,422,835	28,738
Medicago truncatula (Legume)	430,008,013	31,927
Danio rerio (Zebrafish)	1,373,454,788	26,522
Zea mays (Maize)	2,182,786,008	34.337
Rattus norvegicus (Rat)	2,647,915,728	22,228
Mus Musculus (Mouse)	2,728,222,451	22,186
Homo sapiens (Human)	3,099,441,038	20,024

*Data based on NCBI genome database as of Nov. 2022.

DNA elements, see Section 2.4.4) and whole-genome duplication. In terms of genome size, the currently documented range is 112 kbp, coding 137 genes, of *Nasuia deltocephalinicola* (an obligate bacterial endosymbiont living in leafhopper cells) [3] on the lower end, to 149 gigabase pairs (gbp) (number of genes coded still unknown) of *Paris japonica* (a slow-growing flowering plant) [4] on the higher end. Table 2.1 shows the total number of genes in some of the most studied organisms.

2.4.3 Protein-Coding Regions of the Genome

The protein-coding regions are the part of the genome that we foremost study and know most about. The content of these regions directly affects protein synthesis and protein diversity in cells. In prokaryotic cells, functionally related protein-coding genes are often arranged next to each other and regulated as a single unit known as operon. The gene structure in eukaryotic cells is more complicated. The coding sequences (CDS) of almost all eukaryotic genes are not continuous and interspersed among non-coding sequences. The non-coding intervening sequences are called introns (*int* for intervening), while the coding regions are called exons (*ex* for expressed) (see Figure 2.2). During gene transcription, both exons and introns are transcribed. In the subsequent mRNA maturation process, introns are spliced out and exons are joined together for protein translation.

In the human genome, the average number of exons per gene is 8.8. The *titin* gene, coding for a large abundant protein in striated muscle, has 363 exons, the most in any single gene, and also has the longest single exon

(17,106 bp) among all currently known exons. In aggregate the total number of currently known exons in the human genome is around 180,000. With a combined size of 30 Mb, they constitute 1% of the human genome. This collection of all exons in the human genome, or in other eukaryotic genomes, is termed as the exome. Different from the transcriptome, which is composed of all actively transcribed mRNAs in a particular sample, the exome includes all exons contained in a genome. While it only covers a very small percentage of the genome, the exome represents the most important and the best annotated part of the genome. Sequencing of the exome has been used as a popular alternative to whole genome sequencing. While it lacks on coverage, exome sequencing is more cost effective, faster, and easier for data interpretation.

2.4.4 Non-Coding Genomic Elements

While protein-coding genes are the most studied genomic element, they may not necessarily be the most abundant part of the genome. Prokaryotic genomes are usually rich in protein-coding gene sequences, e.g., they account for approximately 90% of the *E. coli* genome. In complex eukaryotic genomes, however, their percentage is lower. For example, only about 1.5% of the human genome codes for proteins (Figure 2.3). Among the non-protein-coding sequences in eukaryotic genomes are introns, regulatory sequences, and other unique non-coding DNA elements. The regulatory sequences are genomic elements that are known to regulate gene expression, including promoters, terminators, enhancers, repressors, and silencers. In comparison, our current understanding of the other unique non-coding DNA elements is the most rudimentary. We know nearly nothing about these elements, with the exception of non-coding RNA genes, which include rRNAs, tRNAs, and other functionally important RNA species, which will be detailed in the next chapter. As mentioned in Chapter 1, rRNAs are key structural components of the ribosome and directly involved in protein translation, while tRNAs transport proper amino acids to the ribosome for protein translation based on the genetic code.

Repetitive sequences occupy more than half of the human genome, and are even more pervasive in some other eukaryotic genomes. For example, in some plants and amphibians, 80% of the genome is composed of repetitive sequences. The percentage of repetitive sequences in prokaryotic genomes is relatively lower but still significant. With respect to their internal structures, some repetitive sequences are tandem repeats, with the basic repeating units connected head-to-tail. In this type of sequence repeats, the length of the repeating units is highly variable, from <10 bp to thousands of base pairs. The other major type of sequence repeats are interspersed repeats, present as a single copy in many genomic loci. These are either transposons, or retrotransposons, that copy themselves via RNA intermediates. Discovered by geneticist Barbara McClintock, transposons (also called transposable

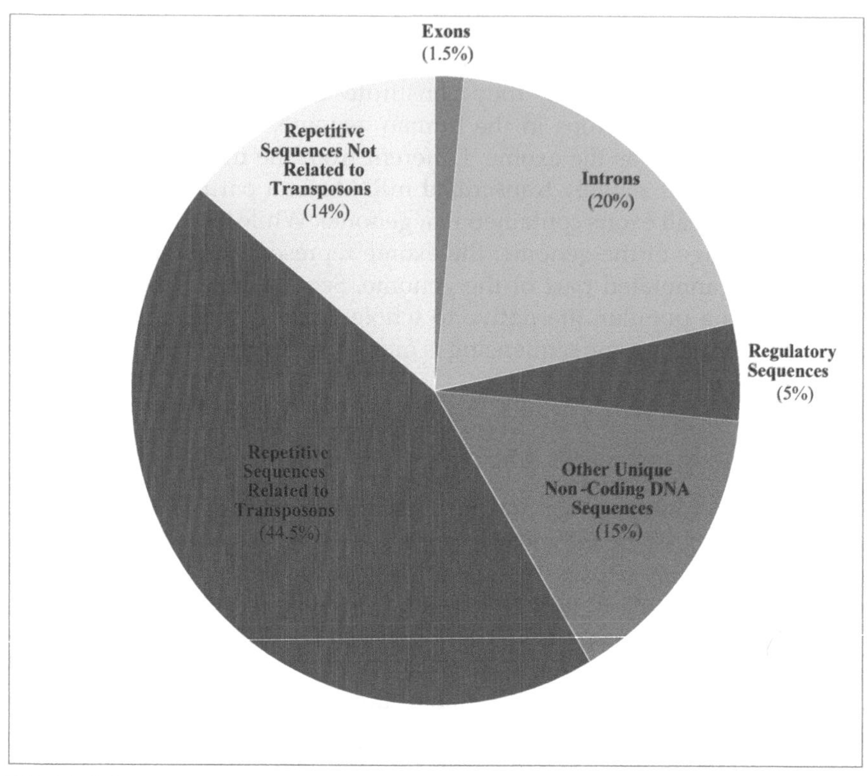

FIGURE 2.3
The composition of the human genome.

elements, or "jumping genes") are DNA sequences that move from one genomic location to another. Repeat sequence units of this type are usually 100 bp to over 10 kb in length, and may appear in over 1 million loci dispersed across the genome.

Many highly repetitive DNA sequences exist in inert parts of chromosomes, such as the centromere and telomere. The centromere, the region where two sister chromatids are linked together before cell division, contains tandem repeat sequences. The telomere, existing at the ends of chromosomes, is also composed of highly repetitive DNA sequences. The telomeric structure protects chromosomal integrity and thereby maintains genomic stability. Besides being essential in maintaining the chromosomal structure, repeat sequences have other functions in the genome, e.g., they play an architectonic role in higher order physical genome structuring [5]. Despite their abundance and function, because sequences associated with repeat regions are not unique, they create a major hurdle for assembling a genome *de novo* from sequencing reads, or mapping reads originated from these regions to a pre-assembled genome.

2.5 DNA Packaging, Sequence Access, and DNA-Protein Interactions

2.5.1 DNA Packaging

In the nucleoid of prokaryotic cells, multiple proteins fold and condense genomic DNA into a supercoiled structure to make it fit into the rather limited space. While being generally condensed, parts of the DNA need to be exposed to allow sequence access for transcription by related proteins factors. While these processes have been studied in prokaryotic cells, DNA packaging and sequence accessing are better studied and understood in eukaryotic cells. In these cells, because of their much larger genome size, genomic DNA is condensed in the nucleus to a much higher degree. For instance, the total length of human genomic DNA is about 2 meters when fully stretched out, but the diameter of the human cell nucleus is only 6 μm. Bound to specific proteins called histones, eukaryotic DNA is packaged in the form of chromatin, in which the positively charged histones bind to the negatively charged DNA molecules through electrostatic interactions. This packaging process involves compacting DNA at different levels. At the first level, DNA wraps around a protein complex composed of eight histone subunits to form the basic structure of nucleosome. Each nucleosome contains around 200 nucleotide pairs and has a diameter of 11 nm. At the second level the nucleosome structure is compacted into a fiber structure. This fiber, with a diameter of 30 nm, is the form most chromatin takes in the interphase between two cell divisions. Prior to cell division, this chromatin fiber is further condensed by two additional levels into chromosome, the extremely condensed form that we can observe under a light microscope.

2.5.2 Sequence Access

Since different DNA sequences in the genome are constantly being transcribed, instead of being permanently locked into the compacted form, DNA sequences at specific loci need to be dynamically exposed to allow transcriptional access to protein factors such as transcription factors and coactivators. Furthermore, DNA replication and repair also require chromatin unpackaging. This unpackaging of the chromatin structure is carried out through two principal mechanisms. One is through histone modification, such as acetylation of lysine residues on histones by histone acetyltransferases, which reduces the positive charge on histones and therefore decreases the electrostatic interactions between histones and DNA. Deacetylation by histone deacetylases, on the other hand, restricts DNA access and represses transcription. The other unpackaging mechanism is through the actions of chromatin remodeling complexes. These large protein complexes consume ATP and use the released energy to expose DNA sequences for transcription through nucleosomal repositioning, nucleosomal eviction, or local unwrapping.

2.5.3 DNA-Protein Interactions

While DNA is the carrier of the code of life, the DNA code cannot be executed without DNA-interacting proteins. Nearly all of the processes mentioned above, including DNA packaging/unpackaging, transcription, repair, and replication, rely on such proteins. Besides histones, examples of these proteins include transcription factors, RNA polymerases, DNA polymerases, and nucleases (for DNA degradation). Many of these proteins, such as histones and DNA/RNA polymerases, interact with DNA regardless of their sequence or structure. Some DNA-interacting proteins bind to DNA of special structure/conformation, e.g., high-mobility group (HMG) proteins that have high affinity for bent or distorted DNA. Some other DNA-interacting proteins bind only to regions of the genome that have certain characteristics such as having damage, the examples of which are DNA repair enzymes such as BRCA1, BRCA2, RAD51, RAD52, and TDG.

The most widely studied DNA-interacting proteins are transcription factors, which bind to specific DNA sequences. Through binding to their specific recognition sequences in the genome, transcription factors regulate transcription of gene targets that contain such sequences in their promoter region. Since they bind to more than one gene locations in the genome, transcription factors regulate the transcription of a multitude of genes in a coordinated fashion, usually as a response to certain internal or external environmental change. For instance, NRF2 is a transcription factor that is activated in response to oxidative stress. Upon activation, it binds to a short segment of specific DNA sequence called the anti-oxidant response element (ARE) located in the promoter region of those genes that are responsive to oxidative stress. Through binding to this sequence element in many regions of the genome, NRF2 regulates the transcription of its target genes and thereby elicits coordinated responses to counteract the damaging effects of oxidative stress.

Study of DNA-protein interactions provides insights into how the genome responds to various conditions. For example, determination of transcription factor binding sites, such as those of NRF2, across the genome can unravel what genes might be responsive to the conditions that activate the transcription factors. While such sites can be predicted computationally, only wet-lab experiment can determine where a transcription factor actually binds in the genome under a certain condition. ChIP-seq, or chromatin immunoprecipitation coupled with sequencing, is one application of NGS that is developed to study genomic binding of transcription factors and other DNA-interacting proteins. Chapter 13 will focus on ChIP-seq data analysis.

2.6 DNA Sequence Mutation and Polymorphism

While DNA replication is a high-fidelity process and the nucleus maintains an army of DNA repair enzymes, sequence mutation does happen, although at a very low frequency. In general, the rate of mutation in prokaryotic and eukaryotic cells is at the scale of 10^{-9} per base per cell division. In multicellular eukaryotic organisms, germline cells have lower mutation rate than somatic cells. In these organisms, because most cells, including germline cells, undergo multiple divisions in the organisms' lifetime, the per-generation mutation rate is significantly higher. For example, whole genome sequencing data collected from human blood cell DNA estimates a mutation rate of 1.1×10^{-8} per base per generation, corresponding to around 70 new mutations in each human diploid genome [6]. Depending on the nature of the change, mutations may have deleterious, neutral, or rarely, beneficial effects on the organism. Mutations lead to sequence variation, and are ultimately the basis of genome evolution and diversification for those carried through the germline. Although mutations in somatic cells are not passed on to the next generation, they can lead to diseases, including cancer, and affect the survival of the individual.

There are various forms of DNA mutations, from single nucleotide substitutions, small insertions/deletions (or indels), to structural variations (SVs) that involve larger genomic regions. Among these different types of mutations, single nucleotide substitutions, also called point mutations, are the most common. These substitutions can be either transitions or transversions. Transitions involve the substitution of a purine for the other purine (i.e., A↔G) or a pyrimidine for the other pyrimidine (i.e., C↔T). Transversions, on the other hand, involve the substitution of a purine for a pyrimidine, or vice versa. Theoretically there are more combinations of transversions than transitions, but due to the nature of the underlying biochemical processes transitions actually occur more frequently than transversions. If single nucleotide substitution takes place in a protein-coding region, it may or may not lead to change in amino acid coding. If it causes the substitution of one amino acid for another, it is a missense mutation, which may lead to change of protein function. If it introduces a stop codon and as a result leads to the generation of a truncated protein, it is a nonsense mutation. Both the missense and nonsense mutations are non-synonymous mutations. If it does not change the coded amino acid due to the redundancy in the genetic code, it is a synonymous mutation and has no effect on protein function. Because of its common occurrence, single nucleotide variation (SNV) is the most frequently observed sequence variation. If an SNV is commonly observed in a population, it is called a single nucleotide polymorphism (or SNP). More than 1,000 million SNPs in the human genome have been cataloged at the time of this writing (November 2022). Because of their high density in the genome, SNPs are often used as flagging markers to cover the entire genome in high resolution when scanning for genomic region(s) that are associated with a phenotype or disease of interest.

Besides single nucleotide substitutions, indels are another common type of mutation. Most indels involve small numbers of nucleotides. In protein-coding regions, small indels lead to the shift of ORF (unless the number of nucleotides involved is a multiple of three), resulting in the formation of a vastly different protein product. Indels that involve large regions lead to alterations of genomic structure and are usually considered as a form of SV. Besides large indels, SVs, defined as changes encompassing at least 50 bp [7], also include inversions, translocations, or duplications. Copy number variation (CNV) is a subcategory of SV, usually caused by large indel or segmental duplication. Although they affect larger genomic region(s) and some lead to observable phenotypic changes or diseases, many CNVs, or SVs in general, have no detectable effects. The frequency of SVs in the genome was underestimated previously due to technological limitations. The emergence of NGS has greatly enabled SV detection, which has led to the realization of its wide existence [8].

2.7 Genome Evolution

The spontaneous mutations that lead to sequence variation and polymorphism in a population are also the fundamental force behind the evolution of genomes and eventually the Darwinian evolution of the host organisms. Gradual sequence change and diversification of early genomes, over billions of years, have evolved into the extremely large number of genomes that had existed or are functioning in varying complexity today. In this process, existing DNA sequences are constantly modified, duplicated, and reshuffled. Most mutations in protein-coding or regulatory sequences disrupt the protein's normal function or alter its amount in cells, causing cellular dysfunction and affecting organismal survival. Under rare conditions, however, a mutation can improve existing protein function or lead to the emergence of new functions. If such a mutation offers its host a competitive advantage, it is more likely to be selected and passed on to future generations.

Gene duplication provides another major mechanism for genome evolution. If a genomic region containing one or multiple gene(s) is duplicated resulting in the formation of an SV, the duplicated region is not under selection pressure and therefore becomes substrate for sequence divergence and new gene formation. Although there are other ways of adding new genetic information to a genome such as inter-species gene transfer, DNA duplication is believed to be a major source of new genetic information generation.

Gene duplication often leads to the formation of gene families. Genes in the same family are homologous, but each member has their specific function and expression pattern. As an example, in the human genome there are 339 genes in the olfactory receptor gene family. Odor perception starts with the binding of odorant molecules to olfactory receptors located on olfactory neurons inside the nose epithelium. To detect different odorants, combination of different olfactory receptors that are coded by genes in this family is required. Based on their sequence homology, members of this large family can be even further grouped into different subfamilies [9]. The existence of pseudogenes in the genome is another result of gene duplication. After duplication, some genes may lose their function and become inactive from additional mutation. Pseudogenes may also be formed in the absence of duplication by the disabling of a functional gene from mutation. A pseudogene called *GULO* mapped to the human chromosome 8p21 provides such an example. The functional *GULO* gene in other organisms codes for an enzyme that catalyzes the last step of ascorbic acid (vitamin C) biosynthesis. This gene is knocked out in primates, including human, and becomes a pseudogene. As a result, we have to get this essential vitamin from food. The inactivation of this gene is possibly due to the insertion to the gene's coding sequence of a retrotransposon-type repetitive sequence called Alu element [10].

DNA recombination, or reshuffling of DNA sequences, also plays an important role in genome evolution. Although it does not create new genetic information, by breaking existing DNA sequences and re-joining them DNA recombination changes the linkage relationships between different genes and other important regulatory sequences. Without recombination, once a harmful mutation is formed in a gene, the mutated gene will be permanently linked to other nearby functional genes, and impossible to regroup all the functional genes back together into the same DNA molecule. Through this regrouping, DNA recombination makes it possible to avoid gradual accumulation of harmful gene mutations. Most DNA recombination events happen during meiosis in the formation of gametes (sperm or eggs) as part of sexual reproduction.

2.8 Epigenome and DNA Methylation

Besides the regulatory DNA sequences introduced earlier, chemical modifications of specific nucleotides in the genome, like the acetylation and deacetylation of histones, offer another layer of regulation on genetic activities. Since they provide additional genetic activity regulation, these chemical modifications on DNA and histones constitute the epigenome. Methylation of the fifth carbon on cytosine (5-methylcytosine, or 5mC) is currently the most studied epigenomic modification in many organisms. Enzymatically, this methylation is carried out and maintained by DNA methyltransferases (three identified in mammals: DNMT1, DNMT3A, and DNMT3B). The cytosines that undergo methylation can occur in three different sequence contexts – CpG, CHG, and CHH (H can be A, G, or T) – each involving different pathways [11]. Most methylated cytosines exist in the CpG context, where the methylation reduces gene expression through recruiting gene silencing proteins, or preventing transcription factors from binding to the DNA. The methylation of cytosines in this context also affects nucleosome positioning and chromatin remodeling, as methyl-CpG-binding domain (MBD) proteins that specifically bind to 5mC at CpG sites can recruit histone modifying proteins and those in the chromatin remodeling complex [12]. The effects of cytosine methylation in the CHG and CHH contexts are less clear but currently available data seems to suggest that they may play a regulatory role in repetitive regions [13].

Just like deacetylation counteracts the effects of acetylation in histones, demethylation of cytosines should be similarly important to reverse the effects of 5mC when the methylation is no longer needed. Until recently, the steps involved in the cytosine demethylation process began to be understood. In this process, the 5mC is first oxidized to 5-hydroxymethylcytosine (5hmC), and then to 5-formylcytosine (5fC) and 5-carboxylcytosine (5caC) in mammals. These oxidative conversions are catalyzed by enzyme systems such as the TET family proteins. The subsequent base excision repair of 5fC/5caC by an enzyme called TDG, or 5mC directly by the glycosylase enzyme in plants, completes the DNA demethylation process [14]. Compared to 5mC, the levels of these demethylation intermediate products are detected to be much lower in most cells (except that 5hmC has been found to be relatively abundant in embryonic stem cells and in the brain).

Different from the genome, which is static, the epigenome is dynamic and changes with environmental conditions. These dynamically changing epigenomic modifications regulate gene expression and thereby play important roles in embryonic development, cell differentiation, stem cell pluripotency, genomic imprinting, and genome stability. In accordance with their regulatory functions, these modifications are highly site specific. To study where cytosine methylations take place in the genome, multiple NGS-based approaches, which will be detailed in Chapter 14, have been developed and widely applied to epigenomics studies. Methodological development for the study of cytosine demethylation is currently still at an early stage.

2.9 Genome Sequencing and Disease Risk

The wide accessibility of DNA sequences, largely fueled by the rapid development of new sequencing technologies, has uncovered extensive sequence variation in individual genomes within a population. The extensiveness in sequence variation was not envisioned in early days of genetics, not even when the Human Genome Project was completed in 2003. This has gradually led to a paradigm shift in disease diagnosis and prevention. As a result, the public becomes more aware of the role of individual genomic makeup in disease development and predisposition. In addition, the easier accessibility to our DNA sequence has further prompted us to look into our genome and use that information for preemptive disease prevention. The declining cost of genome sequencing has also enabled the biomedical community to dig deeper into the genomic underpinnings of diseases, by unraveling the linkage between sequence polymorphism in the genome and disease incidence. Below is a brief overview of the major categories of human diseases that have an intimate connection with DNA mutation, polymorphism, genome structure, and epigenomic abnormality.

2.9.1 Mendelian (Single-Gene) Diseases

The simplest form of hereditary diseases is caused by mutation(s) in a single gene, and therefore also called monogenic or Mendelian diseases. For example, sickle-cell anemia is caused by a mutation in the *HBB* gene located on the human chromosome 11. This gene codes for the β subunit of hemoglobin, an important oxygen-carrying protein in the blood. A mutation of this gene leads to the replacement of the sixth amino acid, glutamic acid, with another amino acid valine in the coded protein. This change of a single amino acid causes conformational change of the protein, leading to the generation of sickle-shaped blood cells that die prematurely. This disease is recessive, meaning that it only appears when both copies (or alleles) of the gene carry the mutation. In dominant diseases, however, one mutant allele is enough to cause sickness. Huntington's disease, a neurodegenerative disease that leads to gradual loss of movement control and mental faculties, is such a dominant single-gene disease. It is caused by mutation in a gene called *HTT* on the human chromosome 4, coding for a protein called huntingtin. The involved mutation is an expansion of tri-nucleotide (CAG, codes for the amino acid glutamine) repeats. When the number of CAG repeats is higher than 36, it leads to the production of an abnormally long polyglutamine tract in the huntingtin protein. This confers a dominant deleterious gain of function on the protein, causing neuronal damage and eventually loss in the striatum and cortical regions of the brain.

2.9.2 Complex Diseases That Involve Multiple Genes

Most common diseases, including heart disease, diabetes, hypertension, obesity, and Alzheimer's disease (AD), are caused by multiple genes. In the case of AD, while its familial or early-onset form can be attributed to one of three genes (*APP*, *PSEN1*, and *PSEN2*), the most common form, sporadic

AD, involves a large number of genes [15]. In this type of complex disease, the contribution of each gene is modest, and it is the combined effects of mutations in these genes that predispose an individual to these diseases. Besides genetic factors, lifestyle and environmental factors often also play a role. For example, history of head trauma, lack of mentally stimulating activities, and high cholesterol levels are all risk factors for developing AD. Because of the number of genes involved and their interactions with non-genetic factors, complex multi-gene diseases are more challenging to study than single-gene diseases.

2.9.3 Diseases Caused by Genome Instability

Aside from the gene-centered disease models introduced above, diseases can also occur as consequences of large-scale genomic changes such as rearrangement of large genomic regions, alterations of chromosome number, and general genome instability. For example, when a genome becomes unstable in an organism, it can cause congenital developmental defects, tumorigenesis, premature aging, etc. Dysfunction in genome maintenance, such as DNA repair and chromosome segregation, can lead to genome instability. Fanconi anemia is an example of a disease caused by genome instability, characterized by growth retardation, congenital malformation, bone marrow failure, high cancer risk, and premature aging. The genome instability in this disease is caused by mutations in a cluster of DNA repair genes, and manifested by increased mutation rates, cell cycle disturbance, chromosomal breakage, and extreme sensitivity to reactive oxygen species and other DNA damaging agents.

Cancer, to a large degree, is also caused by genome instability. This can be hinted by the fact that two well-known high-risk cancer genes, *BRCA1* and *BRCA2*, are both DNA damage repair genes. Mutations in the two genes greatly increase the susceptibility to tumorigenesis, such as breast and ovarian cancers. In general, many cancers are characterized by chromosomal aberrations and genome structural changes, involving deletion, duplication, and rearrangement of large genomic regions. The fact that genome instability is intimately related to major aspects of cancer cells such as cell cycle regulation and DNA damage repair also points to the important role of genome instability in cancer development.

2.9.4 Epigenomic/Epigenetic Diseases

Besides gene mutations and genome instability, abnormal epigenomic/epigenetic pattern can also lead to diseases. Examples of diseases in this category include fragile X syndrome, ICF (immunodeficiency, centromeric instability and facial anomalies) syndrome, Rett syndrome, and Rubinstein-Taybi syndrome. In ICF syndrome, for example, the gene *DNMT3B* is mutated leading to the deficiency of DNA methyltransferase 3B. As a result, patients afflicted with this disease invariably have DNA hypomethylation. Cancer, as a genome disease that is caused by more than one genetic/genomic factor, is also characterized by abnormal DNA methylation, including both hypermethylation and hypomethylation. The hypermethylation is commonly observed in the promoter CpG islands of tumor suppressor genes [16], which leads to their suppressed transcription. The hypomethylation is mostly located in highly repetitive sequences, including tandem repeats in the centromere and interspersed repeats. This lowered DNA methylation has been suggested to play a role in promoting chromosomal relaxation and genome instability [17].

References

1. Fraser CM, Gocayne JD, White O, Adams MD, Clayton RA, Fleischmann RD, Bult CJ, Kerlavage AR, Sutton G, Kelley JM *et al.* The minimal gene complement of *Mycoplasma genitalium*. *Science* 1995, 270(5235):397–403.
2. Hutchison CA, 3rd, Chuang RY, Noskov VN, Assad-Garcia N, Deerinck TJ, Ellisman MH, Gill J, Kannan K, Karas BJ, Ma L *et al.* Design and synthesis of a minimal bacterial genome. *Science* 2016, 351(6280):aad6253.
3. Bennett GM, Moran NA. Small, smaller, smallest: the origins and evolution of ancient dual symbioses in a Phloem-feeding insect. *Genome Biol Evol* 2013, 5(9):1675–1688.
4. Pellicer J, Fay MF, Leitch IJ. The largest eukaryotic genome of them all? *Bot J Linn Soc* 2010, 164(1):10–15.
5. Shapiro JA, von Sternberg R. Why repetitive DNA is essential to genome function. *Biol Rev Camb Philos Soc* 2005, 80(2):227–250.
6. Roach JC, Glusman G, Smit AF, Huff CD, Hubley R, Shannon PT, Rowen L, Pant KP, Goodman N, Bamshad M *et al.* Analysis of genetic inheritance in a family quartet by whole-genome sequencing. *Science* 2010, 328(5978):636–639.
7. Mahmoud M, Gobet N, Cruz-Davalos DI, Mounier N, Dessimoz C, Sedlazeck FJ. Structural variant calling: the long and the short of it. *Genome Biol* 2019, 20(1):246.
8. Ebert P, Audano PA, Zhu Q, Rodriguez-Martin B, Porubsky D, Bonder MJ, Sulovari A, Ebler J, Zhou W, Serra Mari R *et al.* Haplotype-resolved diverse human genomes and integrated analysis of structural variation. *Science* 2021, 372(6537):eabf7117.
9. Malnic B, Godfrey PA, Buck LB. The human olfactory receptor gene family. *Proc Natl Acad Sci U S A* 2004, 101(8):2584–2589.
10. Inai Y, Ohta Y, Nishikimi M. The whole structure of the human nonfunctional L-gulono-gamma-lactone oxidase gene—the gene responsible for scurvy—and the evolution of repetitive sequences thereon. *J Nutr Sci Vitaminol* 2003, 49(5):315–319.

11. Law JA, Jacobsen SE. Establishing, maintaining and modifying DNA methylation patterns in plants and animals. *Nat Rev Genet* 2010, 11(3):204–220.
12. Cedar H, Bergman Y. Linking DNA methylation and histone modification: patterns and paradigms. *Nat Rev Genet* 2009, 10(5):295–304.
13. Guo W, Chung WY, Qian M, Pellegrini M, Zhang MQ. Characterizing the strand-specific distribution of non-CpG methylation in human pluripotent cells. *Nucleic Acids Res* 2014, 42(5):3009–3016.
14. Wu H, Zhang Y. Reversing DNA methylation: mechanisms, genomics, and biological functions. *Cell* 2014, 156(1–2):45–68.
15. Shademan B, Biray Avci C, Nikanfar M, Nourazarian A. Application of next-generation sequencing in neurodegenerative diseases: opportunities and challenges. *Neuromolecular Med* 2021, 23(2):225–235.
16. Nishiyama A, Nakanishi M. Navigating the DNA methylation landscape of cancer. *Trends Genet* 2021, 37(11):1012–1027.
17. Pappalardo XG, Barra V. Losing DNA methylation at repetitive elements and breaking bad. *Epigenetics Chromatin* 2021, 14(1):25.

3

RNA: The Transcribed Sequence

3.1 RNA as the Messenger

The blueprint of life is written in DNA, but almost all life processes are executed by proteins. To convert the information coded in the DNA into the wide array of proteins in each cell, segments of DNA sequence in the genome must be copied into mRNAs first. The transcribed nucleotide sequences in the mRNAs are then translated into proteins through an information decoding process carried out by ribosomes. Because of the intermediary role played by mRNAs between DNA and proteins, the composition of mRNAs in a cell or population of cells – the transcriptome – is often used to study cellular processes and functions. Unlike the genome, which is static and the same for every cell in an organism, the transcriptome is dynamically regulated and therefore can be used as a proxy of cellular functional status.

3.2 The Molecular Structure of RNA

Structurally RNA is closely related to DNA and also made of nucleotides. The nucleotides that make up the RNA molecule are slightly different from those of DNA. Instead of deoxyribose, its five-carbon sugar moiety is a ribose. Among the four nucleobases, uracil (U) is used in place of thymine (T), while the rest of three (A, C, and G) are the same. Unlike the double-stranded structure of DNA, RNA molecules are single-stranded, which gives them great flexibility. If intramolecular sequence complementarity exists between two regions of a single RNA molecule, this structural flexibility allows the regions to bend back on each other and form intramolecular interactions.

As a result of its structural flexibility and internal sequence complementarity, an RNA molecule can assume secondary structures, such as hairpins and stem-loops, and tertiary structures depending on its specific

DOI: 10.1201/9780429329180-4

sequence. These structures can sometimes afford them special chemical properties in cells. For example, some non-messenger RNAs can catalyze chemical reactions like protein enzymes, and are therefore called RNA enzymes (or ribozymes, more details later in Section 3.4.1). Some RNA molecules may assume tertiary structures that enable them to bind to other small molecules such as ligands, or large molecules such as RNA-binding proteins. For mRNAs, their structures may also be important for various steps of their life cycle (see next section for details). One example of this is riboswitch, a region in some mRNAs that binds to small molecule ligands such as metabolites or ions, and thereby regulates their transcription, translation, or splicing via changes in RNA structure upon ligand binding [1]. Binding of proteins to mRNA elements like those located in the 3′ untranslated region (UTR) can also induce structural changes of these elements and affect mRNA translation [2]. Transport of mRNAs to specific cellular locations, such as distal dendritic regions of a neuron, also requires the mRNAs to assume specific structures for RNA-binding proteins to bind as a prerequisite of the transport process. To study structures of individual RNAs, computational prediction and experimental approaches such as RNA fingerprinting that uses a variety of chemical and enzymatic probes, have been the classic methods. With the advent of NGS-based RNA sequencing, transcriptome-wide RNA structural mapping is enabled when integrated with these classic approaches [3].

3.3 Generation, Processing, and Turnover of RNA as a Messenger

When a protein is needed in a cell, its coding gene is first transcribed to mRNA, which is then used as the template to translate into the requisite protein. In a prokaryotic cell, mRNA transcription is immediately followed by protein translation. In a eukaryotic cell, the information flow from DNA to protein through mRNA is more complex, because the two steps of transcription and translation are physically separated and eukaryotic genes contain introns that need to be first removed before translation. In the eukaryotic system, initial transcript (also called primary transcript) is first synthesized from DNA template and then processed, including intron removal, to produce mature mRNA in the nucleus. Then the mRNA is transported from the nucleus to the cytoplasm for translation. When they are no longer needed, the mRNAs are degraded and recycled by the cell. It should also be noted that the transcription process generates a number of mRNA copies from a gene, and the number of copies varies from condition to condition and from gene to gene depending on cellular functional status.

```
DNA
5'... A T G A G A A C G T T A G G C ...3' Sense Strand
3'... T A C T C T T G C A A T C C G ...5' Antisense Strand

                  | Transcription
                  v
mRNA
5'... A U G A G A A C G U U A G G C ...3'

                  | Translation
                  v
Peptide
Met - Arg - Thr - Leu - Gly - ...
```

FIGURE 3.1
How the two strands of DNA template match the transcribed mRNA in sequence, and the genetic code in mRNA sequence corresponds to peptide amino acid sequence.

3.3.1 DNA Template

To initiate transcription, a gene's DNA sequence is first exposed through altering its packing state. In order to transcribe the DNA sequence, the two DNA strands in the region are first unwound and only one strand is used as the template strand for transcription. Since it is complementary to the RNA transcript in base pairing (A, C, G, and T in the DNA template are transcribed to U, G, C, and A, respectively, in the RNA transcript), this DNA template strand is also called the antisense or negative (–) strand (Figure 3.1). The other DNA strand has the same sequence as the mRNA (except with T's in DNA being replaced with U's in RNA) and is called the coding strand, sense, or positive (+) strand. It should be noted that either strand of the genomic DNA can be potentially used as the template, and which strand is used as the template for a gene depends on the orientation of the gene along the DNA. It should also be noted that the triplet nucleotide genetic code that determines how amino acids are assembled in proteins refers to the triplet sequence in the mRNA sequence.

3.3.2 Transcription of Prokaryotic Genes

RNA polymerase catalyzes the transcription of RNA from its DNA template. In prokaryotic cells, there is only one type of RNA polymerase. The prokaryotic RNA polymerase holoenzyme contains a core enzyme of five subunits that catalyzes RNA transcription from a DNA template, and another subunit called the sigma factor that is required for initiation of transcription. The sigma factor initiates the process by enabling binding of the core enzyme to the promoter region and guiding it to the transcription start site (TSS). Promoter is the region upstream of the protein-coding sequence of a gene or an operon. Prokaryotic promoters share some core sequence elements, such as the motif centered at 10 nucleotides upstream of TSS with the consensus sequence TATAAT. Once reaching TSS, the sigma factor disassociates from the core enzyme. The core RNA polymerase, unlike DNA polymerase, does not need a primer, but otherwise the enzyme catalyzes the attachment of nucleotides to the nascent RNA molecule one at a time in the 5′→3′ direction. At a speed of approximately 30 nucleotides/second, the RNA polymerase slides through the DNA template carrying the elongating RNA molecule.

Although the attachment of new nucleotides to the elongating RNA is based on base pairing with the DNA template, the new elongating RNA does not remain associated with the template DNA via hydrogen bonding. On the same template multiple copies of RNA transcripts can be simultaneously synthesized by multiple RNA polymerases one after another. During transcript elongation, these polymerases hold onto the template tightly and do not disassociate from the template until stop signal is transcribed. The stop signal is provided by a segment of palindromic sequence located at the end of the transcribed sequence. Right after transcription, the inherent self-complementarity in the palindromic sequence leads to the spontaneous formation of a hairpin structure. Additional stop signal is also provided by a string of four or more uracil residues after the hairpin structure, which forms weak associations with the complementary A's on the DNA template. The hairpin structure pauses further elongation of the transcript, and the weak associations between the U's on the RNA and the A's on the DNA dissociate the enzyme and the transcript from the template.

Regulation of prokaryotic transcription is conferred by promoters and protein factors such as repressors and activators. Promoter strength, that is, the number of transcription events initiated per unit time, varies widely in different operons. For example, in *E. coli*, genes in operons with weak promoters can be transcribed once in 10 minutes, while those with strong promoters can be transcribed 300 times in the same amount of time. The strength of an operon's promoter is based on the host cell's demand for its protein products, and dictated by its sequence. Specific protein factors may also regulate gene transcription. Repressors, the best known among these factors, prevents RNA polymerase from initiating transcription through binding to an intervening sequence between promoter and TSS called operator. Activators exert an opposite effect and induce higher levels of transcription. The sigma factor, being the initiation factor of the prokaryotic RNA polymerase, provides another mechanism for regulation. There are different forms of this factor in prokaryotic cells, each of which mediates sequence specific transcription. Differential use of these sigma factors, therefore, provides another level of transcriptional regulation in prokaryotic cells.

3.3.3 Pre-mRNA Transcription of Eukaryotic Genes

In eukaryotic cells there are three types of RNA polymerases, among which RNA polymerase II transcribes protein-coding genes, while RNA polymerases I and III transcribe rRNA, tRNA, and various types of small

RNAs. Transcription in eukaryotic cells is in general much more sophisticated, because of the highly compressed packaging of chromosomal DNA, the complex structure of eukaryotic genes, and intricate regulation by multiple factors. Prior to transcription, the highly compressed DNA in the chromatin needs to be uncompressed and gene sequence exposed for access by RNA polymerase.

To perform the transcription of protein-coding genes, besides RNA polymerase II, a variety of other proteins in the nucleus are also required, including transcription factors and coactivators. Transcription factors include general and specific transcription factors. General transcription factors, such as TFIIA, TFIIB, and TFIID, are required in all transcription initiation. Their functions are to position the RNA polymerase at the promoter region and unwind the template DNA strands for transcription. Specific transcription factors, which are detailed next, provide key regulatory function to the transcription initiation process. Coactivators bring together all requisite transcription factors to form the transcription initiation complex. Once transcription is initiated, most of the protein factors in the complex are released and the RNA elongation process is carried out by RNA polymerase II in a manner similar to that occurs in prokaryotic cells. The termination of the elongation process in eukaryotic cells is provided by the signal sequence AAUAAA, which also serves as the signal for cleavage of the transcribed RNA to generate the 3′ end and for polyadenylation (see Section 3.3.4). After completion of the transcription process, the transcript contains both exons and introns and is called the primary transcript or pre-mRNA.

During RNA transcript elongation in both the eukaryotic and prokaryotic systems, like in DNA replication by the DNA polymerase, there is a certain probability of introducing mismatched nucleotides and therefore errors. For proofreading, the prokaryotic and eukaryotic RNA polymerases have 3′→5′ exonuclease activity. If a wrong nucleotide is added to the elongating RNA chain, the RNA polymerase will backtrack and correct the error. Because of this activity, the overall error rate of the transcriptional process in both systems is estimated to be 10^{-4}-10^{-5} per base [4]. Although this is higher than the DNA mutation rate, the transcriptional errors are seldom harmful, because there are multiple copies for each transcript, and transcripts carrying premature stop codons are quickly removed by a process called nonsense-mediated decay.

Besides the step of gene sequence exposure through histone modifications and chromatin remodeling, the eukaryotic gene transcription process is mostly regulated at the initiation step through the use of specific transcription factors. As a large group of DNA-interacting proteins (Chapter 2), these transcription factors bind to specific sequence elements in the promoter region of genes, through which they help assemble general transcription factors and the RNA polymerase into the transcription initiation complex. In addition, specific transcription factors may also bind to specific regulatory sequences at distant locations that are called enhancers or *cis*-regulatory modules. Different from transcription factor binding sites in the promoter regions, enhancers function independent of sequence orientation and from a distance as far as megabases away from the regulated gene, and are sometimes embedded in intergenic regions that otherwise have no known function. Having a significant effect on gene transcription, enhancers exert their regulatory function by DNA looping, which brings enhancer and promoter sequences together affecting formation of the transcription initiation complex. The binding of specific transcription factors to enhancers can have a stimulatory, or inhibitory (through the recruitment of repressors), effect on gene transcription. In general, the transcription of a gene is often regulated by multiple specific transcription factors, and the combined signal input from these transcription factors determines whether the gene will be transcribed, and if yes, at what level. A particular transcription factor can also bind to multiple genomic sites, coordinating the transcription of functionally related genes. NGS-based approaches, such as ChIP-seq (Chapter 13), are often used to locate the binding sites of specific transcription factors across the genome.

3.3.4 Maturation of mRNA

In prokaryotic cells, there is no post-transcription RNA processing, and transcripts are immediately ready for protein translation after transcription. In fact, while mRNAs are still being transcribed, ribosomes already bind to the transcribed portions of the elongating mRNAs synthesizing peptides. In eukaryotic cells, however, primary transcripts undergo several steps of processing in the nucleus to become mature mRNAs. These steps are (1) capping at the 5′ end, (2) splicing of exons and introns, and (3) addition of a poly-A tail at the 3′ end.

The first step, adding a methylated guanosine triphosphate cap to the 5′ end of nascent pre-mRNAs, takes place shortly after the initiation of transcription when the RNA chains are still less than 30 nucleotides long. This step is carried out by adding a guanine group to the 5′ end of the transcripts, followed by methylation of the group. This cap structure marks the transcripts for subsequent transport to the cytoplasm, protects them from degradation, and promotes efficient initiation of protein translation. Once formed, the cap is bound by a protein complex called cap-binding complex.

The second step, splicing of exons and introns, is the most complicated among the three steps. As introns are non-coding intervening sequences, they need to be spliced out while exons are retained to generate mature mRNAs. The molecular machinery that carries out the splicing, called the spliceosome, is assembled from as many as 300 proteins and 5 small nuclear RNAs (snRNAs). The spliceosome identifies and removes introns from primary transcripts, using three positions within each intron: the 5′ end (starts with the consensus sequence 5′-GU, serving as the splice donor), the 3′ end (ends with the consensus sequence AG-3′, as the splice acceptor), and the branch point, which starts around 30 nucleotides upstream of the splicing acceptor and contains an AU-rich region. The actual excision of each intron and the concomitant joining of the two neighboring exons are a three-step process: (1) cleavage at the 5′ end splice donor site; (2) attachment of the cleaved splice donor site to the branch point to form a lariat or loop structure; and (3) cleavage at the 3′ end splicing acceptor site to release the intron and join the two exons.

Beyond simply removing introns from primary transcripts, the splicing process also employs differential use of exons, and sometimes even includes some introns, to create multiple mature mRNA forms from the same primary transcript. This differential splicing, also called alternative splicing (Figure 3.2), provides an additional regulatory step in the production of mRNA populations. When it was first reported in 1980, alternative splicing was considered to be an exception rather than the norm. It is now well established that primary transcripts from essentially all multi-exon genes are alternatively spliced [5, 6]. The biological significance of alternative splicing is obvious: by enabling production of multiple mRNAs and thereby proteins from the same gene, it greatly augments protein and consequently functional diversity in an organism without significantly increasing the number of genes in the genome, and offers explanation to the question why more evolved organisms do not contain many more genes in their genomes (Chapter 2, Table 2.1).

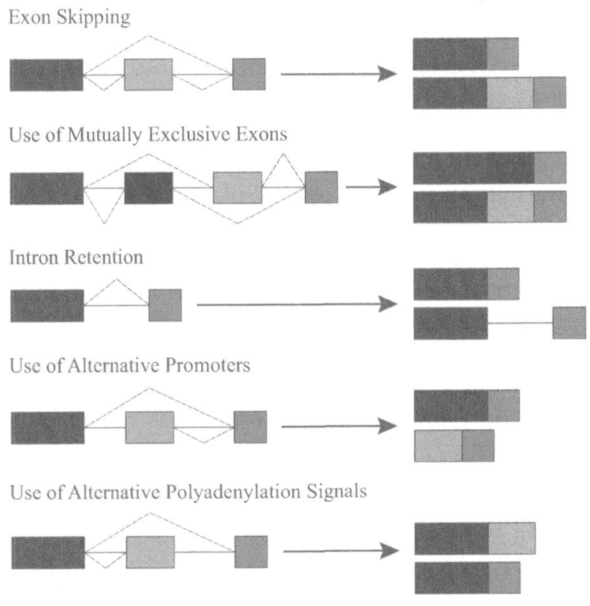

FIGURE 3.2
Varying forms of RNA transcript splicing.

In the third step, once the new primary transcript passes the termination signal sequence, it is bound by several termination-related proteins. One of the proteins cleaves the RNA at a short distance downstream of the termination signal to generate the 3′ end. This is followed by a polyadenylation step that adds 50–200 A's to the 3′ end by an enzyme called poly-A polymerase. This poly-A tail, like the 5′ end cap, increases the stability of the resulting mRNA. This tail is bound and protected by poly(A)-binding protein, which also promotes its transport to the cytoplasm.

Besides these three major constitutive processing steps, some transcripts may undergo additional processing steps. RNA editing, although considered to be rare, is among the best known of these steps. RNA editing refers to the change in RNA nucleotide sequence after it is transcribed. The most common types of RNA editing are conversions from A to I (inosine, read as G during translation), which are catalyzed by enzymes such as ADARs (adenosine deaminases that act on RNA), or from C to U, catalyzed by cytidine deaminases. As a result of these conversions, an edited RNA transcript no longer fully matches the sequence on the template DNA. RNA editing has the potential to change genetic codons, introduce new or remove existing stop codons, or alter splicing sites [7]. Evidence shows that RNA editing and other RNA processing events such as splicing can be coordinated [8].

3.3.5 Transport and Localization

After maturation, mRNAs need to be exported out of the nucleus to the cytoplasm for protein translation. While allowing mature mRNAs to be transported out, the nucleus keeps unprocessed or partially processed transcripts, as well as processed side products like removed introns, inside the nucleus. To move across the nucleus envelope through the nuclear pore complexes, mature mRNAs are packaged into large ribonucleoprotein (RNP) complexes. Once in the cytoplasm, many mRNA species can be used to start synthesizing proteins right away. As the cytoplasm is a crowded place, they may randomly drift in the cytoplasm while translating. Some translations, however, take place at highly localized sites. For example, in neurons some mRNAs are required to be transported to distal dendritic regions for translation. Local protein translation at such target sites has been known to have important biological functions, such as synaptic plasticity that underlies learning and memory [9]. In order to transfer mRNAs to these special locations, the mRNAs bind to special proteins to form mRNA-protein complexes, which are then attached to protein motors to move along cytoskeletal tracks.

3.3.6 Stability and Decay

Steady-state mRNA concentrations, which are the detection target of transcriptomic analysis such as RNA-seq, are determined by not only rates of mRNA production, but also their decay. In general, prokaryotic mRNAs are unstable and quickly degraded by endoribonucleases and exoribonucleases after transcription. As a result, most of them are short-lived and the average prokaryotic mRNA half-life, that is, the amount of time to have half of the mRNAs degraded, is under 10 minutes [10]. This high turnover rate allows prokaryotic cells to respond quickly to environmental changes by altering transcription. In comparison, eukaryotic mRNAs are in general more stable and have a longer average half-life of 7–10 hours [11, 12]. As a general rule, mRNAs for regulatory or inducible proteins, such as transcription factors or stress response proteins, tend to have shorter half-lives (e.g., less than 30 minutes), while those for housekeeping proteins, such as those of metabolism and cellular structure, have long half-lives (e.g., days). The stability and half-lives of mRNAs are also regulatable based on developmental stage or environmental factors. For example, the stability and half-lives of mRNAs of muscle-specific transcription factors, such as myogenin and myoD, are the highest during muscle differentiation, but quickly decline once the differentiation is completed [13].

The regulation of eukaryotic mRNA degradation is not well understood, but has been known to involve interactions between some sequence elements on mRNAs and protein as well as small RNA factors. One example of the mRNA stability regulatory sequences is the AU-rich element, a region on the 3′ untranslated portion of many short-lived mRNAs that, as the name suggests, is rich in adenines and uridines. Many protein factors interact with this element to modulate mRNA turnover, such as the AU-rich binding factor 1 (or AUF1). Small RNAs, including miRNA, siRNA, and piRNA, are also important regulators of mRNA stability and degradation (see Section 3.4.4 for details). P-bodies (for processing bodies), granular structures in eukaryotic cells, are the focal point of mRNA turnover mediated by protein and small RNA factors [14].

Most eukaryotic mRNA decay starts with deadenylation at the 3′ end, that is, removal of the poly-A tail by deadenylases. The deadenylation then leads to mRNA degradation through two alternative mechanisms. One mechanism is through decapping of mRNA at the 5′ end, which leaves the mRNA vulnerable to degradation by 5′→3′ exoribonuclease. The other mechanism is direct 3′→5′ decay from the tail end, which is carried out by a multi-protein complex called the exosome complex. Besides these major deadenylation-dependent mRNA decay pathways, there are also other pathways that do not rely on deadenylation, such as those dependent on 3′-uridylation which adds a uridine-rich tail to the 3′ end [15].

3.3.7 Major Steps of mRNA Transcript Level Regulation

As indicated above, the regulation of both prokaryotic and eukaryotic transcription is mostly applied at the initiation step, and this regulation is heavily dependent on specific protein-DNA interactions. In the prokaryotic system, besides promoter strength, the regulation of transcriptional initiation is provided by protein factors including repressors and activators, both of

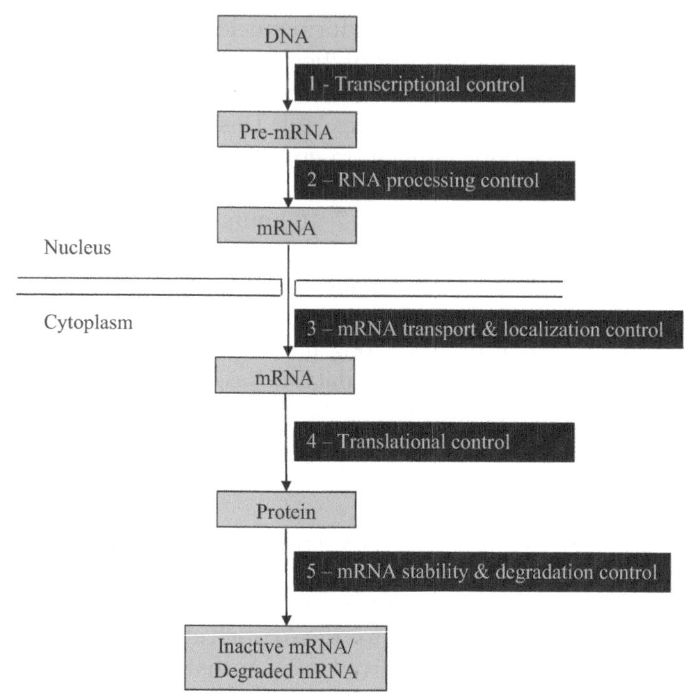

FIGURE 3.3
The regulation of eukaryotic gene expression at multiple levels.

which bind to specific promoter sequences. In the eukaryotic system, specific transcription factors that bind to specific sequences in promoters and/or enhancers offer most of the regulation. In addition, prior to the engagement of transcription initiation complex, gene sequence access is regulated through histone modification and chromatin remodeling. Since the generation of mRNA in the eukaryotic system is a multi-step process, regulatory mechanisms are also applied at subsequent steps (Figure 3.3). During mRNA maturation, regulation of exon and intron splicing leads to generation of alternative splicing variants. Trafficking of mRNAs to localized cellular domains provides additional regulatory mechanism for some genes [16]. Equally important in determining steady-state mRNA levels, mRNA decay is another important but less studied step upon which regulation is also exerted.

3.4 RNA Is More Than a Messenger

Despite its apparent indispensability, mRNAs constitute only about 5% of total cellular RNA. Besides rRNAs and tRNAs, there are also a large number of non-protein-coding RNAs that play important roles in regulating protein-coding genes or carry out essential cellular functions. These non-coding RNAs include microRNAs (or miRNAs), PIWI-interacting RNAs (piRNAs), small interfering RNA (siRNAs), snRNAs, small nucleolar RNAs (snoRNAs), long-non-coding RNAs (lncRNAs), and RNAs that function as catalysts (ribozymes). Some of these non-coding RNAs have been extensively studied, such as the ribozymes, the discovery of which won the 1989 Nobel Prize in Chemistry and has led to the "RNA world" hypothesis. Based on this hypothesis, early life forms were solely based on RNA, whereas DNA and protein evolved later. The rRNAs, tRNAs, and ribozymes are thought by this hypothesis to be evolutionary remnants of the original RNA world [17]. The functional importance and diversity of other non-coding RNAs, such as the many forms of small RNAs and lncRNAs, are still in the process of being fully appreciated because they were discovered more recently. However, because of their wide presence and importance, the 2006 Nobel Prize in Physiology or Medicine was awarded to the discovery of RNA interference (RNAi) by small RNAs. Due to the diverse and important roles that non-coding RNAs play in cells, RNA is not treated as simply a messenger anymore.

3.4.1 Ribozyme

Similar to proteins, RNAs can form complicated three-dimensional structures, and some RNA molecules carry out catalytic functions. These catalytic RNAs are called ribozymes. A classic example of ribozyme is one type of intron called group I intron, which splices itself out of the pre-mRNA that contains it. This self-splicing process, involving two transesterification steps, is not catalyzed by any protein. Group I intron is about 400 nucleotides in length and mostly found in organelles, bacteria, and the nucleus of lower eukaryotes. When a precursor RNA that contains group I intron is incubated in a test tube, the intron splices itself out of the precursor RNA autonomously. Despite variations in their internal sequences, all group I introns share a characteristic spatial structure, which provides active sites for catalyzing the two steps. Another example of ribozyme is the 23S rRNA contained in the large subunit of the prokaryotic ribosome. This rRNA catalyzes the peptide bond formation between an incoming amino acid and the existing peptide chain. Although the large subunit contains over 30 proteins, rRNA is the catalytic component while the proteins only provide structural support and stabilization [18].

Also similar to protein catalysts, the dynamics of the reactions catalyzed by ribozymes follows the same characteristics as those of protein enzyme-catalyzed reactions, which are usually described by the Michaelis–Menten equation. Further similarities of ribozymes to protein enzymes include that ribozyme activity can also be regulated by ligands, usually small molecules, the binding of which leads to structural change in the ribozyme. For instance, a ribozyme may contain a riboswitch, which as part of the ribozyme can bind to a ligand to turn on or off the ribozyme activity.

3.4.2 snRNA and snoRNA

Although group I intron can self-splice, most pre-mRNA introns are not of this type and need the spliceosome for splicing. The spliceosome, even larger than the ribosome in size, contains five snRNAs (U1, U2, U4, U5, and U6), and a large number of proteins. The splicing process heavily depends on the interactions between these snRNAs and pre-mRNAs. For example, to initiate splicing, U1 interacts with the 5′ splice donor site, and for catalysis of the splicing process U2/U5/U6 comprise the active site core. High-resolution structural data have even shown that the spliceosome, like the ribosome, is also a ribozyme [19].

Similarly, snoRNAs are indispensable for pre-rRNA processing. The eukaryotic ribosome contains four rRNAs – 28S, 18S, 5.8S, and 5S – with the first three initially transcribed into a single large rRNA precursor. To generate the three rRNAs, the precursor rRNA needs to be first chemically modified and then cleaved. The chemical modification includes methylation at over 100 nucleotides, and isomerization of uridine at another 100 sites. The snoRNAs are required in this process to identify the specific sites for modification. There are many different types of snoRNAs, each of which can form a complementary region with the precursor rRNA via base pairing. These duplex regions are then recognized as targets for modification.

3.4.3 RNA for Telomere Replication

Located at the tips of a chromosome, telomeres seal the ends of chromosomal DNA. Without telomeres, the integrity of chromosomes would be compromised since DNA repair enzymes would recognize the DNA termini as break points. Inside each telomeric region is a long string of highly repetitive DNA sequences. Normally shortening of telomere length occurs with each chromosome replication, since chromosomal DNA duplication enzymes cannot reach to the very ends of the DNA (the end replication problem). To prevent this problem in germ cells and stem cells, an enzyme called telomerase is responsible for replenishing the telomeric region. The telomerase is a large complex comprised of an RNA component, which serves as a template for the repeat sequence, and a catalytic protein component (reverse transcriptase), which uses the RNA template to synthesize the repetitive telomeric DNA sequence. The telomerase activity is usually turned off or at very low levels in most somatic cells. Therefore, these cells can only divide a certain number of times before reaching senescence due to the gradual shortening of the telomere.

3.4.4 RNAi and Small Non-Coding RNAs

RNAi, as a cellular mechanism that uses small RNAs to silence gene expression, offers an excellent illustration of the significance of non-coding RNAs in the regulation of protein-coding genes. RNAi achieves gene silencing by suppressing mRNA translation, degrading mRNAs, or inhibiting gene transcription [20]. As a native gene regulation mechanism in a wide-range of organisms, RNAi plays an essential role in organismal development and various cellular processes. As viral RNA can activate the RNAi pathway in host cells leading to degradation of the viral RNA, RNAi is also used by plants and some animals to fight viral infections. Furthermore, RNAi can also silence mobile elements in the genome, such as transposons, to maintain genome stability. Currently large amounts of data have established the pervasiveness of small RNA mediated RNAi in many organisms. For example, in the human genome, over 60% of genes are regulated by small non-coding RNAs [21]. Since its discovery, RNAi has been applied as a powerful research tool to silence virtually any gene in the genome in order to decipher their functions. Clinically, small RNAs have been tested as a strategy for gene therapy through turning off faulty genes that underlie many genetic diseases.

RNAi is mediated by three principal groups of small non-coding RNAs: miRNA, siRNA, and piRNA. All these small RNAs induce RNAi through the same basic pathway that involves a ribonucleoprotein complex called the RNA-induced silencing complex (RISC). Below is a more detailed introduction to these three groups of small RNAs and their differences.

3.4.4.1 miRNA

Mature miRNA, at around 22 nucleotides in size, induces gene silencing through mRNA translational repression or decay. The precursor of miRNA is usually transcribed from non-protein-coding genes in the genome (Figure 3.4). The primary transcript, called pri-miRNA, contains internal hairpin structure and is much longer than mature miRNA. For initial processing, the pri-miRNA is first trimmed in the nucleus by a ribonuclease called Drosha that exists as part of a protein complex called the microprocessor, to an intermediate molecule called pre-miRNA, about 70-nucleotide in size. Alternatively, some miRNA precursors originate from introns spliced out from protein-coding transcripts. These precursors, to be processed for the generation of mirtrons (miRNAs derived from introns), bypass the microprocessor complex in the nucleus. For further processing, the pre-miRNA and the mirtron precursor are exported out of the nucleus into the cytoplasm, where they are cleaved by the endoribonuclease Dicer to form double-stranded miRNA. The double-stranded miRNA is subsequently loaded into RISC. Argonaute, the core protein component of RISC, unwinds the two miRNA strands and discard one of them [22]. The remaining strand is used by Argonaute as the guide sequence to identify related mRNA targets through imperfect base pairing

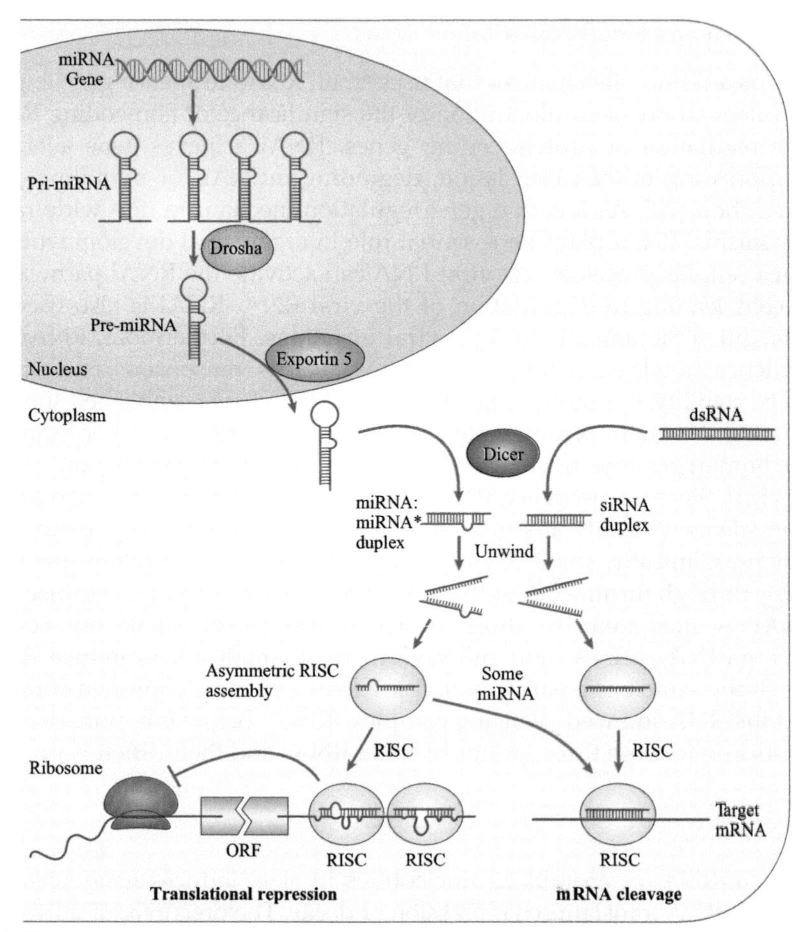

FIGURE 3.4
The generation and functioning of miRNA and siRNA in suppressing target mRNA activity. Genomic regions that code for miRNAs are first transcribed into pri-miRNAs, which are processed into smaller pre-miRNAs in the nucleus by Drosha. The pre-miRNAs are then transported by exportin 5 into the cytoplasm, where they are further reduced to miRNA:miRNA* duplex by Dicer. While both strands of the duplex can be functional, only one strand is assembled into the RNA-induced silencing complex (RISC), which induces translational repression or cleavage of target mRNAs. Long double-stranded RNA can also be processed by Dicer to generate siRNA duplex, which also uses RISC to break down target mRNA molecules. (Adapted by permission from Macmillan Publishers Ltd: *Nature Reviews Genetics*, He L. and Hannon G.J. (2004) MicroRNAs: small RNAs with a big role in gene regulation. *Nature Reviews Genetics* 5, 522–531, ©2004.)

with seed sequence usually located in the 3′-UTR of mRNAs. Through this miRNA-mRNA interaction, RISC induces silencing of target genes through repressing translation of the mRNAs and/or their deadenylation and degradation. Because the base pairing is imperfect, one miRNA can target multiple target genes' mRNAs. Conversely, one mRNA may be targeted by multiple miRNAs.

3.4.4.2 siRNA

While being similar in size and using basically the same system for gene silencing (Figure 3.4), siRNA differs from miRNA in a number of aspects. On origin, siRNA is usually exogenously introduced, such as from viral invasion or artificial injection. But they can also be generated endogenously, e.g., from repeat-sequence-generated transcripts (such as those from telomeres or transposons), or RNAs synthesized from convergent transcription (in which both strands of a DNA sequence are transcribed from the two opposite orientations with corresponding promoters), or other naturally occurring sense-antisense transcript pairs [23]. To generate mature siRNA, exogenously introduced double-stranded RNA, or endogenously transcribed precursor that is transported from the nucleus to the cytoplasm, is cleaved by Dicer. The mature siRNA is then loaded into RISC for silencing target mRNAs by Argonaute. On target mRNA identification, siRNA differs from mRNA in that it has perfect or nearly perfect sequence complementarity with their target. On the mechanism of gene silencing, siRNA usually leads to endonucleolytic cleavage, also called slicing, of the mRNAs.

3.4.4.3 piRNA

As a relatively newer class of small non-coding RNA, piRNAs are between 23 and 31 nucleotides in length, and have functions mostly found in animal germline tissues as a defense mechanism against transposons (or transposable elements, "selfish" DNA elements that have the capability to move around in the genome). While functioning with a similar basic RNAi mechanism, piRNA is different from miRNA and siRNA in two major aspects. One is that its biogenesis does not involve Dicer, and the other is that, for target gene silencing, it specifically interacts with PIWI proteins, a different clade in the Argonaute protein family. The biogenesis of piRNA, independent Dicer activity, starts from transcription of long RNAs from specific loci of the genome called piRNA clusters. With regard to these clusters, it has been found that while their locations in the genome do not show much change in related species, their sequences are not conserved even in closely related species, indicating that they are derived from invading transposable elements serving as an adaptive genome immunity mechanism. After transcription, the long piRNA precursor is transported out of the nucleus to the cytoplasm for processing into mature piRNA. To induce target gene silencing, mature piRNA is loaded into RISC that contains PIWI, which uses the piRNA sequence as guide to silence target mRNAs by slicing. Besides this post-transcriptional silencing, piRNA-loaded mature RISC can also be transported into the nucleus, where it finds and silences target mRNAs that are still in the process of being transcribed. This transcription-level gene silencing is achieved through interactions with other protein factors in the nucleus, and histone modification that alters chromatin structure and gene access. The currently best-known function of piRNAs, through post-transcriptional and transcriptional gene silencing, is to repress transposon activity and thereby maintain genome stability in germline cells. Transposon mobilization in the absence of this repression can lead to DNA damage disrupting germ line development. Non-transposon gene targets of piRNAs have also been reported such as those related to development.

3.4.5 Long Non-Coding RNAs

Some non-coding RNAs, unlike the small RNAs, are rather long with an average length of over 200 nucleotides in their mature form. These RNAs, called long non-coding RNAs, have been discovered more recently and therefore less studied. The biogenesis of lncRNAs is somewhat similar to that of mRNAs, as many of them are transcribed by RNA polymerase II and subject to splicing, capping at the 5′ end, and polyadenylation at the 3′ end. Unlike mRNAs, however, they are usually shorter with a median length of ~600 nucleotides, have fewer exons, and are generally expressed at levels lower than those of mRNAs. Furthermore, their expression displays higher tissue and developmental stage specificity than mRNAs, and are mostly localized in the nucleus rather than transported to the cytoplasm.

Although they are relatively new, evidence on their importance in regulating fundamental cellular functions is rapidly accumulating. They have been known to control many steps of gene activity, including chromatin remodeling, transcriptional regulation, mRNA processing, stability, localization, and translation [24, 25]. For example, some lncRNAs, such as Xist and HOTAIR, repress gene transcription at target genomic sites through interacting with chromatin remodeling protein complexes [26, 27]. A class of lncRNAs that was discovered by NGS is transcribed from enhancer regions of protein-coding genes. These transcripts, called eRNAs (or enhancer RNAs), have been shown to affect transcription of protein-coding genes that are regulated by the enhancers [28]. In general, lncRNAs regulate gene activity via binding to transcription factors, repressing promoter activity, and interacting with mRNA-binding proteins and splicing factors. In addition, lncRNAs can directly interact with mRNAs and thereby influence their stability and translation [29, 30]. Because of their functional importance, it is not surprising that abnormal lncRNA expression can lead to diseases such as cancer and aging-related neurodegenerative disorders [31, 32].

3.4.6 Other Non-Coding RNAs

Deep sequencing of the cellular transcriptome has led to the discovery of other non-coding RNAs. For example, circular RNAs, or circRNAs, exist in many species and cell types. Unlike linear RNAs, which include all the RNA species introduced so far, circRNAs have their 5′ and 3′ ends joined forming a loop structure. This structure makes them less vulnerable to attacks from RNases and expectedly more stable. Because their widespread existence was only unveiled with the use of RNA-seq, the functions of most circRNAs are still being investigated. Among currently established functions are the roles they play in sequestering miRNA and RNA-binding proteins from their targets, and regulating transcription, splicing, and translation events [33]. Besides the major non-coding RNAs introduced in this chapter, there are also other classes of non-coding RNA species in cells that perform a remarkable array of functions [34]. It is highly possible that new classes of non-coding RNAs will continue to be discovered through RNA sequencing.

3.5 The Cellular Transcriptional Landscape

Traditionally protein-coding mRNA transcripts used to be the major targets of transcriptional studies and as a result were often mis-regarded as the major component of a transcriptome. However, with the evolution of transcriptomics technologies and as a result the discovery of the wide variety of non-coding RNAs, it has been gradually realized that protein-coding transcripts only constitute a minor fraction of a cell's transcribed sequences. Large-scale studies on the landscape of cellular transcription, as carried out by consortia including the FANTOM (Functional Annotation Of the Mammalian genome) and ENCODE (Encyclopedia of DNA Elements), have revealed that the majority of the genome is transcribed, and a large proportion of the transcriptome is non-coding RNAs [35, 36]. For example, after studying the transcriptional landscape of 15 human cell lines, encompassing RNA populations isolated from different cellular sub-compartments including the cytosol and the nucleus, the ENCODE consortium found that the transcription of the genome is pervasive and 75% of genomic sequences, including those located in gene-poor regions, are present in transcripts. Many of the transcripts come from intronic and intergenic regions that are not characterized currently and therefore novel. The complex cellular RNA landscape adds further evidence that RNA is not simply a messenger between DNA and protein.

References

1. Bedard AV, Hien EDM, Lafontaine DA. Riboswitch regulation mechanisms: RNA, metabolites and regulatory proteins. *Biochim Biophys Acta Gene Regul Mech* 2020, 1863(3):194501.

2. Ray PS, Jia J, Yao P, Majumder M, Hatzoglou M, Fox PL. A stress-responsive RNA switch regulates VEGFA expression. *Nature* 2009, 457(7231):915–919.
3. Xu B, Zhu Y, Cao C, Chen H, Jin Q, Li G, Ma J, Yang SL, Zhao J, Zhu J *et al.* Recent advances in RNA structurome. *Sci China Life Sci* 2022, 65(7):1285–1324.
4. Imashimizu M, Oshima T, Lubkowska L, Kashlev M. Direct assessment of transcription fidelity by high-resolution RNA sequencing. *Nucleic Acids Res* 2013, 41(19):9090–9104.
5. Wang ET, Sandberg R, Luo S, Khrebtukova I, Zhang L, Mayr C, Kingsmore SF, Schroth GP, Burge CB. Alternative isoform regulation in human tissue transcriptomes. *Nature* 2008, 456(7221):470–476.
6. Pan Q, Shai O, Lee LJ, Frey BJ, Blencowe BJ. Deep surveying of alternative splicing complexity in the human transcriptome by high-throughput sequencing. *Nat Genet* 2008, 40(12):1413–1415.
7. Keegan LP, Gallo A, O'Connell MA. The many roles of an RNA editor. *Nat Rev Genet* 2001, 2(11):869–878.
8. Bratt E, Ohman M. Coordination of editing and splicing of glutamate receptor pre-mRNA. *RNA* 2003, 9(3):309–318.
9. Pfeiffer BE, Huber KM. Current advances in local protein synthesis and synaptic plasticity. *J Neurosci* 2006, 26(27):7147–7150.
10. Rustad TR, Minch KJ, Brabant W, Winkler JK, Reiss DJ, Baliga NS, Sherman DR. Global analysis of mRNA stability in *Mycobacterium tuberculosis*. *Nucleic Acids Res* 2013, 41(1):509–517.
11. Sharova LV, Sharov AA, Nedorezov T, Piao Y, Shaik N, Ko MS. Database for mRNA half-life of 19 977 genes obtained by DNA microarray analysis of pluripotent and differentiating mouse embryonic stem cells. *DNA Res* 2009, 16(1):45–58.
12. Yang E, van Nimwegen E, Zavolan M, Rajewsky N, Schroeder M, Magnasco M, Darnell JE, Jr. Decay rates of human mRNAs: correlation with functional characteristics and sequence attributes. *Genome Res* 2003, 13(8):1863–1872.
13. Figueroa A, Cuadrado A, Fan J, Atasoy U, Muscat GE, Munoz-Canoves P, Gorospe M, Munoz A. Role of HuR in skeletal myogenesis through coordinate regulation of muscle differentiation genes. *Mol Cell Biol* 2003, 23(14):4991–5004.
14. Kulkarni M, Ozgur S, Stoecklin G. On track with P-bodies. *Biochem Soc Trans* 2010, 38(Pt 1):242–251.
15. Labno A, Tomecki R, Dziembowski A. Cytoplasmic RNA decay pathways – enzymes and mechanisms. *Biochim Biophys Acta* 2016, 1863(12):3125–3147.
16. Willis DE, Twiss JL. Regulation of protein levels in subcellular domains through mRNA transport and localized translation. *Mol Cell Proteomics* 2010, 9(5):952–962.
17. Jeffares DC, Poole AM, Penny D. Relics from the RNA world. *J Mol Evol* 1998, 46(1):18–36.
18. Cech TR. Structural biology. The ribosome is a ribozyme. *Science* 2000, 289(5481):878–879.
19. Zhang L, Vielle A, Espinosa S, Zhao R. RNAs in the spliceosome: insight from cryoEM structures. *Wiley Interdiscip Rev RNA* 2019, 10(3):e1523.
20. Wilson RC, Doudna JA. Molecular mechanisms of RNA interference. *Annu Rev Biophys* 2013, 42:217–239.
21. Friedman RC, Farh KK, Burge CB, Bartel DP. Most mammalian mRNAs are conserved targets of microRNAs. *Genome Res* 2009, 19(1):92–105.

22. Kawamata T, Tomari Y. Making RISC. *Trends Biochem Sci* 2010, 35(7):368–376.
23. Carthew RW, Sontheimer EJ. Origins and mechanisms of miRNAs and siRNAs. *Cell* 2009, 136(4):642–655.
24. Liu X, Hao L, Li D, Zhu L, Hu S. Long non-coding RNAs and their biological roles in plants. *Genomics Proteomics Bioinformatics* 2015, 13(3):137–147.
25. Derrien T, Johnson R, Bussotti G, Tanzer A, Djebali S, Tilgner H, Guernec G, Martin D, Merkel A, Knowles DG *et al.* The GENCODE v7 catalog of human long noncoding RNAs: analysis of their gene structure, evolution, and expression. *Genome Res* 2012, 22(9):1775–1789.
26. Gupta RA, Shah N, Wang KC, Kim J, Horlings HM, Wong DJ, Tsai MC, Hung T, Argani P, Rinn JL *et al.* Long non-coding RNA HOTAIR reprograms chromatin state to promote cancer metastasis. *Nature* 2010, 464(7291):1071–1076.
27. Zhao J, Sun BK, Erwin JA, Song JJ, Lee JT. Polycomb proteins targeted by a short repeat RNA to the mouse X chromosome. *Science* 2008, 322(5902):750–756.
28. Li W, Notani D, Ma Q, Tanasa B, Nunez E, Chen AY, Merkurjev D, Zhang J, Ohgi K, Song X *et al.* Functional roles of enhancer RNAs for oestrogen-dependent transcriptional activation. *Nature* 2013, 498(7455):516–520.
29. Yoon JH, Abdelmohsen K, Srikantan S, Yang X, Martindale JL, De S, Huarte M, Zhan M, Becker KG, Gorospe M. LincRNA-p21 suppresses target mRNA translation. *Mol Cell* 2012, 47(4):648–655.
30. Gong C, Maquat LE. lncRNAs transactivate STAU1-mediated mRNA decay by duplexing with 3′ UTRs via Alu elements. *Nature* 2011, 470(7333):284–288.
31. Yarmishyn AA, Kurochkin IV. Long noncoding RNAs: a potential novel class of cancer biomarkers. *Front Genet* 2015, 6:145.
32. Ni YQ, Xu H, Liu YS. Roles of Long Non-coding RNAs in the development of aging-related neurodegenerative diseases. *Front Mol Neurosci* 2022, 15:844193.
33. Nisar S, Bhat AA, Singh M, Karedath T, Rizwan A, Hashem S, Bagga P, Reddy R, Jamal F, Uddin S *et al.* Insights into the role of CircRNAs: biogenesis, characterization, functional, and clinical impact in human malignancies. *Front Cell Dev Biol* 2021, 9:617281.
34. Cech TR, Steitz JA. The noncoding RNA revolution—trashing old rules to forge new ones. *Cell* 2014, 157(1):77–94.
35. Carninci P, Kasukawa T, Katayama S, Gough J, Frith MC, Maeda N, Oyama R, Ravasi T, Lenhard B, Wells C *et al.* The transcriptional landscape of the mammalian genome. *Science* 2005, 309(5740):1559–1563.
36. Djebali S, Davis CA, Merkel A, Dobin A, Lassmann T, Mortazavi A, Tanzer A, Lagarde J, Lin W, Schlesinger F *et al.* Landscape of transcription in human cells. *Nature* 2012, 489(7414):101–108.

Part II

Introduction to Next-Generation Sequencing (NGS) and NGS Data Analysis

4

Next-Generation Sequencing (NGS) Technologies: Ins and Outs

4.1 How to Sequence DNA: From First Generation to the Next

The sequence of nucleotides in a DNA molecule can be determined in multiple ways. Early in the 1970s, biochemists (Drs. Walter Gilbert and Frederick Sanger) devised different methods to sequence DNA. Dr. Gilbert's method is based on chemical procedures that break down DNA specifically at each of the four bases. Dr. Sanger's method, on the other hand, takes advantage of the DNA synthesis process. In this process, a new DNA chain is synthesized base by base using sequence information on the template (Chapter 2). The use of chemically modified nucleotides, that is dideoxynucleotides, as irreversible DNA chain terminators in Dr. Sanger's method randomly stops the synthesis process at each base position, so a series of new DNA chains of various lengths that differ by one base are produced (Figure 4.1). Determining the lengths in single-base resolution of specifically broken DNA molecules in Dr. Gilbert's method, or new DNA chains that are randomly terminated at each of the four nucleotides in Dr. Sanger's method, enabled sequencing of the template DNA. Over the years the Sanger method was further developed. The integration of automation into the process reduced human involvement and improved efficiency. The use of fluorescently labeled terminators, instead of the radioactively labeled terminators that were used initially, made it safer to operate and sequence detection more robust. The improved separation of DNA chains with the use of capillary electrophoresis, instead of slab gels, enabled high-confidence basecalls. All these developments had made the Sanger method widely adopted and become the method of choice for the Human Genome Project. Even today, this method is still widely used for single- or low-throughput DNA sequencing. With the coming of NGS, this method has become the synonym of first-generation sequencing.

While it is robust in sequencing individual DNA fragments, the Sanger method cannot easily achieve high-throughput, which is the key to lowering

DOI: 10.1201/9780429329180-6

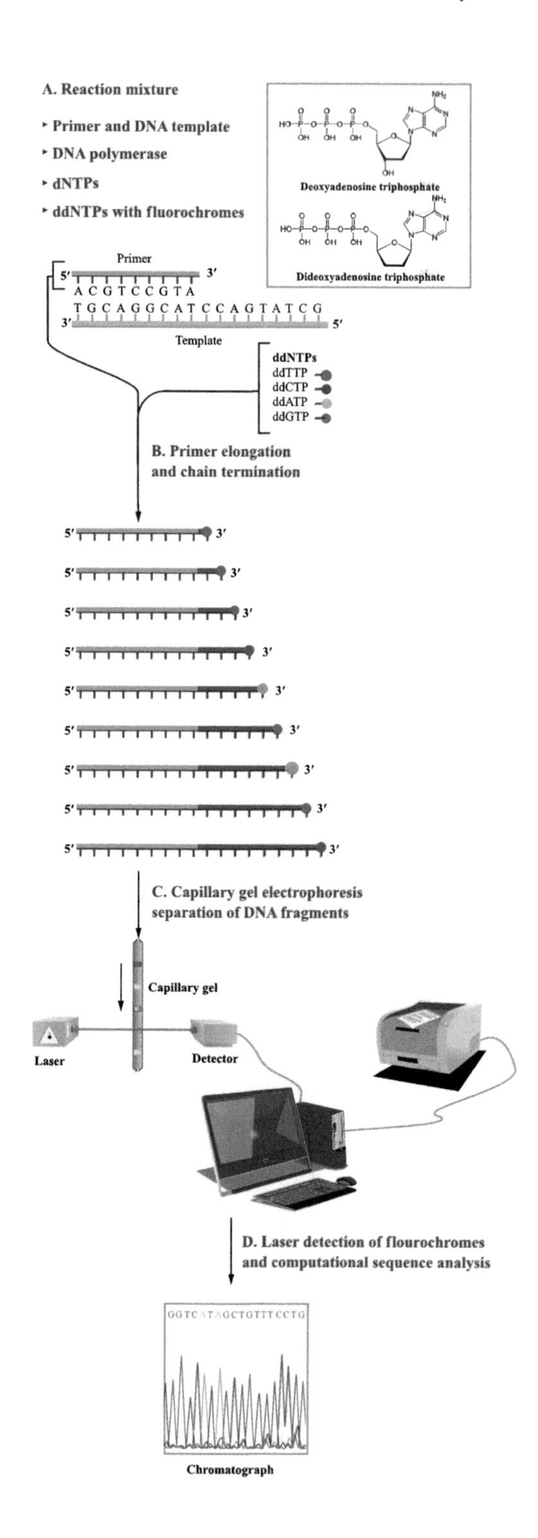
A. Reaction mixture
Primer and DNA template
DNA polymerase
dNTPs
ddNTPs with fluorochromes
Deoxyadenosine triphosphate
Dideoxyadenosine triphosphate
Primer
5′ 3′
A C G T C C G T A
T G C A G G C A T C C A G T A T C G
3′ 5′
Template
ddNTPs
ddTTP
ddCTP
ddATP
ddGTP
B. Primer elongation and chain termination
C. Capillary gel electrophoresis separation of DNA fragments
Capillary gel
Laser
Detector
D. Laser detection of flourochromes and computational sequence analysis
Chromatograph

FIGURE 4.1
The Sanger sequencing method as originally proposed. This method involves a step for new DNA strand synthesis using the sequencing target DNA as template, followed by sequence deduction through resolution of the newly synthesized DNA strands. In the first step (A), the new strand synthesis reaction mixture contains denatured DNA template, primer, DNA polymerase, and dNTPs. Besides the dNTPs, the Sanger method is characterized by the use of dideoxynucleotides (ddG, ddA, ddT, and ddC; the inset illustrates the structural difference between ddATP and dATP) that are labeled with different fluorochromes. The DNA polymerase in the reaction mixture incorporates dideoxynucleotides into the elongating DNA strand along with regular nucleotides, but once a dideoxynucleotide is incorporated, the strand elongation terminates. In this sequencing scheme, the ratio of these dideoxynucleotides to their regular counterparts is controlled so that the polymerization can randomly terminate at each base position. The end product is a population of DNA fragments with different lengths, with the length of each fragment dependent on where the dideoxynucleotide is incorporated. These fragments are then separated using capillary electrophoresis, in which smaller fragments migrate faster than larger ones and as a result pass through the laser detector sooner. The fluorochrome labels they carry enable computational deduction of the specific sequence of the original DNA. (Image adapted from https://commons.wikimedia.org/w/index.php?curid=23264166 by Estevezj. Used under the Creative Commons Attribution-Share Alike 3.0 Unported (CC BY-SA 3.0) license (https://creativecommons.org/licenses/by-sa/3.0/deed.en).)

sequencing cost, largely due to the segregation of its DNA synthesis process and the subsequent DNA chain separation/detection process. Its principle of sequencing-by-synthesis, however, becomes the basis of many NGS technologies, including Illumina's reversible terminator sequencing, Pacific Biosciences' single-molecule real-time (SMRT) sequencing, ThermoFisher's Ion Torrent semiconductor sequencing, and the discontinued 454/Roche's pyrosequencing. Different from the first-generation method, these technologies use nucleotides with reversible terminator or other cleavable chemical modifications, or regular unmodified nucleotides, so the new DNA strand synthesis is not permanently terminated and therefore can be monitored as or after each base is incorporated.

Not all NGS technologies are based on the principle of sequencing-by-synthesis. For example, Oxford Nanopore sequencing and the discontinued SOLiD sequencing from Life Technologies use nanopore sensing and sequencing-by-ligation, respectively. Despite the differences in how different NGS technologies work in principle, there is one common denominator among them that separate them from first-generation sequencing, which is their massive data throughput by sequencing millions to billions of DNA molecules simultaneously. Besides the ingenuity in the development of new sequencing chemistries or detection schemes to be detailed next, the success of NGS technologies in achieving extremely high throughput is also due to modern engineering and computing feats. Advancements in microfluidics and microfabrication make signal detection from micro-volume of sequencing reaction possible. Developments in modern optics and imaging technology enable tracking of sequencing reactions in high resolution, high fidelity, and high speed. Some NGS platforms also rely on the decades of progress in the semiconductor industry or more recent but rapid development in nanopore technology (such as the Ion Torrent and Nanopore platforms, respectively). High-performance computing makes it possible to process and deconvolve the torrent of signals recorded from millions of these processes.

As different NGS technologies employ different mechanisms and implementation strategies, in the next section the specifics of some of the most adopted NGS platforms at the time of writing (early 2022) are detailed. As NGS technologies continue to evolve, new platforms will appear while some current technologies become obsolete. While an overview of NGS platforms usually becomes outdated fairly soon, the guiding principles on the analysis of NGS data introduced in this book will remain.

4.2 Ins and Outs of Different NGS Platforms

4.2.1 Illumina Reversible Terminator Short-Read Sequencing

4.2.1.1 Sequencing Principle

The Illumina NGS platform is by far the most popular and has generated the largest amounts of NGS data. At the core of the Illumina sequencing technology is the employment of fluorescently labeled nucleotides with reversible terminator [1]. As mentioned previously, this method is based on the same basic principle of sequencing-by-synthesis as the Sanger method; but unlike the Sanger method, after the incorporation of each of these specially modified nucleotides, the terminator moiety they carry only temporarily prevents the new DNA strand from extending. After optical detection of the incorporated nucleotide based on its specific fluorescent label, the terminator moiety is cleaved and thereby the new strand synthesis resumes for the next cycle of nucleotide incorporation. For simultaneous detection of nucleotide incorporation in millions to billions of sequencing reactions, dATP, dCTP, dGTP, and dTTP are labeled with different fluorescent labels so each nucleotide can be detected by the different fluorescence signal they emit. The fluorescent labels and the reversible terminator moiety are attached to the nucleotides via the same chemical bond, so both of them can be cleaved off in a single reaction after each nucleotide incorporation and detection cycle to prepare for incorporation of the next nucleotide.

4.2.1.2 Implementation

The sequencing reaction in an Illumina NGS system takes place in a flow cell (Figure 4.2). The fluidic channels in the flow cell, often called lanes, are

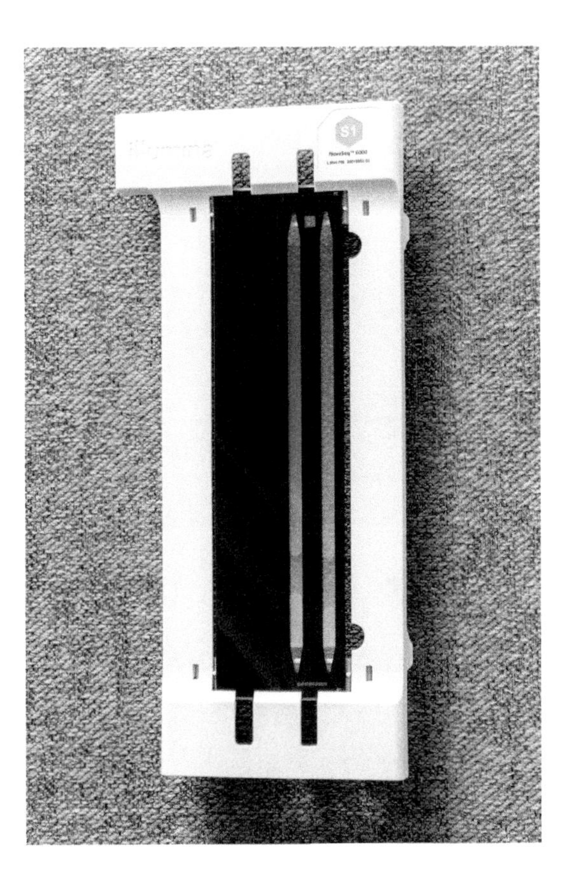

FIGURE 4.2
An Illumina sequencing flow cell. It is a special glass slide that contains fluidic channels inside (called lanes). Sequencing libraries are loaded into the lanes for massively parallel sequencing after template immobilization and cluster generation. In each step of the sequencing process, DNA synthesis mixture, including DNA polymerase and modified dNTPs, is pumped into and out of each of the lanes through their inlet and outlet ports located at the two ends.

where sequencing reactions take place and sequencing signals are collected through scanning. The top and bottom surface of each lane is covered with a lawn of oligonucleotide sequences that are complementary to the anchor sequences in Illumina adapters. When sequencing libraries, prepared from DNA through fragmentation and adapter ligation, are loaded into each of the lanes, DNA templates in the libraries bind to these oligonucleotide sequences and become immobilized onto the lane surface (Figure 4.3). After immobilization, each template molecule is clonally amplified through an isothermal process called "bridge amplification," through which up to 1,000 identical copies of the template are generated in close proximity (<1 micron in diameter) forming a cluster. During sequencing, these clusters are basic detection units, which generate enough signal intensity for basecalling.

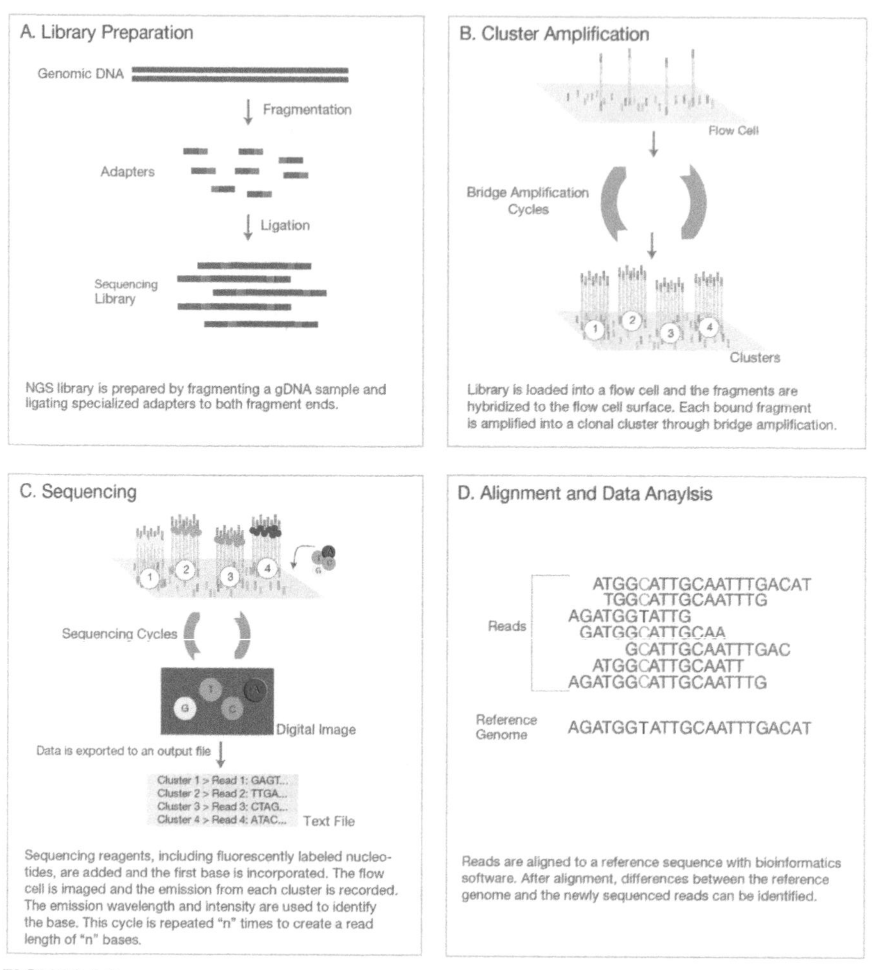

FIGURE 4.3

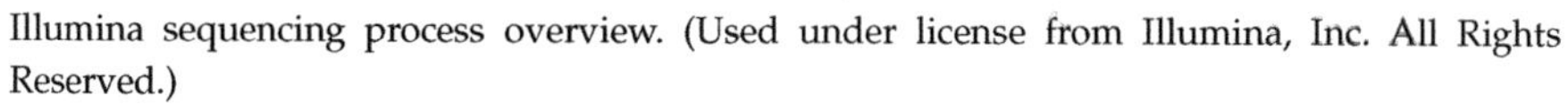

Illumina sequencing process overview. (Used under license from Illumina, Inc. All Rights Reserved.)

Under ideal conditions the simultaneous incorporation of nucleotides to the many identical copies of sequencing templates in a single cluster is expected to be in synchronization from step to step and therefore remain in phase. In reality, a small percentage of templates loses sync with the majority of molecules in the same cluster, leading to either falling behind (called phasing), due to incomplete removal of the terminator as well as missing a cycle, or being one or several bases ahead (pre-phasing), due to incorporation of nucleotides with no terminators. The existence of phasing and pre-phasing in a cluster leads to increased background noise and decreased basecall quality. When more and more sequencing cycles are conducted, this problem becomes worse. This is why platforms that are based on clonal amplification (which also include the Ion Torrent platform to be detailed later) have declining basecall quality scores toward the end. Eventually the decrease in basecall quality reaches a threshold beyond which the quality scores become simply unacceptable. The gradual loss of synchronicity is a major determinant of read length for these platforms.

4.2.1.3 Error Rate, Read Length, Data Output, and Cost

The overall error rate of the Illumina sequencing method is below 1%, which makes it one of the most accurate NGS platforms currently available. The most common type of errors is single nucleotide substitution. On read length, all available Illumina sequencers available at the time of writing can produce reads of at least 150 bases. Some sequencers can generate reads of up to 250 bases (NovaSeq 6000) or 300 bases (MiSeq) in length. Besides reading from one end of a DNA template (i.e., single-end sequencing), Illumina sequencers can also be read from both ends of the DNA fragment (called paired-end sequencing). Besides doubling the total number of sequence reads, paired-end sequencing also has the advantage of facilitating subsequent alignment to a reference genome (for details see next chapter) or genome assembly (Chapter 12), thereby reducing the limitation caused by the relatively short read length.

With regard to data output, the different sequencers/flow cells offered by Illumina have different output levels (see Table 4.1), with the maximum of 3 Tb (Terabases) being achieved on the NovaSeq 6000 using the S4 flow cell. Sequencing run time depends on read length and is typically in the range of less than one day to a couple of days. On sequencing cost, depending on sequencer and flow cell type, each Gb data costs US$4.80–41.67 on production-scale sequencers (i.e., NovaSeq 6000 and NextSeq 2000). This cost calculation is based on the current list price of sequencing reagents divided by corresponding data output, so other costs, including library preparation reagents, personnel time, sequencer depreciation, and service contract, are not included. It should also be noted that the numbers provided here are as of early 2022 and will change with future system updates.

4.2.1.4 Sequence Data Generation

There are three steps in the Illumina sequence data generation process. Firstly, raw images captured after each cycle are analyzed to locate clusters and report signal intensity, coordinates, and noise level for each cluster. This step is conducted by the instrument control software. The output from this step is fed into the next step of basecalling by the instrument Real-Time Analysis (RTA) software, which uses cluster signal intensity and noise level to make basecalls and quality score calculation. This step also filters out low-quality reads. In the third step, the basecall files, or bcl files, are converted to

FASTQ files, which contain raw reads. Since multiple samples are typically sequenced together in a multiplex fashion, demultiplexing of the sequence data is also performed in the third step. This is typically performed using Illumina's bcl2fastq tool, but other tools such as IlluminaBasecallsToFastq [2] can also be used. The demultiplexed FASTQ files in a compressed format are what an end user typically receives from an NGS facility after the completion of a run.

4.2.2 Pacific Biosciences Single-Molecule Real-Time (SMRT) Long-Read Sequencing

4.2.2.1 Sequencing Principle

The Pacific Biosciences' SMRT sequencing platform is usually regarded as the third-generation sequencing technology, as it is sensitive enough to sequence single DNA molecules and therefore can bypass amplification [3]. In addition, this platform generates much longer reads than the Illumina NGS platform, with the current read length reaching 25 kb and beyond. While it is also based on the principle of sequencing-by-synthesis, different from the Illumina method, SMRT sequencing uses nucleotides that carry distinct fluorescent labels linked to their end phosphate group but no terminator group. When a nucleotide is incorporated into an elongating DNA strand, with the cleavage of the end phosphate group (actually a pyrophosphate group as mentioned earlier) the fluorescent label is simultaneously released, which enables real-time signal detection. As this process does not involve a separate step of fluorescent label releasing and detection, the sequence-detecting signal is recorded continuously as a movie for up to 30 hours instead of using scanner images.

4.2.2.2 Implementation

At the core of PacBio sequencing is the SMRT cell, which carries millions of wells, technically called zero-mode waveguides (or ZMWs), for simultaneous sequencing of millions of DNA templates (the current version, as of early 2022, has 8 million of ZMWs). ZMWs are essentially holes tens of nanometers in diameter microfabricated in a metal film of 100 nm thickness, which is in turn deposited onto a glass substrate. Because the diameter of a ZMW is smaller than the wavelength of visible light, and the natural behavior of visible light passing through such a small opening from the glass bottom, only the bottom 30 nm of the ZMW is illuminated. Having a detection volume of only 20 zetpoliters (10^{-21} L), this detection scheme greatly reduces background noise and enables detection of the light of different wavelengths emitted from nucleotide incorporation into a new DNA strand.

While the SMRT platform performs single-molecule sequencing, the standard library prep protocol still requires DNA samples at the μg level to start (for lower DNA input amplification is needed). The library prep process includes fragmentation of DNA into desired length, end repair/A-tailing, and ligation of a hairpin loop adapter. This leads to the formation of a circular structure called SMRTbell (Figure 4.4). To prepare for sequencing, a SMRTbell template is annealed to a sequencing primer, and a DNA polymerase enzyme molecule is subsequently bound to the template/primer structure. The template-primer-polymerase complex is then immobilized to the bottom of a SMRT cell prior to sequencing.

The currently available PacBio SMRT sequencers (Sequel II/IIe) have two sequencing modes called continuous long reads (CLR) and circular consensus sequencing (CCS). With CLR, the DNA polymerase continues to advance along a template until it stops, thereby producing long reads in one pass. With CCS, the DNA polymerase goes through the SMRTbell structure multiple times and traverses both strands of the template in order to generate a consensus read (Figure 4.4).

4.2.2.3 Error Rate, Read Length, Data Output, and Cost

The CCS mode significantly improves sequencing accuracy. For example, with ten passes on a template the accuracy can reach 99.9%, i.e., Q30 [4]. In PacBio terms, reads generated from the CCS mode are high-fidelity (or HiFi) reads. In comparison, the accuracy of CLR reads is lower at around 90%. In general, the most common error type from SMRT sequencing is indels, with most of them in homopolymer regions. While more accurate, CCS reads are relatively shorter, currently in the range of 15–25 kb. In comparison, the length of CLR reads are typically 30–60 kb with reported maximum at over 200 kb. In terms of data output, CLR generates 150–250 Gb data in one run, while CCS produces 10–30 Gb. Based on the current list price of PacBio sequencing reagents, the per Gb cost of CLR reads approaches that of Illumina high-end sequencers, while CCS reads are still more costly due to the multiple passes needed to generate consensus sequences (see Table 4.1).

4.2.2.4 Sequence Data Generation

Primary data processing to generate raw basecalls is carried out on the sequencer. This includes processing of the movie to extract sequencing signals, basecalling from the extracted traces and pulses, and quality check of the basecalls. To generate CLR reads, adapter sequence needs to be trimmed from original polymerase-generated reads. To generate CCS reads, adapter sequence similarly needs to be removed first to generate subreads. Each subread contains sequence that corresponds to one pass of a DNA template from one direction. Subsequent post-processing of the subreads collapses multiple consecutive subreads to create a consensus sequence. All these data generation steps take place on the sequencer.

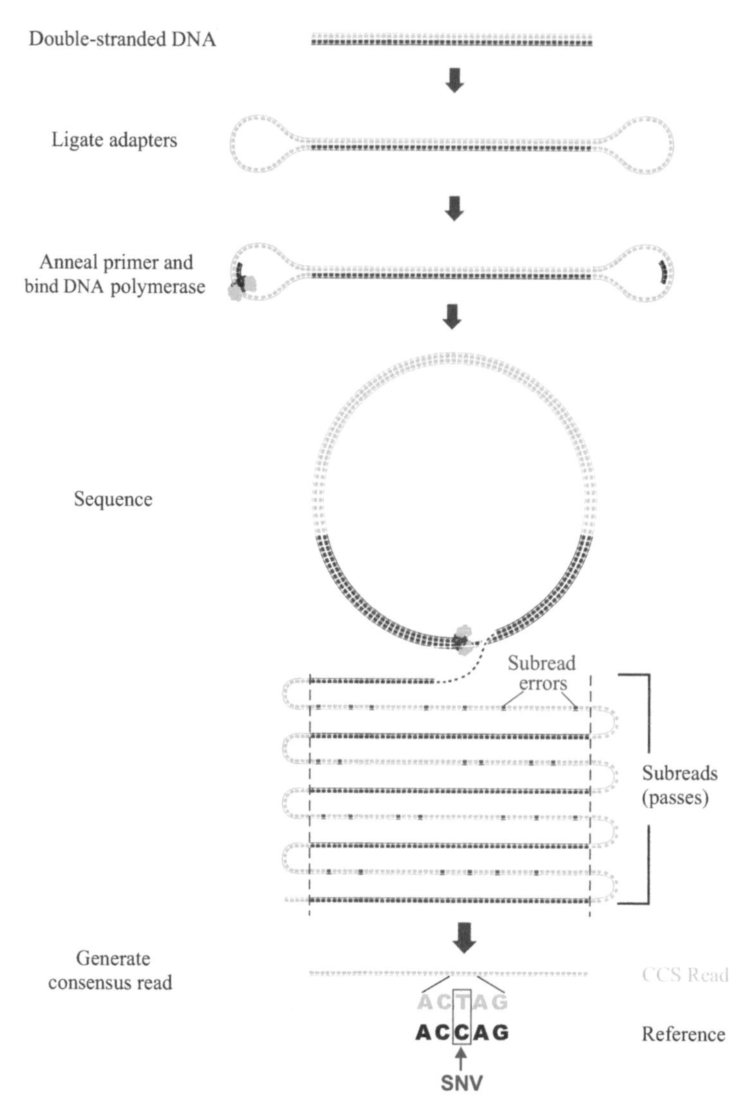

FIGURE 4.4
PacBio sequencing library preparation and sequencing. The library prep process mostly involves ligation of hairpin loop adapters to create the SMRTbell structure. A SMRTbell template can be sequenced using either circular consensus sequencing (CCS, shown here) or continuous long reads (CLR) mode. In the CCS mode, a template undergoes multiple passes to produce error-prone subreads in each pass followed by generation of accurate consensus reads. (Adapted by permission from Springer Nature Customer Service Centre GmbH: Springer Nature, *Nature Biotechnology*, Accurate circular consensus long-read sequencing improves variant detection and assembly of a human genome, Aaron M. Wenger et al., Copyright 2019.)

4.2.3 Oxford Nanopore Technologies (ONT) Long-Read Sequencing

4.2.3.1 Sequencing Principle

Like the PacBio platform, the Oxford Nanopore platform is also a single-molecule sequencing platform, but the reads it produces can be even longer. Instead of using sequencing by synthesis, it detects the order of nucleotides off a strand of DNA (or RNA) using physical measurement when the molecule passes through a bio-nanopore (Figure 4.5). To achieve the detection, a voltage is first applied across both sides of the pore, which is embedded in an electrically insulating membrane immersed in an electrolytic solution. Because DNA/RNA is negative charged, the electric field drives their movement through the pore. During this process, the nanopore serves as a biosensor to detect the subtle change in ionic current caused by the passing of nucleotides. Because different nucleotides result in different change patterns in this process, the collected electrical signal is then decoded by basecalling algorithms to reveal the underlying nucleotide sequence. It should be noted that the detected ionic current change is emitted from 5–9 nucleotides on a

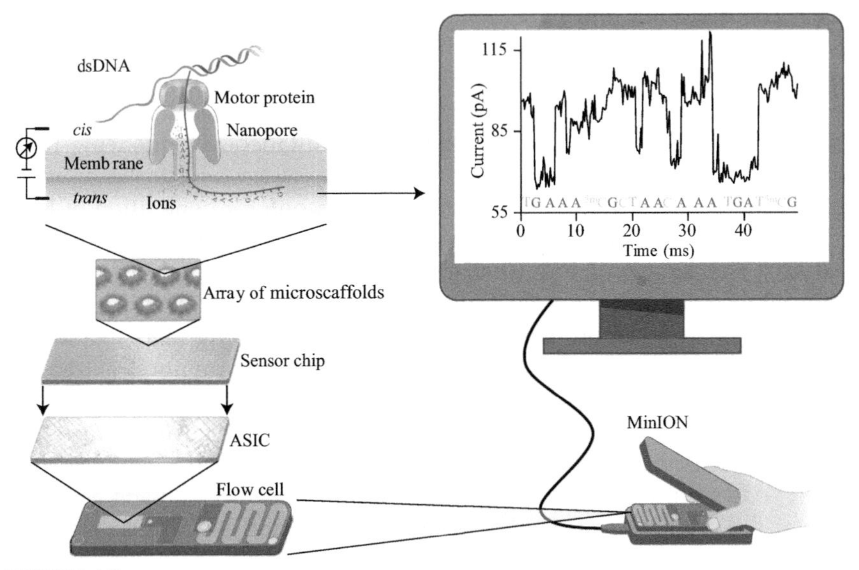

FIGURE 4.5
Nanopore sequencing. Illustrated here is sequencing with a MinION flow cell which contains 512 channels with 4 nanopores in each channel. The electrically insulating membrane that carries the nanopores is supported by an array of microscaffolds which is underlain by a senor chip. There are electrodes on the sensor chip that correspond to the individual channels, and electrical signals from the electrodes are recorded by the application-specific integrated circuit or ASIC. The recorded signals are then analyzed to make basecalls. (Adapted by permission from Springer Nature Customer Service Centre GmbH: Springer Nature, *Nature Biotechnology*, Nanopore sequencing technology, bioinformatics and applications, Yunhao Wang et al., Copyright 2021.)

strand of DNA/RNA (the actual number of nucleotides detected varies with the version of the nanopore ONT uses), not individual nucleotides [5]. To achieve reliable detection, the speed at which the DNA/RNA strand passes through the pore is important, as it needs to be slowed down to a rate that allows the electrical current change to be reliably recorded. This speed control is achieved by a motor protein situated at the orifice of the pore, and the speed at the time of writing is usually in the range of 200–450 bases per second depending on the version.

4.2.3.2 Implementation

ONT currently offers three main devices at different data throughput levels: MinION, GridION, and PromethION. MinION, at the lower end, is a USB drive-sized device that holds one flow cell. The GridION as a level up can hold up to five flow cells. The flow cells used in MinION and GridION is of the same type that contains 2,048 nanopores in 512 channels. The PromethION at the high end has the capacity to run up to 48 flow cells simultaneously. The flow cell used in PromethION has more capacity with 12,000 nanopores in 3,000 channels. For both the MinION/GridION and PromethION flow cells, in each channel only one pore can perform sequencing at a time. Besides these flow cells, for commonly conducted, smaller-scale sequencing, ONT also offers a flow cell dongle called Flongle, which provides an adapter for MinION/GridION to allow use of smaller and lower-cost flow cells. The Flongle adapter has 126 channels allowing simultaneous sequencing from 126 nanopores.

ONT provides two modes of sequencing, with one for generation of long reads (currently defined as below 100 kb) and the other for ultra-long reads (≥100 kb). The length of input DNA determines which mode to use. The sample library prep process for long-read sequencing involves fragmentation and/or size selection (optional), end repair, A-tailing, and ligation of sequencing adapters. Ultra-long sequencing library prep requires extraction of ultra-high-molecular-weight DNA. While the steps involved in ultra-long sequencing library prep may continue to evolve, the current procedure includes a transposition step that cleaves the template and attaches tags to the cleaved ends simultaneously, a subsequent step to add sequencing adapters to the tagged ends, and lastly an overnight elution of the DNA library prior to loading into a flow cell.

4.2.3.3 Error Rate, Read Length, Data Output, and Cost

The ONT platform has the capability to sequence the entire length of DNA/RNA fragments brought to it. The longest read length achieved so far by ONT is over 4 Mb. Typically for the long-read sequencing mode, the length is 10–100 kb, and for ultra-long read sequencing it is 100–300 kb. With the newest nanopore (R10.4) available at the time of writing, the raw read error rate is at 1% to achieve 99% (Q20) accuracy. Homopolymer error is the most common error type in ONT sequencing. In terms of data output, the MinION/GridION can typically produce 10–20 Gb data (30 Gb maximum at the current moving speed of 250 bases/second) from each flow cell. The PromethION has a throughput of 50–100 Gb (170 Gb maximum) per flow cell. With a top loading capacity of 48 flow cells, the data output from the PromethION can exceed that of PacBio Sequel II and Illumina NovaSeq 6000. The cost of sequencing on the MinION/GridION platforms is US$45–90 per Gb, and US$13–40 per Gb on the PromethION. Again this calculation is based on the list price per flow cell at the time of writing divided by the typical data output on each platform.

4.2.3.4 Sequence Data Generation

ONT uses its MinKNOW software for device control, raw data collection, and data processing to generate basecalls. For onboard real-time basecalling, its proprietary algorithm called Guppy is used. Unlike the Illumina or PacBio platforms where basecalling is more mature, basecalling from nanopore sequencing signal is still under constant improvement by ONT and independent groups and significant progress has been made [6]. ONT sequencers store the collected raw electrical signal in FAST5 files and can provide sequence reads after basecalling in FASTQ files. The raw data in FAST5 files can be used for independent basecalling, calling of DNA/RNA base modifications (e.g., methylation), and downstream analyses using other software tools. As the accuracy of sequence reads is heavily dependent on the basecall algorithm employed, reprocessing of raw data in FAST5 files over time can lead to discovery of new information. Storing the raw data files, however, does increase demands for storage space significantly.

4.2.4 Ion Torrent Semiconductor Sequencing

4.2.4.1 Sequencing Principle

Developed before the ONT platform, the Ion Torrent semiconductor sequencing system is the first NGS platform that does not rely on chemically modified nucleotides, fluorescence labeling, and the time-consuming step of image scanning, thereby achieving much faster speed, lower cost, and smaller equipment footprint. The Ion Torrent platform sequences DNA through detecting the H^+ ion released after each nucleotide incorporation in the sequencing-by-synthesis process. When a nucleotide is incorporated into a new DNA strand, the chemical reaction catalyzed by DNA polymerase releases a pyrophosphate group and a H^+ (proton). The release of H^+ leads to pH change in the vicinity of the reaction, which can be detected and used to determine the nucleotide incorporated in the last cycle. As the change in pH value is not nucleotide-specific, to determine DNA sequence, each of the substrate nucleotides (dATP, dGTP, dCTP, and dTTP) is added to the reaction in order at different times. A detected pH change after the introduction of a nucleotide suggests that the template strand contains its complementary base at the last position.

4.2.4.2 Implementation

The library construction process in this technology is similar to other NGS technologies, involving ligation of platform-specific primers to DNA shotgun fragments. The library fragments are then clonally amplified by emulsion PCR onto the surface of 3-micron diameter beads. The microbeads coated with the amplified sequence templates are then deposited into an Ion chip. Each Ion chip has a liquid flow chamber that allows influx and efflux of native nucleotides (introduced one at a time), along with DNA polymerase and buffer that are needed in the sequencing-by-synthesis process. For measuring possible pH change associated with each introduction of nucleotide, there are millions of pH microsensors that are manufactured on the chip bottom by the employment of standard processes used in the semiconductor industry.

4.2.4.3 Error Rate, Read Length, Date Output, and Cost

The overall accuracy of the Ion Torrent platform is over 99%. The major type of errors is indels caused by homopolymers. When the DNA template contains a homopolymeric region, i.e., a stretch of identical nucleotides (such as TTTTT), the signal in pH change is stronger and proportional to the number of nucleotides contained in the homopolymer. For example, if the template contains two Ts, the influx of dATPs will generate a pH change signal that is about twice as strong as that generated for a single T. Accordingly the signal for 3 Ts will be 1.5-fold that of 2 Ts, and signal for 6 Ts will be reduced to 1.2 fold that of 5 Ts. Therefore, with the increase in the total number of the repeat base, there is a gradual decrease in signal strength ratio that reduces the reliability of calling the total number of the base correctly. It is estimated that the current error rate for calling a 5-base homopolymer is 3.5%.

There are currently (as of early 2022) three sequencing systems in the Ion Torrent family: GeneStudio S5, Genexus, and PGM Dx. Of these systems, the GeneStudio S5 can produce reads up to 600 bases in length and total output of up to 25 Gb per chip. The Ion PGM Dx system on the other end of the spectrum generates 200 base reads at a total output of up to 1 Gb. The Genexus system is more than just a sequencer, as it provides an integrated workflow starting from nucleic acid extraction, library prep, template preparation, to sequencing and reporting. Genexus uses its GX5 chip for sequencing, and the cost listed on Table 4.1 is for the sequencing step only. GeneStudio S5

TABLE 4.1

Comparison of Current NGS Platforms

Platform	Read length	Run time	Data Output per Flow Cell	Cost per Gb (US$)	Common Error Type	Error Rate
Illumina Reversible Terminator Short-Read Sequencing						
NovaSeq 6000	Up to 250 bases from each end	~13–44 hours	Up to 3,000 Gb	$5–30	Single nucleotide substitution	<1%
NextSeq 2000	Up to 150 bases from each end	~11–48 hours	Up to 360 Gb	$8–42	See above	See above
MiSeq	Up to 300 bases from each end	~4–55 hours	Up to 15 Gb	$114–259	See above	See above
Pacific Biosciences Single Molecule Real-Time (SMRT) Long-Read Sequencing						
PacBio Sequel II/IIe	CLR: typically 30–60 kb, maximum >200 kb; CCS: 15–25 kb, maximum >25 kb	Up to 30 hours	CLR: typically 150–250 Gb; CCS: 10–30 Gb	CLR: US$5–8 per Gb; CCS: US$38–113 per Gb	Indels (mostly at homopolymers)	CLR: 8–13% CCS: <1%
Oxford Nanopore Technologies (ONT) Sequencing						
MinION/ GridION	20 bases – 4.2 Mb, typically 20–200 bp for short reads, 50–100 kb for long reads, over 100 kb for ultra-long reads	Up to 72 hours	Typically 10–20 Gb (30 Gb max)	US$45–90 per Gb	Long homopolymers	<1%
PromethION	See above	See above	Typically 50–100 Gb (170 Gb max)	US$13–40 per Gb	See above	See above
Ion Torrent Semiconductor Sequencing						
Ion GeneStudio S5	200, 400, 600 bases	~3–21.5 hours	Up to 25 Gb per chip	US$37–698 per Gb	Indels (mostly at homo-polymers)	<1%
Genexus	Up to 400 bases	~14–31 hours	24 Gb	US$56 per Gb	See above	<1%
Ion PGM Dx	200 bases	4.4 hours	0.6–1 Gb	US$941 per Gb	See above	<1%

employs five chip types: Ion 510, 520, 530, 540, and 550, with the 550 chip producing the most data (20–25 Gb) and being the most cost effective and the 510 having the least amount of data (0.3–1 Gb) and being the least cost effective. The Ion 318 Dx chip used on the PGM Dx system, similar to the 510 chip in throughput, generates 600 Mb to 1Gb data and is suitable for running molecular diagnostic tests that do not require a lot of data.

4.2.4.4 Sequence Data Generation

Sequencer operation and basecalling on the GeneStudio S5 and Ion PGM Dx are provided by the Torrent Suite Software. The Genexus system has its own software for managing the workflow and generating basecalls. The basecalling process used by the software is similar. Raw voltage measurement signal from pH change is first saved in DAT files for each cycle. After a run all DAT files are condensed into a single WELLS file, which is then used as input to generate basecalls. The WELLS file can be saved and used for reanalysis. Basecalls are reported in an unmapped BAM file. Besides generating raw sequence reads, the software also provides additional analytic functions, including read trimming and filtering, read mapping to a reference genome, and other tertiary analyses through the use of plugins.

4.3 A Typical NGS Workflow

Despite the differences in how different NGS technologies work in principle, the overall NGS workflow is similar. Sequencing genomic DNA, or RNA transcripts, using these NGS technologies all involves multiple steps (Figure 4.6). The early steps in this process are to construct sequencing libraries from DNA or RNA molecules extracted from biological samples of interest. As they are usually too large to be directly handled by most NGS technologies, especially those that produce short reads, the extracted DNA or RNA molecules often need to be broken into smaller fragments first. This fragmentation can be achieved with different techniques, including enzymatic treatment, acoustic shearing, sonication, or chemical shear (typically for RNA). The fragmentation step is usually followed by a size selection step to collect fragments in a certain target range. If ultra-long reads need to be obtained from the long-read platforms, a special extraction procedure is required to retain high-molecular-weight DNA.

A key step in the sequencing library construction process is the ligation of adapters to the two ends of DNA fragments. For RNA fragments, they are usually converted to complementary DNA (cDNA) first before adding the adapters. The adapters are artificial sequences that contain multiple components including universal sequencing primer sequence(s) that initiate

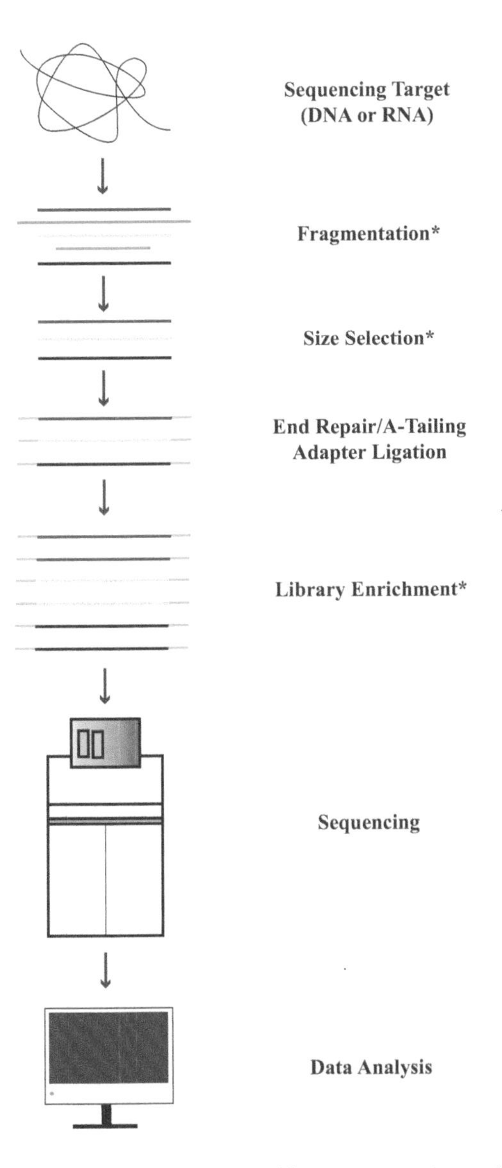

FIGURE 4.6
The general workflow of an NGS experiment. For library construction, only core steps shared by the different sequencing platforms are shown. The steps marked with asterisks are not used in some library construction protocols. Adapters ligated to sequencing targets are specific to each platform. There are other library construction strategies or procedures, such as use of non-ligation or target sequence capture, that are not shown here.

sequencing reactions on each fragment, and indexing (or "barcode") sequences to differentiate multiple samples when they are sequenced together. While they generally serve similar functions in different NGS platforms, the actual standard adapter sequences are specific to each platform. It is also possible to design custom adapters to meet special needs as long as key adapter sequence elements essential for a platform are in place. Prior to adapter ligation, the two ends of the DNA (or cDNA) fragments need to be prepared in an end repair step. After adapter ligation, the sequencing DNA templates in the resultant library may need to be enriched through a PCR amplification step using common sequences in the adapters. Alternatively, the constructed library may be sequenced PCR-free without enrichment on most of the platforms.

4.4 Biases and Other Adverse Factors That May Affect NGS Data Accuracy

Just as a certain level of erroneous basecalls is inherent to an NGS platform, the multiple steps that lead to the generation of sequence calls are not immune to biases. Different from errors, biases affect accurate representation of the original DNA or RNA population leading to higher (or lower) representation of some sequences than expected. The major source of biases in NGS is the molecular steps involved in the library construction and the sequencing process itself. Besides biases, there are also other potential factors that may lead to the generation of inaccurate sequencing signals. Detailed next are the various potential biases and other adverse factors during sequencing library construction and sequencing that may affect NGS data accuracy. It should be noted that while it is impossible to avoid them altogether, being aware of their existence is the first step toward minimizing their influence through careful experimental design and data analysis, and developing more robust analytic algorithms.

4.4.1 Biases in Library Construction

Biases in DNA fragmentation and fragment size selection. The initial step of library construction, i.e., DNA fragmentation, is usually assumed to be a random process and not dependent on sequence context. This has been shown to be not the case [7]. For example, sonication and nebulization cause DNA strand breaks after a C residue more often than expected. After DNA fragmentation, the size selection process may also introduce bias. For example, if gel extraction is employed for this process, the use of high gel melting temperature favors recovery of fragments with high GC content.

Ligation biases. After fragmentation and size selection, double-stranded DNA fragments are usually adenylated, after end repair, at the two 3′-ends generating 3′-dA tails that facilitate subsequent ligation of adapters that carry

5′-dT overhangs and thereby avoid self-ligation of DNA fragments or adapters. This AT overhang-based adapter ligation process, however, tend to be biased against DNA fragments that start with a T [8]. The sequencing of large RNA species, such as mRNAs or long non-coding RNAs, is also affected by this bias, as cDNA molecules reverse transcribed from these species are also subjected to the same adapter ligation process. Small RNA sequencing is not affected by this bias, as the ligation of adapters in small RNA sequencing library preparation is carried out prior to the reverse transcription step. The small RNA adapter ligation step, however, introduces a different type of bias, which affects some small RNAs in a sequence-specific manner. Sequence specificity underlies small RNA secondary and tertiary structure, which is also affected by temperature, concentration of cations, and destabilizing organic agents (such as DMSO) in the ligation reaction mixture. The efficiency of small RNA adapter ligation is influenced by their secondary and tertiary structure [9].

PCR biases. After adapter ligation, the DNA library is usually enriched by PCR for sequencing on most of the current NGS platforms. PCR, based on the use of DNA polymerases, is known to be biased against DNA fragments that are extremely GC- or AT-rich [10]. This can lead to variation in the coverage of different genomic regions and under-representation of those regions that are GC- or AT-rich. While optimization of PCR conditions can ameliorate this bias to some degree especially for high-GC regions, this bias can only be eliminated via adoption of a PCR-free workflow. To achieve this, Illumina provides PCR-free options. For single-molecule sequencing carried out on the PacBio SMRT and ONT platforms, PCR amplification is typically not required unless the input amount of starting DNA/RNA is low.

4.4.2 Biases and Other Factors in Sequencing

Like PCR, the sequencing-by-synthesis process carried out by most current NGS systems is also based on the use of DNA polymerases, which introduces similar coverage bias against genomic regions of extreme GC or AT content. As the use of DNA polymerases is at the core of these technologies, it is difficult to eradicate this bias completely. This bias should be kept in mind though when sequencing genomes or genomic regions of extremely high GC or AT content (>90%). Besides this enzymatic procedure, other aspects of the sequencing process, including equipment operation and adjustment, image analysis, and basecalling, may also introduce biases as well as artifacts. For example, air bubbles, crystals, dust, and lint in the buffers could obscure existing clusters (or beads) and lead to the generation of artificial signals. Misalignment of the scanning stage, or even unintended light reflections, can cause significant imaging inaccuracies. Unlike some of the inherent biases mentioned above, these artifacts can be minimized or avoided by experienced personnel.

The sequencing signal processing and basecalling steps may also introduce bias. For example, on the Illumina platform, basecalling results can be affected by color crosstalk, spatial crosstalk, as well as phasing and pre-phasing [11, 12]. On MiSeq, for instance, four images are generated from four detection channels after each cycle, which need to be overlaid to extract signal intensities for basecalling. This procedure is complicated by three factors: 1) signals from the four channels are not totally independent, as there is crosstalk between A and C, and between G and T channels, due to the overlapping in the emission spectra of their fluorescent labels; 2) neighboring clusters may partially overlap leading to spatial crosstalk between adjacent clusters; and 3) signals from a particular cycle are also dependent on signals from the cycles before and after, due to phasing and pre-phasing. While the Illumina's proprietary software is efficient at dealing with these factors for basecalling, there are other commercial and open-source tools that employ different algorithms for these tasks and generate varying results [13]. The algorithms these methods use (including the Illumina method) make different assumptions on signal distribution, which may not strictly represent the collected data, and therefore introduce method-specific bias to basecalling.

4.5 Major Applications of NGS

4.5.1 Transcriptomic Profiling (Bulk and Single-Cell RNA-Seq)

NGS has replaced microarray as the major means of detecting transcriptomic profiles and changes. The transcriptomic profile of a biological sample (such as a cell, tissue, or organ) is determined by and reflects on its developmental stage, internal, and external functional conditions. By sequencing existing RNA species in the transcriptome, NGS provides answers to key questions such as what genes are active and at what activity levels. RNA-seq at single-cell level interrogates cellular heterogeneity and reveals different cell types and states in a mixed population of cells or a tissue. Transcriptomic studies are almost always comparative studies, contrasting one tissue/stage/condition with another. Besides gene-level analysis, RNA-seq can also be used to study different transcripts derived from the same gene through alternative splicing. As an integral part of the transcriptome, small RNAs can be similarly studied by NGS. Analysis of bulk RNA-seq data generated from large and small RNA species is covered in Chapters 7 and 9, respectively. Due to its uniqueness single-cell RNA-seq data analysis is covered in a dedicated chapter (Chapter 8).

4.5.2 Genetic Mutation and Variation Identification

Detecting and cataloging genetic mutation or variation among individuals in a population is a major application of NGS. Existing NGS studies have already shown that severe diseases such as cancer and autism are associated with novel somatic mutations. Projects such as the 1000 Genomes Project have revealed the great amount of genetic variation in a population that accounts for individual differences in physical traits, disease predisposition, and drug response. Chapter 10 focuses on data analysis techniques in a research setting on how to identify mutations and various types of variations, and test their associations with traits or diseases. Chapter 11 focuses on the clinical application of NGS to identify actionable variants to guide bedside decision making.

4.5.3 *De Novo* Genome Assembly

Sanger sequencing used to be regarded as the golden standard for *de novo* genome assembly, but more and more genomes, including large complex genomes, have been assembled with NGS reads alone. Technological advancements in the NGS arena, including the gradual increases of read length in short-read technologies and the maturation of long-read technologies, have contributed to this trend. The development of new algorithms for NGS-based genome assembly is another force behind this progress. Chapter 12 focuses on how to use these algorithms to assemble a new genome from NGS reads.

4.5.4 Protein-DNA Interaction Analysis (ChIP-Seq)

The normal functioning of a genome depends on its interaction with a multitude of proteins. Transcription factors, for example, are among some of the best-known DNA-interacting proteins. Many of these proteins interact with DNA in a sequence- or region-specific manner. To determine which regions of the genome these proteins bind to, the bound regions can be first captured by a process called chromatin immunoprecipitation (or ChIP) and then sequenced by NGS. ChIP-seq can be applied to study how certain conditions, such as a developmental stage or disease, affect the binding of protein factors to their affinity regions. ChIP-seq data analysis is covered in Chapter 13.

4.5.5 Epigenomics and DNA Methylation Study (Methyl-Seq)

Chemical modifications of certain nucleotides and histones provide an additional layer of genome modulation beyond the regulatory mechanism embedded in the primary nucleotide sequence of the genome. These modifications and the modulatory information they provide constitute the epigenome. NGS-based epigenomics studies have revealed how monozygotic twins display difference in certain phenotypes, and how changes in epigenomic profile may lead to diseases such as cancer. Cytosine methylation is a major form of epigenomic change. Chapter 14 covers analysis of DNA methylation sequencing data.

4.5.6 Metagenomics

To study a community of microorganisms like the microbiome in the gut or those in a bucket of seawater, where extremely large but unknown numbers of species are present, a brutal force approach that involves the study of all genomes contained in such a community is metagenomics. Recently the field of metagenomics has been greatly fueled by the development of NGS technologies. By quickly sequencing everything in a metagenome, researchers can get a comprehensive profile of the makeup and functional state of a microbial community. Compared to NGS data generated from a single genome, the metagenomics data is much more complicated. Chapter 15 focuses on metagenomics NGS data analysis.

References

1. Bentley DR, Balasubramanian S, Swerdlow HP, Smith GP, Milton J, Brown CG, Hall KP, Evers DJ, Barnes CL, Bignell HR *et al.* Accurate whole human genome sequencing using reversible terminator chemistry. *Nature* 2008, 456(7218):53–59.
2. Picard toolkit (https://broadinstitute.github.io/picard/)
3. Eid J, Fehr A, Gray J, Luong K, Lyle J, Otto G, Peluso P, Rank D, Baybayan P, Bettman B *et al.* Real-time DNA sequencing from single polymerase molecules. *Science* 2009, 323(5910):133–138.
4. Wenger AM, Peluso P, Rowell WJ, Chang PC, Hall RJ, Concepcion GT, Ebler J, Fungtammasan A, Kolesnikov A, Olson ND *et al.* Accurate circular consensus long-read sequencing improves variant detection and assembly of a human genome. *Nat Biotechnol* 2019, 37(10):1155–1162.
5. Jain M, Fiddes IT, Miga KH, Olsen HE, Paten B, Akeson M. Improved data analysis for the MinION nanopore sequencer. *Nat Methods* 2015, 12(4):351–356.
6. Rang FJ, Kloosterman WP, de Ridder J. From squiggle to basepair: computational approaches for improving nanopore sequencing read accuracy. *Genome Biol* 2018, 19(1):90.
7. Poptsova MS, Il'icheva IA, Nechipurenko DY, Panchenko LA, Khodikov MV, Oparina NY, Polozov RV, Nechipurenko YD, Grokhovsky SL. Non-random DNA fragmentation in next-generation sequencing. *Sci Rep* 2014, 4:4532.
8. Seguin-Orlando A, Schubert M, Clary J, Stagegaard J, Alberdi MT, Prado JL, Prieto A, Willerslev E, Orlando L. Ligation bias in Illumina next-generation DNA libraries: implications for sequencing ancient genomes. *PLoS One* 2013, 8(10):e78575.

9. Hafner M, Renwick N, Brown M, Mihailovic A, Holoch D, Lin C, Pena JT, Nusbaum JD, Morozov P, Ludwig J *et al.* RNA-ligase-dependent biases in miRNA representation in deep-sequenced small RNA cDNA libraries. *RNA* 2011, 17(9):1697–1712.
10. Aird D, Ross MG, Chen WS, Danielsson M, Fennell T, Russ C, Jaffe DB, Nusbaum C, Gnirke A. Analyzing and minimizing PCR amplification bias in Illumina sequencing libraries. *Genome Biol* 2011, 12(2):R18.
11. Wang B, Wan L, Wang A, Li LM. An adaptive decorrelation method removes Illumina DNA base-calling errors caused by crosstalk between adjacent clusters. *Sci Rep* 2017, 7:41348.
12. Whiteford N, Skelly T, Curtis C, Ritchie ME, Lohr A, Zaranek AW, Abnizova I, Brown C. Swift: primary data analysis for the Illumina Solexa sequencing platform. *Bioinformatics* 2009, 25(17):2194–2199.
13. Cacho A, Smirnova E, Huzurbazar S, Cui X. A Comparison of base-calling algorithms for illumina sequencing technology. *Brief Bioinform* 2016, 17(5): 786–795.

5

Early-Stage Next-Generation Sequencing (NGS) Data Analysis: Common Steps

In general, NGS data analysis is divided into three stages. In the primary analysis stage, bases are called based on deconvolution of the optical or physicochemical signals generated in the sequencing process. Regardless of sequencing platforms or applications, the basecall results are usually stored in the standard FASTQ format. Each FASTQ file contains a massive number of reads, i.e., sequence readouts of DNA fragments sampled from a sequencing library. In the secondary analysis stage, reads in the FASTQ files are quality checked, preprocessed, and then mapped to a reference genome. The data quality check or control (QC) step involves examining a number of sequence reads quality metrics. Based on data QC result, the NGS sequencing files are preprocessed in order to filter out low-quality reads, trim off portions of reads that have low-quality basecalls, and remove adapter sequences or other artificial sequences (such as PCR primers) if they exist. Subsequent mapping (or aligning) of the preprocessed reads to a reference genome aims to determine where in the genome the reads come from, the critical information required for most tertiary analysis (except *de novo* genome assembly). The stage of tertiary analysis is highly application-specific and detailed in the chapters of Part III. This chapter focuses on steps in the primary and secondary stages, especially on reads QC, preprocessing, and mapping, which are common and shared among most applications (Figure 5.1).

5.1 Basecalling, FASTQ File Format, and Base Quality Score

The process of basecalling in the primary stage from fluorescence images, movies, or physicochemical measurements is carried out with platform-specific, proprietary algorithms. For example, Illumina uses Bustard, a statistical model-based basecaller. ONT currently employs Guppy, based on the use of a deep learning approach called RNNs (or recurrent neural networks). Besides these onboard basecallers that come with the sequencers, other stand-alone basecalling algorithms have also been developed towards the goal of further improving accuracy [1, 2]. Increasing basecalling accuracy is

DOI: 10.1201/9780429329180-7

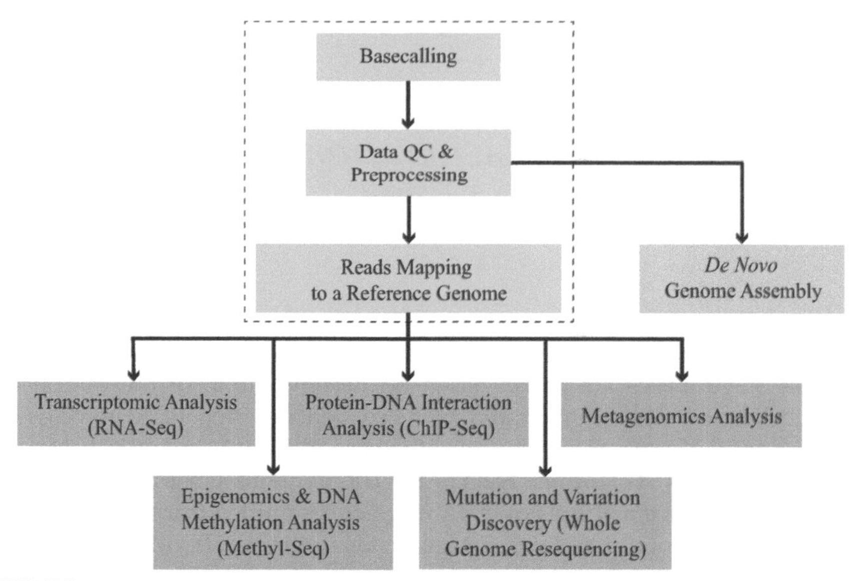

FIGURE 5.1
General overview of NGS data analysis. The steps in the dashed box are common steps conducted in primary and secondary analysis.

especially important for the long-read sequencing platforms, with the ONT platform serving as a good example with active basecalling algorithm development. Multiple machine learning-based basecallers have been developed by ONT, including Albacore, Scappie, Flappie, and Bonito, besides Guppy. At the time of writing, Guppy offers faster speed than the others while maintains relatively high accuracy, while Bonito, as the latest iteration of basecallers from ONT, uses another deep learning approach called CNN (convolutional neural networks) to achieve even better accuracy than Guppy but at a slower speed. Other open-source basecalling algorithms developed by the community include DeepNano [3], Nanocall [4], Chiron [5], and causalcall [6]. As a result of these algorithmic development efforts, significant progress has been made and basecalling accuracy has been significantly increased.

Most end users do not usually intervene in the basecalling process but rather focus on analysis of the basecalling results. Regardless of the sequencing platform, basecalling results are usually reported in the universally accepted FASTQ format. In file size, a typical compressed FASTQ file is usually in the multi-GB range and may contain millions to billions of reads. In a nutshell, the FASTQ format is a text-based format, containing the sequence of each read along with the confidence score of each base. Figure 5.2 shows an example of one such read sequence reported in the FASTQ format.

The confidence (or quality) score, as a measure of the probability of making an erroneous basecall, is an essential component of the FASTQ format. The

```
@HISEQ:131:C5NWFACXX:1:1101:3848:2428 1:N:0:CGAGGCTGCTCTCTAT
CTTTTATCAGACATATTTCTTAGGTTTGAGGGGGAATGCTGGAGATTGTAATGGGTATGGAGACATATCATATAAGTAATGCTAGG
GTGAGTGGTAGGAAG
+
BB7FFFFB<F<FBFBBFBFBFFFIFFFFIIIFF<FBFFFBFIFFBFFFIFFFBFB07<BFFF7BBFFFBFFFFFF<BFBFBBBBBB
B'77B<770<BBBBB
```

FIGURE 5.2
The FASTQ sequence read report format. Shown here is one read generated from an NGS experiment. A FASTQ file usually contains millions to billions of such reads, with each containing several lines as shown here. Line 1, starting with the symbol '@,' contains sequence ID and descriptor. Line 2 is the read sequence. Line 3 (optional) starts with the '+' symbol, which may be followed by the sequence ID and description. Line 4 lists confidence (or quality) scores for each corresponding base in the read sequence (Line 2). For Illumina-generated FASTQ files, the sequence ID in Line 1 in this example basically identifies where the sequence was generated. This information includes the equipment ("HISEQ" in the above example), sequence run ID ("131"), flow cell ID ("C5NWFACXX"), flow cell lane ("1"), tile number within the lane ("1101"), x/y-coordinates of the sequence cluster within the tile ("3848" and "2848," respectively). The ensuing descriptor contains information about the read number ("1" is for single read here; for paired-end read it can be 1 or 2), whether the read is filtered ("N" here means it is not filtered), control number ("0"), and index (or sample barcode) sequence ("CGAGGCTGCTCTCTAT").

NGS basecall quality score (Q-score) is similar to the Phred score used in Sanger sequencing and is calculated as:

$$Q = -10 \times \log_{10} P_{Err}$$

where P_{Err} is the probability of making a basecall error. Based on this equation, a 1% chance of incorrectly calling a base is equivalent to a Q-score of 20, and Q30 means a 1/1000 chance of making a wrong call. Usually for a basecall to be reliable, it has to have a Q-score of at least 20. High-quality calls have Q-scores above 30, usually up to 40. For better visualization of Q-scores associated with their corresponding basecalls, they are usually encoded with ASCII characters. While there have been different encoding scheme versions (e.g., Illumina 1.0, 1.3, and 1.5), currently the NGS field has mostly settled on the use of the same encoding scheme used by Sanger sequencing (Figure 5.3). In the FASTQ example shown in Figure 5.2, the first base, C, has an encoded Q-score of B, i.e., 33.

To come up with the P_{Err}, a control lane or spike control is usually used to generate a basecall score calibration table in Illumina sequencing for lookup. A precomputed calibration table can also be used in the absence of a control lane and spike control. Because each platform calibrates their Q-scores differently, if they are to be compared with each other or analyzed in an integrated fashion, their Q-scores need to be recalibrated. To carry out the recalibration, a subset of reads is used that maps to regions of the reference genome that contain no SNPs, and any mismatch between the reads and the reference

```
ASCII Character: ! " # $ % & ' ( ) * +  ,  -  .  /  0  1  2  3  4  5  6  7  8  9  :
  Quality Score: 0 1 2 3 4 5 6 7 8 9 10 11 12 13 14 15 16 17 18 19 20 21 22 23 24 25

 ;  <  =  >  ?  @  A  B  C  D  E  F  G  H  I  J
26 27 28 29 30 31 32 33 34 35 36 37 38 39 40 41
```

FIGURE 5.3
Encoding of basecall quality scores with ASCII characters. ASCII stands for American Standard Code for Information Interchange, and an ASCII code is the numerical representation of a character in computers (e.g., the ASCII code of the letter 'B' is 66). In this encoding scheme, the ASCII character codes equal to Q-scores plus 33. Current major NGS platforms, including Illumina (after ver. 1.8), use this encoding scheme for Q-score representation.

sequence is considered a sequencing error. Based on the rate of mismatch at each base position of the reads, a new calibration table is constructed, which is then used for recalibration. Even without cross-platform NGS data comparison and integration, NGS data generated on the same platform can still be recalibrated post mapping (see next) using the same approach, which often leads to improved basecall quality scores.

5.2 NGS Data Quality Control and Preprocessing

After NGS data generation, the first step should be data quality check. While this step does not directly generate biological insights, it is nonetheless essential and should be carried out carefully. Doing so will avoid production of non-sensical or even erroneous results in later steps and unnecessary consumption of computational resources and time. In this process, the following metrics of data quality need to be examined:

1) *Q-Scores*: These can be examined in different ways. On a per-base basis, this can be conducted by examining quality scores across all base positions of all reads, from the first sequenced base to the last. As a general trend, for platforms based on sequencing-by-synthesis, base positions covered at early phases of a sequencing procedure tend to have higher Q-values than those sequenced later in the procedure. The Q-scores for even the late-phase base positions, however, should still have a median value of at least 20. If there is a significant Q-score drop in the late phase, the affected base positions need to be examined closely and low-quality bases should be trimmed off from affected reads. In addition, increased percentage of N calls also helps determine loss of basecall quality (an N is called when the basecalling algorithm cannot call any of the four bases with confidence). Another way of inspecting Q-scores is by plotting the average Q-score of each read and examining

their distribution pattern. For a successful run, the majority of reads should have average Q-score of over 30, and only a very small percentage of reads have average Q-score below 20.

2) *Percentage of each base at each position*: If reads are obtained from a sequencing library constructed from randomly generated DNA fragments, the chance of observing each of the four bases at each base position should be constant. Therefore, when plotting the percentages of each base across all base positions, the plots for A, C, G, and T should be roughly parallel to each other, and the overall percentage shown in each plot should reflect the overall frequency of each base in the target library. If the plots deviate significantly from being parallel, this indicates problem(s) in the library construction process, such as existence of over-represented sequences in the library (such as rRNA in an RNA-seq library), or non-random fragmentation.
3) *Read length distribution*: For platforms that produce reads of varying length (such as the PacBio and ONT platforms), the distribution of read length should also be examined. In combination with the distribution of Q-scores, this determines the total amount of useful data a run generates. In addition, with data quality and total volume being equal, a run that produces longer reads is more advantageous in terms of sequence alignment or assembly than one with more relatively short reads.

Besides examining reads quality and length distribution, other QC metrics should also be examined, such as existence of artificial sequences including adapters and PCR primers, or duplicated sequences based on sequence identity (sequence duplication can also be checked based on reference genome mapping result). After inspecting sequence data quality, filtering should be performed to remove low-quality reads. Furthermore, low-quality basecalls, e.g., bases at the 3′ end that have Q-scores below 20, as well as artificial sequence contaminants, should also be trimmed off if they exist. While some platforms (e.g., Illumina) perform sequence filtering by default prior to FASTQ file generation, if the distribution of sequence Q-scores is found to be unsatisfactory after examination, additional filtering/trimming may need to be performed. Execution of these preprocessing tasks is a requirement for high-quality downstream analysis.

The most commonly used NGS data QC software includes FastQC [7], NGS QC Toolkit [8], and fastp [9]. These toolkits have functional modules to examine per-read and per-base Q-scores, base frequency distribution, read length distribution, and existence of duplicated sequences and artificial sequences. FastQC is written in Java and has a user-friendly interface on most operating systems including Windows. Developed in C/C++ and employing multi-threading for parallel processing, fastp aims to achieve fast speed for QC, as well as other preprocessing steps including adapter trimming, quality filtering, per-read quality pruning, etc. More recent NGS QC tools, such as seqQscorer [10], apply machine learning approaches in an attempt to achieve better understanding of quality issues and automated quality control. Tools such as FQC Dashboard [11] and MultiQC [12] serve as aggregators of QC results from other tools (e.g., FastQC) and present them in a single report. After data QC, to perform stand-alone preprocessing tasks such as adapter trimming and read filtering, tools such as cutadapt [13] and Trimmomatic [14] are often used.

Some of the QC tools mentioned above, including FastQC and fastp, can be used for both short and long reads. There are also tools such as NanoQC (part of NanoPack) [15] specifically designed for long-read QC. Besides NanoQC, NanoPack has a set of utilities for trimming, filtering, summarization, visualization, etc. PycoQC [16] is another tool that provides interactive QC metrics for ONT data. For PacBio long reads, SequelTools [17] provides QC data, as well as other tasks such as reads filtering, summarization, and visualization.

5.3 Read Mapping

After the data is cleaned up, the next step is to map, or align, the reads to a reference genome if it is available, or conduct *de novo* assembly. As shown in Figure 5.1, most NGS applications require read mapping to a reference genome prior to conducting further analysis. The purpose of this mapping process is to locate origins of the reads in the genome. Compared to searching for the location(s) of a single or a small number of sequences in a genome by tools such as BLAST, simultaneous mapping of millions of NGS reads, sometimes very short, to a genome is not trivial. Further challenge comes from the fact that any particular genome from which NGS reads are derived deviates from the reference genome at many sites because of polymorphism and mutation. As a result any algorithm built for this task needs to accommodate such sequence deviations. To further complicate the situation, sequencing errors are often indistinguishable from true sequence deviations.

5.3.1 Mapping Approaches and Algorithms

The mapping of NGS reads to a reference genome is not a new task in itself. As indicated above, before the advent of NGS, a number of sequence alignment algorithms already existed, the best known of which is BLAST. These aligners use hash tables and seed-and-extend methods to perform the computationally intensive process of aligning an individual query sequence against a sequence database (such as GenBank). To use these methods to align the millions of NGS reads to a reference genome, however, creates a scalability problem, as they simply cannot scale up to the data volume, and scale down to the short length of many NGS reads (as short read carries less information). As a result, new generations of algorithms have been devised for the mapping of NGS reads, either through optimizing the previous methods or introducing new approaches.

Mapping of NGS reads can be separated into three steps (Figure 5.4). The first step is to index the reference genome sequence in computer memory to boost subsequent searching speed. The use of hash table is one way to achieve this, while another approach is to use suffix-tree based techniques (to be detailed next). The second step is the so-called "global positioning," with the goal of determining possible matching location(s) in the indexed reference genome of seed sequences extracted from a read. The third step is extensive alignment between the entire read and location(s) matched in Step 2 to verify the alignment and in the meantime produce alignment information such as sequence variation and their type.

Hash table-based reference genome indexing is often used for fast lookup of matching locations of exact subsequences (or *k*-mers) from reads. A hash table is a data structure that stores associative data, in this case, *k*-mers and their associated genomic locations. After genomic sequences are indexed by hashing, *k*-mers are then extracted from each read and used as seeds to search the hash table to identify their possible locations in the genome (Figures 5.4 and 5.5). Aligners based on this approach include minimap2 [18], SOAP (Short Oligonucleotide Alignment Program) [19], MAQ (Mapping and Assembly with Qualities) [20], Illumina's Isaac Genome Alignment Software [21], and Novoalign (Commercial). Among these aligners, minimap2 uses minimizers [22] to reduce the amount of computer memory needed to store the hash table in order to further increase speed. It should be noted that while most aligners use hash table to index the reference genome, some aligners (such as MAQ) create a hash table from NGS reads instead, and in such cases *k*-mers extracted from the reference genome are used to look up matches in the reads.

In the seed-and-extend approach used by BLAST, the seed used is consecutive sequence designed to locate near-exact matches, which is not ideal for aligning sequences that contain variations especially indels. To increase alignment robustness, NGS reads aligners have migrated from the use of consecutive exact-match seeds to nonconsecutive (or spaced) seeds. By allowing space between seeds, the chance of finding a match is increased. In SOAP and Novoalign, for example, to perform alignment using spaced seeds, the reference genome sequence is first cut into equal-sized small fragments and saved in a big hash table in memory. The NGS reads are then cut in a similar fashion into subsequences, which are searched against the reference genome (Figure 5.5a). Computationally, these aligners are memory and processor intensive and therefore not very fast.

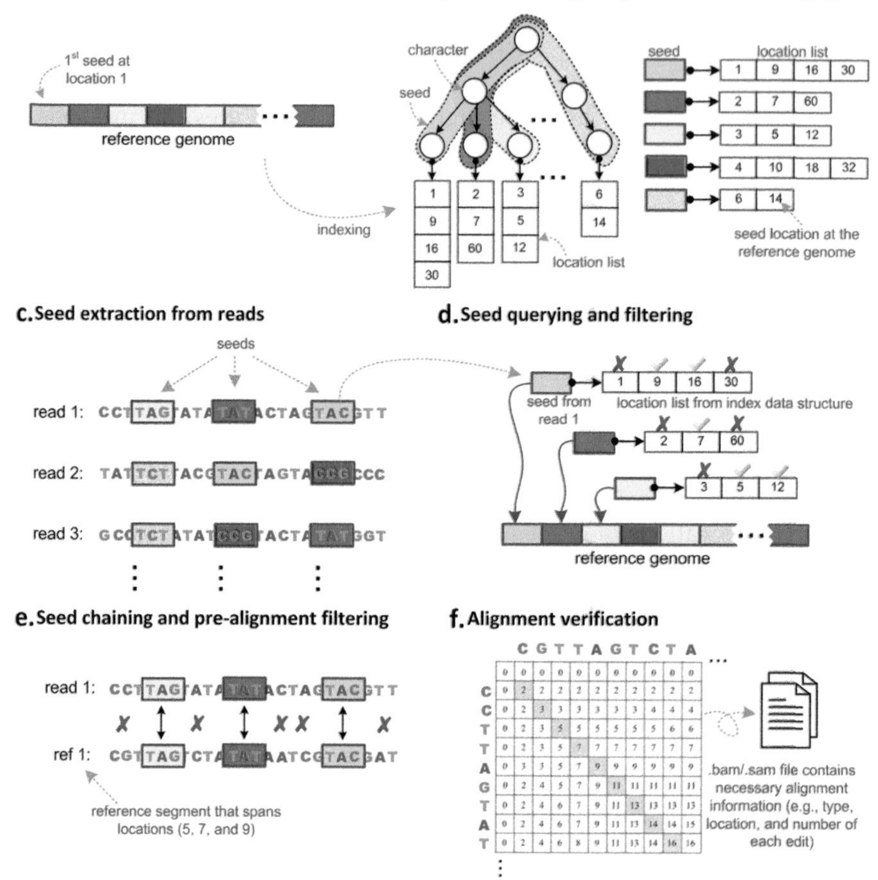

FIGURE 5.4
Major steps in mapping NGS reads. In the first stage, the reference genome is indexed. This is achieved through extracting seed sequences from the reference genome (a) and subsequently the seed sequences are indexed using suffix tree or hash table (b). In the second stage, seed sequences are extracted from reads (c), which are then used to search the indexed reference genome for possible matching locations (d). In the example shown, each seed extracted from read 1 is searched to locate their potential locations in the indexed genome. Based on their adjacency some of the locations are excluded (red X) as such locations are unlikely to span the read. In the last stage, the adjacent seeds are chained and the gap sequences between the seeds are inspected for mismatches (red X), based on which pre-alignment filters determine whether to accept the alignment between the read and the genomic region (e). In the last step, the alignment is subject to verification to generate alignment result including sequence differences and their locations. (From Alser, M., Rotman, J., Deshpande, D. et al. Technology dictates algorithms: recent developments in read alignment. *Genome Biol* 22, 249 (2021). https://doi.org/10.1186/s13059-021-02443-7. Used under the terms of the Creative Commons Attribution 4.0 International License, http://creativecommons.org/licenses/by/4.0/, © 2021 Alsher et al.)

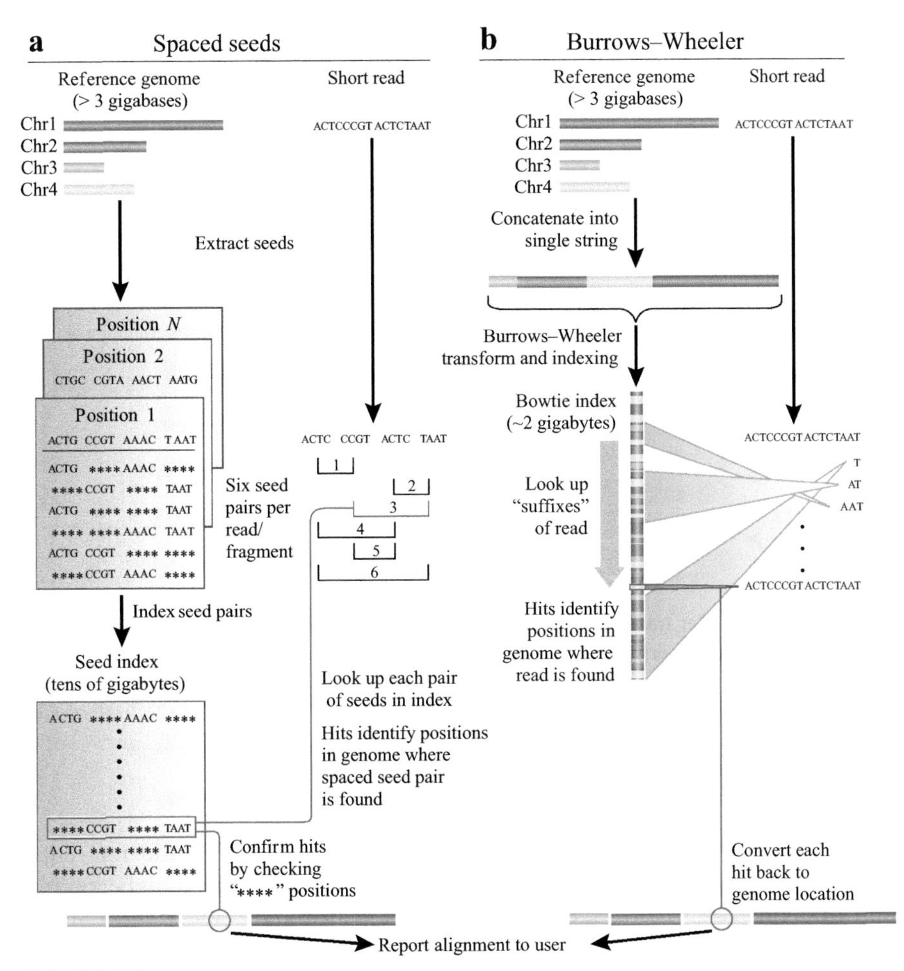

FIGURE 5.5
NGS read mapping approaches. Panel (a) shows the approach based on spaced seed indexing. In this illustration, spaced seeds extracted from the reference genome sequence are indexed by a hash table. Panel (b) shows the approach based on Burrows–Wheeler transform (BWT). In this example, the algorithm Bowtie performs mapping by looking up reads base by base, from right to left, against the transformed and indexed genome. (Adapted by permission from Macmillan Publishers Ltd: *Nature Biotechnology*, How to map billions of short reads onto genomes, C Trapnell and SL Salzberg, copyright 2009.)

To reduce demands on computational resources, another approach is through the use of Burrows–Wheeler transform (or BWT). BWT is an algorithm that applies reversible transformation to a block of text (genomic nucleotide sequence in the case of NGS data) to enable lossless data compression [23]. The transformation is achieved through a text reordering process that can be efficiently implemented using the suffix array data structure

```
(a)
              acaacg$    $acaacg
              $acaacg    aacg$ac
              g$acaac    acaacg$
acaacg$ →     cg$acaa →  acg$aca → gc$aaac
              acg$aca    caacg$a
              aacg$ac    cg$acaa
              caacg$a    g$acaac

(b)
        aac          aac          aac
      $acaacg      $acaacg      $acaacg
      aacg$ac      aacg$ac      aacg$ac
      acaacg$      acaacg$      acaacg$
      acg$aca      acg$aca      acg$aca
      caacg$a      caacg$a      caacg$a
      cg$acaa      cg$acaa      cg$acaa
      g$acaac      g$acaac      g$acaac
```

FIGURE 5.6
How Burrows–Wheeler transform (BWT) works for aligning NGS reads. Panel (a) shows the BWT procedure for a short example sequence 'acaacg.' Panel (b) shows how to use BWT to identify the locations of read sequences prefixed by 'aac.' (Adapted by permission from Springer Nature Customer Service Centre GmbH: Springer Nature, *Genome Biology*, Ultrafast and memory-efficient alignment of short DNA sequences to the human genome, B Langmead, C Trapnell, M Pop, and SL Salzberg, copyright 2009.)

(Figure 5.6a). After the transformation, efficient indexing of the reference genome sequence can be achieved using an algorithm proposed by Ferragnia and Manzini [24]. Figure 5.6(b) provides an example of how such a suffix array is used for lookup of *k*-mers in the indexed genome. The combined use of BWT, suffix array, and FM-index effectively reduces the amount of computer memory needed to store an indexed reference genome for fast mapping. As an example, using this approach the indexed human genome only takes 2–3 GB of computer memory instead of over 50 GB used by the spaced-seed indexing approach, and the run time is cut from hours to minutes. BWT is employed by algorithms such as BWA (Burrows–Wheeler Alignment) [25], Bowtie/Bowtie 2 [26, 27], and SOAP2 [28]. Figure 5.5b shows an example how this approach works in the case of Bowtie.

After global position(s) of seed sequences from NGS reads are located in the reference genome based on the use of either suffix tree or hash table, adjacent seeds are chained and the gaps between them are evaluated for mismatches. If this filtering process is passed and the read and the genomic region are matched, pairwise alignment is then used to verify the alignment and generate alignment results in SAM or BAM formats (see next) (Figure 5.4). This pairwise alignment procedure can be performed using different techniques, among which Smith–Waterman, Hamming Distance, and Needleman–Wunsch algorithms are often used. The Smith–Waterman algorithm is used in commonly used tools such as STAR [29]. Smith–Waterman is a local alignment procedure based on dynamic programming (DP) [30]. DP was first introduced by Needleman and Wunsch in 1970 to alignment of DNA and protein sequences towards the goal of producing an alignment for two sequences that achieves maximum alignment score. The Needleman–Wunsch algorithm itself, however, is used for global alignment [31]. Alignment methods such as Novoalign use Needleman–Wunsch. Hamming Distance [32] is a non-DP-based measure of dissimilarity between two sequences, i.e., the number of locations at which nucleotides are different. The HD-based method is used by RMAP [33], Bowtie [34], and mrsFAST [35].

5.3.2 Selection of Mapping Algorithms and Reference Genome Sequences

Table 5.1 lists some of the often-used mapping algorithms. When selecting aligners, factors including accuracy, speed, and computer memory requirement need to be considered. As these factors are usually conflicting, some aligners put more emphasis on accuracy while others stress speed and memory efficiency. If speed and memory efficiency are more important, Bowtie2 is recommended. If higher accuracy is preferred, hash table-based tools such as Novoalign, Stampy [36], and SHRiMP2 [37] are often used. BWA strikes a balance between speed and sensitivity. Most of these aligners are initially developed to map very short reads, such as those of 35 nucleotides from early Illumina sequencers. With the gradual increase in read length, these aligners have been adapted accordingly. For instance, BWA-MEM is an adaptation of the original BWA algorithm for aligning longer Illumina short reads [38].

For aligning much longer reads such as those from the PacBio and ONT platforms, aligners designed to handle long sequences, such as minimap2, GraphMap [39], BLASR [40], LAST [41], NGMLR [42], or Winnowmap/Winnowmap2 [43, 44] should be used. In general, mapping of long reads follows a similar seed-chain-extend approach as used for short-read aligners. For effective mapping, long reads are typically broken into shorter subsequences that are then used as seeds for finding exact matches in the reference genome. One of the challenges for long read mapping, however, is the significant number of short seeds extracted from each long read. To counter this problem and achieve fast mapping speed, a new approach based on the use of a minimum set of representative seeds is often used. This minimizer approach provides a quick method for sampling and summarization of *k*-mers in a long sequence, based on which the similarity of two long sequences can be assessed more readily. Based on this approach, if two long sequences share identical subsequences of sufficient length, the same *k*-mer (or minimizer) will be selected for the two subsequences [22, 45]. By adopting

TABLE 5.1
Commonly Used Alignment Methods

Name	Description	Reference
Minimap2	A general-purpose alignment tool for both long and short reads. Uses hash table for reference genome indexing. Achieves fast speed through use of minimizers. Splice-aware and can be used for long RNA-seq reads	[18]
BWA-MEM2	The often-used algorithm in the BWA package designed for short reads. Employs suffix array lookup of seed sequences and Smith–Waterman-based extended alignment	[48]
Bowtie2	A short-read aligner based on the use of BWT and FM for reference genome indexing, and Smith–Waterman or Needleman–Wunsch for local or global alignment	[27]
SOAP2	Uses BWT compression to index the reference genome to achieve high speed for short-read alignment	[28]
Stampy	A short-read aligner that has high sensitivity in mapping reads that contain variation(s) or diverge from the reference sequence. Uses fast hashing to build reference index and a statistical model for alignment	[36]
NGMLR	A mapper designed for PacBio and ONT long reads. Splits long reads to shorter anchor sequences for lookup using hashing and then deploys Smith–Waterman for final alignment	[42]
GraphMap	A long-read aligner that uses spaced seeds for hashing-based index construction and lookup, and then performs graph-based mapping and progressive refinement to achieve alignment of long but error-prone reads	[39]
LAST	Implements the standard seed-and-extend approach but with the use of adaptive seeds instead of fixed-length seeds	[41]

minimizers, localizing subsequences shared in two long sequences becomes more computationally tractable. To use minimap2 as an example, the step of indexing a reference genome is achieved through indexing minimizers from a reference genome in a hash table. To map long reads, minimizers from the query reads are used as seeds to find exact match(es) in the reference genome. Co-localized seeds are then linked together as chains for the extension step.

Besides mapping algorithms, selection of reference genome sequences, when multiple reference genome sequences are available, also affects mapping result. By the design of most current mappers, reads that are more similar to the selected reference sequence align better than those that deviate more from the reference. If the deviation is sufficiently large, it might be discarded as a mismatch. As a result, the use of different reference genome sequences can introduce a "reference bias." The use of any one particular reference genome invariably introduces this bias, as a single reference genome simply cannot accommodate sequence variations and polymorphisms that are naturally present in a population or species. This bias should be kept in mind, especially when the genetic background of the source organism is different from the reference genome. In this situation, comparison of mapping results from the use of different references can help select a reference that is more appropriate. More recent mapping strategies to counter the reference bias include use of a "major-allele" reference genome, in which the most common allele is used for each variant site [46]; multiple reference genomes, which is used by the Reference Flow method [47]; or graph-based reference genome (see Chapter 16, Section 16.6).

5.3.3 SAM/BAM as the Standard Mapping File Format

Mapping results generated from the various algorithms are usually stored in the SAM or BAM file format. SAM, standing for Sequence Alignment/Map, has a tab-delimited text format. It is human readable and easy to examine but relatively slow to parse. BAM, being the compressed binary version of SAM, is smaller in size and faster to parse. Due to their wide use, SAM/BAM have become the *de facto* standard for storing read mapping results. The basic structure of a SAM/BAM file is straightforward, containing a header section (optional) and an alignment section. The header section, if it exists, provides generic information about the SAM/BAM file and is placed above the alignment section. Each line in the header section starts with the symbol "@." For the alignment section there are 11 mandatory fields (listed in Table 5.2). An example of the SAM/BAM format is presented in Figure 5.7.

In the example shown in Figure 5.7, the header section contains two lines. The first line has the two-letter record type code HD, signifying it as the header line, which is always the first line if present. This record has two tags: VN, for format version, and SO, for sorting order (in this case the alignments are sorted by coordinate). The second line is for SQ, i.e., the reference sequence dictionary. It also has two tags SN and LN, for reference sequence name and

TABLE 5.2

Mandatory Fields in the SAM/BAM Alignment Section

Col	Field	Type	Description
1	QNAME	String	Query sequence read (or template) NAME
2	FLAG	Integer	bitwise FLAG
3	RNAME	String	Reference sequence NAME
4	POS	Integer	leftmost mapping POSition on the reference sequence
5	MAPQ	Integer	MAPping Quality
6	CIGAR	String	CIGAR string
7	RNEXT	String	Reference name of the NEXT read (For paired-end reads)
8	PNEXT	Integer	Position of the NEXT read (For paired-end reads)
9	TLEN	Integer	observed Template LENgth
10	SEQ	String	segment SEQuence
11	QUAL	String	ASCII of Phred-scaled base QUALity+33

A

```
Coor        12345678901234  5678901234567890123456789012345
Ref         TACGATCGAAGGTA**ATGACATGCTGGCATGACCGATACCGCGACA

+r001/1           CGAAGGTACTATGA*ATG
+r002            cggAAGGTA*TATGA
+r003                        TGACAT..............TACCG
-r001/2                                              ACCGCGACA
```

B

```
@HD VN:1.6 SO:coordinate
@SQ SN:ref LN:45
r001   99 ref  7 30 8M2I4M1D3M = 37  39 CGAAGGTACTATGAATG *
r002    0 ref  9 30 3S6M1P1I4M *  0   0 CGGAAGGTATATGA    *
r003    0 ref 16 30 6M14N5M    *  0   0 TGACATTACCG       *
r001  147 ref 37 30 9M         =  7 -39 ACCGCGACA         * NM:i:1
```

FIGURE 5.7
The SAM/BAM format for storing NGS reads alignment results. The alignment shown in panel (a) is captured by the SAM format shown in panel (b). In panel (a), the reference sequence is shown on the top with the corresponding coordinates. Among the sequences derived from it, r001/1 and r001/2 are paired reads. The bases in lower cases in r002 do not match the reference and as a result are clipped in the alignment process. The read r003 represents a spliced alignment. In panel (b), the SAM format contains 11 mandatory fields that are explained in more detail on Table 5.2.

reference sequence length, respectively. For the alignment section, while most of the fields listed in Table 5.1 are self-explanatory, some fields may not be so clear at first glance. The FLAG field uses a simple decimal number to track the status of 11 flags used in the mapping process, such as whether there are multiple segments in the sequencing (like r001 in the example) or if the SEQ is reverse complemented. To check on the status and meaning of these flags, the decimal number needs to be converted to its binary counterpart. For the POS field, SAM uses a 1-based coordinate system, that is, the first base of the reference sequence is counted as 1 (instead of 0). The MAPQ is the mapping quality score, which is calculated similarly to the Q-score introduced earlier ($MAPQ = -10 \times \log10(P_{MapErr})$). The CIGAR (or Concise Idiosyncratic Gapped Alignment Report) field describes in detail how the SEQ maps to the reference sequence, with the marking of additional bases in the SEQ that are not present in the reference, or missing reference bases in the SEQ. In the example above, the CIGAR field for r001/1 shows a value of "8M2I4M1D3M," which means the first eight bases matching the reference, the next two bases being insertions, the next four matching the reference, the next one being a deletion, and finally the last three again being matches. For more details (such as those on the different FLAG status) and full specification of the SAM/BAM format, refer to the documentation from the SAM/BAM Format Specification Working Group. It should be noted that BAM may also be used to store unaligned raw reads as an "off-label" use of the format. For example, the

PacBio platform outputs raw sequencing reads in unaligned BAM format. Besides the SAM/BAM format, an alternative format to store aligned reads is CRAM [49] designed by the European Bioinformatics Institute. CRAM files are smaller than equivalent BAM files, because CRAM uses a reference-based compression scheme, i.e., only bases that are different from the reference sequence are stored.

5.3.4 Mapping File Examination and Operation

After carrying out the mapping process, the mapping results reported in SAM/BAM files should be closely examined. Firstly, summary statistics, such as the percentage of aligned reads, especially uniquely mapped reads, should be generated. Currently the mapping rates are still far from 100%. Even under ideal conditions, most aligners find unique genomic position match for 70–75% of sequence reads. This inability to locate the genomic origin of a significant number of reads can be attributed to multiple factors, including the existence of repetitive sequences in most genomes, the relatively short length and therefore limited positioning information of most short NGS reads, algorithmic limitation, sequencing error, and DNA sequence variation and polymorphism in a population. Mapping performance improves with increasing read length and better-designed algorithms from active developments in this area.

Secondly, reads that map to multiple genomic locations, often called multireads, usually do not contribute to subsequent analysis and therefore should be filtered out. The ambiguity in the mapping of multireads is due to the aforementioned sequence deviation caused by polymorphism and mutation, sequencing error, and existence of highly similar sequences in the genome such as those from duplicated genes. Inclusion of these reads in downstream analysis may lead to biased or erroneous results. For most experiments, these reads should be excluded from further analysis. As filtering of multireads usually removes a significant number of reads, which may lead to potential loss of information, there are some algorithms (such as BM-Map [50]) that are designed to reuse multireads by probabilistically allocating them to competing genomic loci.

Thirdly, besides multireads, duplicate reads should also be identified and filtered out for many experiments. In a diverse non-enriched sequencing library, because of the randomness of the fragmentation process the chance of getting identical fragments is extremely low. Even with a PCR step to enrich DNA fragments, the chance of generating duplicate reads is still very low (usually < 5%), as the number of cycles in the PCR process is limited and the subsequent sequencing process is a random sampling of the DNA library (to varying depth). Existence of excessive numbers of duplicate reads therefore suggests PCR over-amplification. Duplicate reads can be detected based on sequence identity, but due to sequencing error this tends to underestimate the number of duplicate reads. It is more appropriate, therefore, to detect

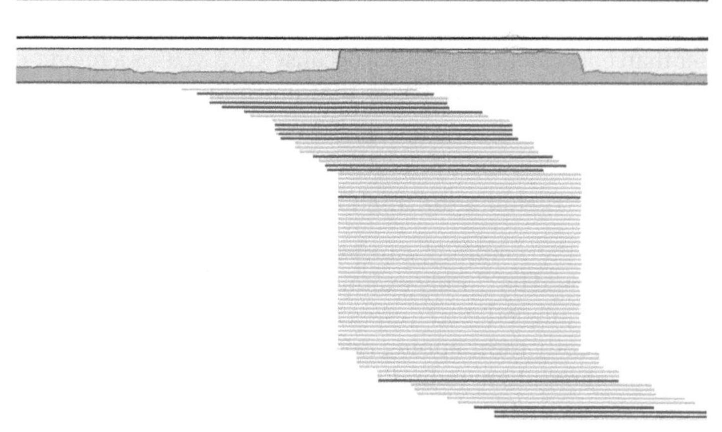

FIGURE 5.8
Detection of duplicate reads after the mapping process. Depth of coverage of the reference genomic region is shown on the top. Mapped reads, along with a set of duplicate reads that map to the same area, are show underneath. The green and red colors denote the two DNA strands. (Generated with CLC Genomics Workbench and used with permission from CLC Bio.)

duplicate reads after the mapping step (Figure 5.8). As technical duplicates caused by PCR over-amplification and true biological duplicates are indistinguishable, researchers should exert caution when making decisions on whether to remove duplicate reads from further analysis. While removing duplicate reads can lead to increased performance in subsequent analysis in many cases (such as variant discovery), in circumstances that involve less complex or mostly enriched sequencing targets, including those from an extremely small genome, or those used in RNA-seq or ChIP-seq, removing them can lead to loss of true biological information.

Furthermore, a variety of other steps can also be conducted to operate SAM/BAM files. These steps are usually provided by SAMtools and Picard, two widely used packages for operating SAM/BAM files. These operations include:

- SAM and BAM interconversion. SAMtools can also convert other alignment file formats to SAM/BAM
- Merging of multiple BAM files into a single BAM file
- Indexing of SAM/BAM files for fast random access
- Sorting reads alignment using various criteria, e.g., genomic coordinates, lanes, libraries, or samples
- Additional reads alignment filtering, such as removing paired reads that only one of the pair maps to the reference genome

```
ref 181 A 24  ,.$.....,,.,.,...,,,.,..^+. <<<+;<<<<<<<<<<<<=<;<;7<&
ref 182 C 23  ,.....,,.,.,...,,,.,..A <<<;<<<<<<<<<3<=<<<;<<+
ref 183 A 23  ,.$....,,.,.,...,,,.,...    7<7;<;<<<<<<<<<=<;<;<<6
ref 184 G 23  ,$.....,,.,.,...,,,.,....^l.  <+;9*<<<<<<<<<<=<<:;<<<<
ref 185 G 22  ...T,,.,.,...,,,.,....  33;+<<7=7<<7<&<<1;<<6<
ref 186 C 22  ....,,.,.,.A.,,,.,..G.  +7<;<<<<<<<<&<=<<:;<<&<
ref 187 G 23  ....,,.,.,...,,,.,....^k.   %38*<<;<7<<7<=<<<;<<<<<
ref 188 A 23  C..T,,.,.,...,,,.,.... ;75&<<<<<<<<<=<<<9<<:<<
```

FIGURE 5.9
The pileup file format as generated from SAMtools. A pileup file shows how sequenced bases in mapped reads align with the reference sequence at each genomic coordinate. The columns are (from left to right): chromosome (or reference name), genomic coordinate (1-based), reference base, total number of reads mapped to the base position, read bases, and their call qualities. In the read bases column, dot signifies match to the reference base, comma to the complementary strand, and 'AGCT' are mismatches. Additionally, the '$' symbol marks the end of a read, while '^' marks the start of a read and the character after the '^' represents mapping quality.

- Generation of a pileup format file (Figure 5.9) to show matching (or mismatching) bases from different reads at each genomic coordinate (SAMtools)
- Simple visualization using a text-based viewer for close examination of read alignment in a small genomic region (SAMtools)

SAMtools and Picard are very versatile in handling and analyzing SAM/BAM files. In fact, the steps mentioned earlier, that is, generation of alignment summary statistics and removal of multireads and duplicate reads, can be directly conducted with these tools. For example, both SAMtools and Picard have utilities to detect and remove duplicate reads called markdup and markduplicates, respectively. These utilities mark reads that are mapped to the same starting genomic locations as duplicates.

Lastly, in terms of examining mapping results, nothing can replace direct visualization of the mapped reads in the context of the reference genome. While a text-based alignment viewer, such as that provided by SAMtools, offers a simple way to examine a small genomic region, direct graphical visualization of mapping results by overlaying mapped read sequences against the reference genome provides a more intuitive way of examining the data and looking for patterns. This visualization process serves multiple purposes, including additional data QC, experimental procedure validation, and mapping pattern recognition. Commonly used visualization tools include Integrative Genomics Viewer (IGV) [51], Artemis [52], SeqMonk [53], JBrowse [54], and Tablet [55]. The UCSC and Ensembl genome browsers also provide visualization options by adding customized BAM tracks. Post-mapping data QC tools such as Qualimap 2 [56] also provide visual summaries on key metrics including overall coverage across the reference genome.

5.4 Tertiary Analysis

After the sequence read mapping step, subsequent analyses vary greatly with application. For example, the workflow for RNA-seq data analysis is different from that for mutation and variant discovery. Therefore, it is not possible to provide a "typical" workflow for all NGS data analyses in this chapter beyond the common steps of data QC, preprocessing, and read mapping. Chapters in Part III provide details on application-specific tertiary analytic steps and commonly used tools.

References

1. Cacho A, Smirnova E, Huzurbazar S, Cui X. A comparison of base-calling algorithms for Illumina sequencing technology. *Brief Bioinform* 2016, 17(5): 786–795.
2. Wick RR, Judd LM, Holt KE. Performance of neural network basecalling tools for Oxford Nanopore sequencing. *Genome Biol* 2019, 20(1):129.
3. Boza V, Brejova B, Vinar T. DeepNano: deep recurrent neural networks for base calling in MinION nanopore reads. *PLoS One* 2017, 12(6):e0178751.
4. David M, Dursi LJ, Yao D, Boutros PC, Simpson JT. Nanocall: an open source basecaller for Oxford Nanopore sequencing data. *Bioinformatics* 2017, 33(1):49–55.
5. Teng H, Cao MD, Hall MB, Duarte T, Wang S, Coin LJM. Chiron: translating nanopore raw signal directly into nucleotide sequence using deep learning. *GigaScience* 2018, 7(5):giy037.
6. Zeng J, Cai H, Peng H, Wang H, Zhang Y, Akutsu T. Causalcall: nanopore basecalling using a temporal convolutional network. *Front Genet* 2019, 10:1332.
7. FastQC: A Quality Control Tool for High Throughput Sequence Data [Online] (www.bioinformatics.babraham.ac.uk/projects/fastqc/)
8. Patel RK, Jain M. NGS QC Toolkit: a toolkit for quality control of next generation sequencing data. *PLoS One* 2012, 7(2):e30619.
9. Chen S, Zhou Y, Chen Y, Gu J. fastp: an ultra-fast all-in-one FASTQ preprocessor. *Bioinformatics* 2018, 34(17):i884–i890.
10. Albrecht S, Sprang M, Andrade-Navarro MA, Fontaine JF. seqQscorer: automated quality control of next-generation sequencing data using machine learning. *Genome Biol* 2021, 22(1):75.
11. Brown J, Pirrung M, McCue LA. FQC Dashboard: integrates FastQC results into a web-based, interactive, and extensible FASTQ quality control tool. *Bioinformatics* 2017, 33(19):3137–3139.
12. Ewels P, Magnusson M, Lundin S, Kaller M. MultiQC: summarize analysis results for multiple tools and samples in a single report. *Bioinformatics* 2016, 32(19):3047–3048.

13. Martin M. Cutadapt removes adapter sequences from high-throughput sequencing reads. *EMBnet J* 2011, 17(1):10–12.
14. Bolger AM, Lohse M, Usadel B. Trimmomatic: a flexible trimmer for Illumina sequence data. *Bioinformatics* 2014, 30(15):2114–2120.
15. De Coster W, D'Hert S, Schultz DT, Cruts M, Van Broeckhoven C. NanoPack: visualizing and processing long-read sequencing data. *Bioinformatics* 2018, 34(15):2666–2669.
16. Leger A, Leonardi T. pycoQC, interactive quality control for Oxford Nanopore Sequencing. *J Open Source Softw* 2019, 4(34):1236.
17. Hufnagel DE, Hufford MB, Seetharam AS. SequelTools: a suite of tools for working with PacBio Sequel raw sequence data. *BMC Bioinformatics* 2020, 21(1):429.
18. Li H. Minimap2: pairwise alignment for nucleotide sequences. *Bioinformatics* 2018, 34(18):3094–3100.
19. Li R, Li Y, Kristiansen K, Wang J. SOAP: short oligonucleotide alignment program. *Bioinformatics* 2008, 24(5):713–714.
20. Li H, Ruan J, Durbin R. Mapping short DNA sequencing reads and calling variants using mapping quality scores. *Genome Res* 2008, 18(11):1851–1858.
21. Raczy C, Petrovski R, Saunders CT, Chorny I, Kruglyak S, Margulies EH, Chuang HY, Kallberg M, Kumar SA, Liao A *et al.* Isaac: ultra-fast whole-genome secondary analysis on Illumina sequencing platforms. *Bioinformatics* 2013, 29(16):2041–2043.
22. Roberts M, Hayes W, Hunt BR, Mount SM, Yorke JA. Reducing storage requirements for biological sequence comparison. *Bioinformatics* 2004, 20(18):3363–3369.
23. Burrows M, Wheeler D. A block-sorting lossless data compression algorithm. In: *Digital SRC Research Report: 1994*: Citeseer; 1994.
24. Ferragina P, Manzini G. Opportunistic data structures with applications. In: *Proceedings 41st annual symposium on foundations of computer science: 2000*: IEEE; 2000: 390–398.
25. Li H, Durbin R. Fast and accurate long-read alignment with Burrows–Wheeler transform. *Bioinformatics* 2010, 26(5):589–595.
26. Langmead B, Salzberg SL. Fast gapped-read alignment with Bowtie 2. *Nat Methods* 2012, 9(4):357–359.
27. Langmead B, Wilks C, Antonescu V, Charles R. Scaling read aligners to hundreds of threads on general-purpose processors. *Bioinformatics* 2019, 35(3):421–432.
28. Li R, Yu C, Li Y, Lam TW, Yiu SM, Kristiansen K, Wang J. SOAP2: an improved ultrafast tool for short read alignment. *Bioinformatics* 2009, 25(15):1966–1967.
29. Dobin A, Davis CA, Schlesinger F, Drenkow J, Zaleski C, Jha S, Batut P, Chaisson M, Gingeras TR. STAR: ultrafast universal RNA-seq aligner. *Bioinformatics* 2013, 29(1):15–21.
30. Smith TF, Waterman MS. Identification of common molecular subsequences. *J Mol Biol* 1981, 147(1):195–197.
31. Needleman SB, Wunsch CD. A general method applicable to the search for similarities in the amino acid sequence of two proteins. *J Mol Biol* 1970, 48(3):443–453.
32. Hamming RW. Error detecting and error correcting codes. *Bell Syst Tech J* 1950, 29(2):147–160.

33. Smith AD, Xuan Z, Zhang MQ. Using quality scores and longer reads improves accuracy of Solexa read mapping. *BMC Bioinformatics* 2008, 9:128.
34. Langmead B, Trapnell C, Pop M, Salzberg SL. Ultrafast and memory-efficient alignment of short DNA sequences to the human genome. *Genome Biol* 2009, 10(3):R25.
35. Hach F, Hormozdiari F, Alkan C, Hormozdiari F, Birol I, Eichler EE, Sahinalp SC. mrsFAST: a cache-oblivious algorithm for short-read mapping. *Nat Methods* 2010, 7(8):576–577.
36. Lunter G, Goodson M. Stampy: a statistical algorithm for sensitive and fast mapping of Illumina sequence reads. *Genome Res* 2011, 21(6):936–939.
37. David M, Dzamba M, Lister D, Ilie L, Brudno M. SHRiMP2: sensitive yet practical SHort Read Mapping. *Bioinformatics* 2011, 27(7):1011–1012.
38. Li H. Aligning sequence reads, clone sequences and assembly contigs with BWA-MEM. *arXiv:13033997*, 2013.
39. Sovic I, Sikic M, Wilm A, Fenlon SN, Chen S, Nagarajan N. Fast and sensitive mapping of nanopore sequencing reads with GraphMap. *Nat Commun* 2016, 7:11307.
40. Chaisson MJ, Tesler G. Mapping single molecule sequencing reads using basic local alignment with successive refinement (BLASR): application and theory. *BMC Bioinformatics* 2012, 13:238.
41. Kielbasa SM, Wan R, Sato K, Horton P, Frith MC. Adaptive seeds tame genomic sequence comparison. *Genome Res* 2011, 21(3):487–493.
42. Sedlazeck FJ, Rescheneder P, Smolka M, Fang H, Nattestad M, von Haeseler A, Schatz MC. Accurate detection of complex structural variations using single-molecule sequencing. *Nat Methods* 2018, 15(6):461–468.
43. Jain C, Rhie A, Zhang H, Chu C, Walenz BP, Koren S, Phillippy AM. Weighted minimizer sampling improves long read mapping. *Bioinformatics* 2020, 36(Suppl_1):i111–i118.
44. Jain C, Rhie A, Hansen NF, Koren S, Phillippy AM. Long-read mapping to repetitive reference sequences using Winnowmap2. *Nat Methods* 2022.
45. Zheng H, Kingsford C, Marcais G. Improved design and analysis of practical minimizers. *Bioinformatics* 2020, 36(Suppl_1):i119–i127.
46. Shukla HG, Bawa PS, Srinivasan S. hg19KIndel: ethnicity normalized human reference genome. *BMC Genomics* 2019, 20(1):459.
47. Chen NC, Solomon B, Mun T, Iyer S, Langmead B. Reference flow: reducing reference bias using multiple population genomes. *Genome Biol* 2021, 22(1):8.
48. Vasimuddin M, Misra S, Li H, Aluru S. Efficient architecture-aware acceleration of BWA-MEM for multicore systems. In: *2019 IEEE International Parallel and Distributed Processing Symposium (IPDPS): 2019*: IEEE; 2019: 314–324.
49. Hsi-Yang Fritz M, Leinonen R, Cochrane G, Birney E. Efficient storage of high throughput DNA sequencing data using reference-based compression. *Genome Res* 2011, 21(5):734–740.
50. Yuan Y, Norris C, Xu Y, Tsui KW, Ji Y, Liang H. BM-Map: an efficient software package for accurately allocating multireads of RNA-sequencing data. *BMC Genomics* 2012, 13 Suppl 8:S9.
51. Thorvaldsdottir H, Robinson JT, Mesirov JP. Integrative Genomics Viewer (IGV): high-performance genomics data visualization and exploration. *Brief Bioinform* 2013, 14(2):178–192.

52. Carver T, Harris SR, Berriman M, Parkhill J, McQuillan JA. Artemis: an integrated platform for visualization and analysis of high-throughput sequence-based experimental data. *Bioinformatics* 2012, 28(4):464–469.
53. SeqMonk (www.bioinformatics.babraham.ac.uk/projects/seqmonk/)
54. Buels R, Yao E, Diesh CM, Hayes RD, Munoz-Torres M, Helt G, Goodstein DM, Elsik CG, Lewis SE, Stein L *et al.* JBrowse: a dynamic web platform for genome visualization and analysis. *Genome Biol* 2016, 17:66.
55. Milne I, Bayer M, Cardle L, Shaw P, Stephen G, Wright F, Marshall D. Table—next generation sequence assembly visualization. *Bioinformatics* 2010, 26(3):401–402.
56. Okonechnikov K, Conesa A, Garcia-Alcalde F. Qualimap 2: advanced multi-sample quality control for high-throughput sequencing data. *Bioinformatics* 2016, 32(2):292–294.

6

Computing Needs for Next-Generation Sequencing (NGS) Data Management and Analysis

The gap between our ability to pump out NGS data and our capability to extract knowledge from these data is getting broader. To manage and process the tsunami of NGS data for deep understanding of biological systems, significant investment in computational infrastructure and analytical power is needed. How to gauge computing needs and build a system to meet the needs, however, poses serious challenges to small research groups and even large research organizations. To meet this unprecedented challenge, the NGS field can borrow solutions from other "big data" fields such as high-energy particle physics, climatology, and social media. For biologists without much training in bioinformatics, while getting expert help is needed, having a good understanding of the various aspects of NGS data management and analysis is beneficial for years to come.

6.1 NGS Data Storage, Transfer, and Sharing

NGS has itself become a major producer of big data in scientific research. With the continuous drop in sequencing cost, the speed at which NGS data is pumped out will only pick up. This translates into a concomitant increase in the demand for more data storage, access, and processing power. Compared to files generated from other biological assays, such as gel pictures or even microarray data files, NGS files are much larger. For an individual lab, a single typical run generates data at the level of tens to thousands of gigabytes (GB) in compressed FASTQ format. After aligning to a reference genome, the processed files increase in size appreciably. Further analysis leads to the generation of more and more files and propagation in data volume. To accommodate raw and processed files from multiple runs, tens of terabytes (TB) of storage space or more is required. Storing and archiving these files are no trivial task. To make the situation even worse, the raw sequencing

DOI: 10.1201/9780429329180-8

signal intensity files in formats such as scanned images or movies are in the scale of TB from a single run (this amount is not counted in the data volume mentioned above). As these raw signal files accumulate, they can easily overwhelm most data storage systems. While these raw images files can be retained long term, newer sequencing systems process them on-the-fly and delete them by default once they have been analyzed to alleviate the burden of storing them. Oftentimes it is easier and more economical to rerun the samples in case of data loss rather than archiving these huge raw signal files.

Due to the huge size of most NGS files, transferring them from one place to another is non-trivial. For a small-sized project to transfer sequencing files from a production server to a local storage space, download via FTP or HTTP might be adequate if a fast network connection is available. As for network speed, a 1 Gbps network is essential while a 10/100 Gbps network offers improved performance for high traffic conditions. When the network speed is slow or the amount of data to be transferred is too large, the use of external hard drive might be the only option. When the data reaches the lab, for fast local file reading, writing, and processing, they need to be stored in a hard drive array inside a dedicated workstation or server.

For a production environment, such as an NGS core facility or a large genome center, which generates NGS data for a large number of projects, enterprise-level data storage system, such as DAS (Directly Attached Storage), SAN (Storage Area Network), or NAS (Network Attached Storage), is required to provide centralized data repositories with high reliability, access speed, and security. To avoid accidental data loss, these data storage systems are usually backed up, mirrored, or synced to data servers distributed at separate locations. For large-scale collaborative projects that involve multiple sites and petabytes to exabytes of data, the processes of data transfer and sharing pose more challenges, which prompts the development of high-capacity and high-performance platforms such as Globus.

Data sharing among collaborating groups creates additional technical issues beyond those dealt with by individual labs. A centralized data repository might be preferred over simple data replication at multiple sites to foster effective collaboration and timely discussion. Along with data sharing come also the issues of data access control and privacy for data generated from patient-oriented studies. In a broader sense, NGS data sharing with the entire life science community also increases the value of a research project. For this reason, many journals enforce a data sharing policy that requires deposition before publication of sequence read data and processed data into a publicly accessible database (such as the NCBI's Sequence Read Archive [SRA] or the European Nucleotide Archive [ENA]). To facilitate data interpretation and potential meta-analysis, relevant information about such an experiment must also be deposited with the data. Some organizations, such as the Functional Genomics Data Society, have developed guidelines on what information should be deposited with the data. For example, the MINSEQE

(Minimum Information about a high-throughput Nucleotide SEQuencing Experiment) guidelines specify the following information to be provided with the sequence read data and processed data: (1) description of the biological system, samples, and experimental variables; (2) experimental summary and sample data relationships; and (3) essential experimental and data processing protocols. Archiving NGS data and associated information for the community is a huge undertaking and requires sizeable investment in maintaining and growing the requisite infrastructure and expert support. The NCBI SRA repository was shut down in 2011 due to high costs and government budgetary constraints. However, because its vital importance to the community, NIH resumed its support to SRA later that year.

6.2 Computing Power Required for NGS Data Analysis

Processing the large volume of NGS data requires a lot of computing power. The question of how much computing power is needed is dependent on the type of analysis to be performed. For example, *de novo* assembly of a large genome requires much more computing power than resequencing for variant discovery, or transcriptomic analysis for the identification of differentially expressed genes. Therefore, to determine the computing power needed for a project, a lab, or an organization, the type(s) of NGS work to be performed need to be analyzed first. If the work will require intensive computation, or involve development and optimization of new algorithms and software tools, a high-performance cluster may be needed. On the other hand, if the work will use established workflow that does not require highly intensive computation, a powerful workstation may suffice. It is also advisable that the computer system be scalable to accommodate increases in future computing needs due to unforeseeable change of future research projects or further development of high-throughput genomics technologies.

For a small-sized project, the most basic system needed for NGS data analysis can be simply a 64-bit computer with 8 GB of RAM and two 2-GHz quad-core processors. With such a computer, basic mapping to a reference genome can be performed on obtained sequence reads. This basic setup allows handling of one dataset at a time. For simultaneous processing of multiple datasets or projects, high-performance computing (HPC) systems with more memory and CPU cores are needed. The number of cores an HPC system needs depends on the number of simultaneous tasks to be run at one time. For each task, the number of cores that is needed relies on the nature of the task and the algorithm that carries it out.

Besides the number of CPU cores, the amount of memory a system has also heavily affects its performance. Again memory needs depend on the number and complexity of jobs to be processed, for example, read mapping to a small genome may need only a few GB of memory while *de novo* assembly of a large genome may require hundreds of GB or even TB-level memory. The current estimation is that for each CPU core the amount of memory needed should not be less than 3 GB. In an earlier implementation of *de novo* assembly of the human genome using the SOAPdenovo pipeline (to be detailed in Chapter 12), a standard supercomputer with 32 cores (eight AMD quad-core 2.3 GHz CPUs) and 512 GB memory was used [1]. As a more recent example of the computing power needed for *de novo* genome assembly, a server with 64 cores (eight Intel Xeon X6550 8-core 2.00 GHz CPUs) and 2 TB RAM is used by a Swedish team [2]. For *de novo* assembly of small genomes such as those of microbes, a machine that contains at least 8 CPU cores, 256 GB of RAM, and a fast data storage system can get a job completed in a reasonable time frame. By current estimation, an 8-core workstation with 32 GB RAM and 10 TB storage can work for many projects that do not conduct *de novo* genome assembly.

The amount of time needed to complete a job varies greatly with the complexity of the job and accessible computing power. As a more concrete example, running the deep learning-based WGS variant calling tool DeepVariant (see Chapter 10 for details) needs 24–48 hours when using the minimum setting of a 8-core computer with 16 GB RAM, but the processing time is reduced by more than half when using a graphics processing unit (GPU) with 4 GB dedicated video RAM and CUDA support for parallel computing [3]. To map an RNA-seq dataset of 80 million 75-bp reads to the human genome using Bowtie on a computer equipped with 32 cores and 128 GB RAM, it took <2 hours and even less time in subsequent steps including normalization and differential expression statistical tests [4]. In a small RNA-seq study, with a 32-core and 132 GB memory workstation, processing 20 multiplex barcoded samples with a total of 160 million reads took a little over 2 hours for sample de-multiplexing, and about the same amount of time for read mapping to the host genome and small RNA annotation databases [5].

6.3 Cloud Computing

As clearly demonstrated above, NGS data storage, transfer, and sharing are no trivial tasks. And one limitation of a locally built computing system is its scalability. With the rates at which NGS technologies advance and sequencing costs drop being faster than those of development in the computer hardware industry, the gap between NGS data generation and our ability to handle and analyze them will only widen. To narrow this gap and speed up NGS data processing, the NGS community has embraced a trend from the long-existing model of local computing to cloud computing. Companies such as Amazon, Microsoft, and Google have been building mega-scale cloud computing clusters and data storage systems for end users to use over the Internet. Compared to local computing, cloud computing enables access to supercomputing and mass data storage capabilities without the need to build and maintain a local workstation, server, or HPC cluster.

At the core of cloud computing is virtualization technology, which allows an end user to create a virtual computer system on demand with the flexibility of specifying the number of CPU cores, memory size, disk space, as well as operating system that are required for a job. With this technology, multiple virtual computer systems can be run simultaneously on the same physical cloud server. The adoption of cloud computing for NGS data processing has demonstrated the advantages of this "supercomputing-on-demand" model, which include flexibility, scalability, and oftentimes cost savings. The flexibility and scalability offered by cloud computing allow a researcher to conduct NGS data analysis using supercomputing capabilities that previously only existed in large genome centers. Cost savings are achieved as the user only needs to pay for the time used by the user-configured computing instance.

Another advantage of using the cloud is on data sharing among researchers and projects. By providing a single centralized data storage, the cloud enables different groups located in different geographical locations to have access to the same dataset and share analytical results. Furthermore, with cloud computing the task of bringing software tools to the "big" NGS data can be realized more readily. In contrast to the large sizes of NGS data files, the software and scripts designed to process them are much smaller. Therefore, it is much easier and more efficient to download and install them to wherever the data is stored, rather than moving or replicating the high volumes of NGS data to where the tools are installed. By directly storing production data in the cloud, the burden of data transfer is greatly reduced; by coupling data and tools in the same place, optimal performance can be achieved.

While cloud computing enables users to off-load the hassle and cost of running and maintaining a local computing system, it does have downsides that need to be considered. One of the practical barriers of moving to the cloud is the speed of data transfer into and out of the cloud. It may take a week to upload 100 GB of data to the cloud using low-speed Internet connections. The question of whether to run analysis in the cloud is heavily dependent on the amount of data to be transferred and the computational complexity of the analytical steps. As a general rule, it is only worthwhile to upload data to the cloud for processing when the analytical task requires more than 10^5 CPU cycles per byte of data [6]. So for projects that deal with large amounts of data but do not involve a lot of highly intensive computational steps, more time may be spent on data transfer to the cloud rather than data processing. Other potential factors include data security, cost ineffectiveness under some circumstances, availability of analytical tools in the cloud environment, and network downtime. While users can access their data from anywhere on the

TABLE 6.1

Current Providers of Cloud Computing That Can Be Used for NGS Data Analysis

Provider	URL
Amazon Elastic Compute Cloud	http://aws.amazon.com/ec2/
Google Cloud	https://cloud.google.com
Microsoft Azure	http://azure.microsoft.com/
Rackspace	www.rackspace.com
IBM Cloud	www.ibm.com/cloud

Internet, the convenience also means the possibility of data security being breached or compromised. Some heavy users may find cloud computing not as cost effective as running a local server. While more and more tools are becoming available in the cloud, users still need to use due diligence to make sure that the tools they need are available. For users at places that suffer frequent network outage, cloud computing can be problematic as all cloud-based operations are dependent on Internet traffic.

Despite the potential downsides, cloud computing has been proven to be a viable approach for NGS data analysis. Table 6.1 is a list of some of the current cloud computing providers that can be used for NGS applications. To illustrate how cloud computing can be deployed for analyzing NGS data, below is an example on the conduct of reads alignment using the Amazon Elastic Compute Cloud (or EC2) Cloud. As the first step, input data files (FASTQ files and a reference genome file) are uploaded from a local computer to a "bucket" in the Amazon Simple Storage Service (S3). This cloud storage bucket, which is also used to hold program scripts and output files, can be created with the Amazon Web Services (AWS) Management Console, a unified interface to access all Amazon cloud resources. To initiate alignment, a workflow must be defined first using the Console's "create workflow" function. To define the workflow, the input sequence read files, the aligner script, and the saving location for alignment output files are specified. In the meantime, the number of Amazon EC2 instances required for the job, which determines memory and processor allocation, is also configured. After the configuration the job is submitted through the Management Console. When the instances are finished, alignment output files are deposited into the pre-specified file location in the S3 cloud storage.

6.4 Software Needs for NGS Data Analysis

To configure a cloud instance or set up a local workstation or server, operating system and software need to be selected and installed. While some NGS analysis software (such as CLC Genomics Workbench) can operate in the Windows environment, most tools only operate in the Unix (or Linux) environment. Therefore, Unix or Linux is usually the operating system installed on such a machine. Installing software in Unix or Linux is not as straightforward as in Windows, as uncompiled software source code downloaded from a developer site needs to be compiled first before being installed to a particular distribution of the operating system. If the reader is not familiar with the Unix/Linux environment and the command line interface it uses, an introductory book or web-based tutorial is suggested.

One approach to reducing the barrier of using tools developed for the Unix/Linux environment is to access them through a "bridging" system, such as Galaxy [7], that provides a more user-friendly interface to the command line tools. Developed by the Nekrutenko lab at Penn State and the Taylor lab at Johns Hopkins University, the Galaxy system provides a mechanism to deploy these tools via the familiar web browser interface, making them accessible to users regardless of the operating system they use. The Galaxy system is highly extensible, with latest tools being constantly wrapped for execution through the web interface. Besides providing a user-friendly interface, such a system also allows creation of data analysis workflow from different tools, which enables fast deployment of multiple tools in tandem, achievement of consistency and reproducibility, and sharing of analytical procedures with other researchers. Galaxy can be accessed through a publicly available server (e.g., usegalaxy.org), installed on a local instance, or in the cloud. With a public server, the user does not need to maintain a local server, but the usable storage space assigned to each account is usually limited and the computing resource is shared with many other users. Creating a local Galaxy instance in Unix/Linux or Mac OS takes some effort and the user does need to provide maintenance, but the user has more control on storage space, computing power, and selection and installation of tools from the entire collection of genomics tools that are made available through the Galaxy Tool Shed (currently has close to 10,000 tools). The Galaxy team has made it very easy to install a local instance by offering detailed and easy-to-follow instructions. Galaxy is also available in the cloud via dedicated instances in AnVIL, a cloud-based platform for data analysis, storage, and management built on the Google Cloud Platform (GCP) and supported by the U.S. National Human Genome Research Institute [8]. Besides providing a highly scalable computing environment, the Galaxy cloud instance in AnVIL also offers protection to private datasets while facilitates collaboration.

There are also other community projects that provide alternative platforms to facilitate user access to various NGS and other genomics analysis tools. Bioconductor, an open-source and open-development software project, is among the best known of these projects. This large-scale project is based on R, a programming language and software environment designed for statistical computing and graphics. With the goal of providing tools for the analysis and comprehension of high-throughput genomics data, the Bioconductor software library contains over 2,200 software packages as the time of this writing, many of which are designed or can be used to process NGS data. The R environment and the Bioconductor library can be installed in all major operating systems including Windows. The Bioconductor project web portal (www.bioconductor.org) and the R project site (www.r-project.org) provide detailed information and tutorials for the installation and use of these packages. Each tool is well documented with actual use examples provided.

Identifying, installing, and maintaining suitable NGS analysis software from an ever-growing number of tools for a local Unix/Linux workstation, a local Galaxy instance, or a local Bioconductor R library are not trivial. New software tools are developed and introduced constantly, while many existing ones are updated from time to time. To evaluate candidate packages and identify appropriate tools for installation and use, it is better to employ multiple test datasets, not just using those from computer simulation but also those from real-world biological samples. In addition, almost all tools have adjustable parameters, which should be set equivalently to facilitate performance comparison. Also in terms of performance, earlier NGS software usually does not take advantage of high-performance parallel computing. To increase performance and take full advantage of the multiple cores or nodes in an HPC system, more recent algorithms tend to use threading or message passing interface (MPI) to spread the work across multiple processes. Therefore, when evaluating NGS tools, it also helps to examine if these types of parallel processing are employed to take advantage of the power of multi-core computing architecture.

6.4.1 Parallel Computing

Parallelization, a computation term that describes splitting of a task into a number of independent subtasks, can significantly increase the processing speed of highly parallelizable tasks, which include many NGS data analysis steps. For example, although millions of reads are generated from a sequencing run, mapping of these reads to a reference genome is a process that is "embarrassingly parallel," as each read is mapped independently to the reference. As parallel computing can be efficiently carried out by GPUs since rendering of each pixel on a computer screen is also a highly parallel process, the integration of GPUs with CPUs in heterogeneous computing systems can increase throughput 10- to 100-fold, and turn individual computers into mini-supercomputers. While these systems can be applied to various aspects of NGS data analysis, many NGS analytical tools have yet to take full advantage of the power of parallel computing in such systems.

Parallelization is also an important factor in determining how increase in the number of CPU (or GPU) cores might affect actual NGS data processing performance. If a step is highly parallelizable, and the algorithm designed for it employs parallelization, then an increase in core number will most likely lead to improved performance. On the contrary, if the step is not readily parallelizable, or even if the task is parallelizable but the algorithm deployed does not use parallelization, simply having more cores may not lead to improvement in performance.

6.5 Bioinformatics Skills Required for NGS Data Analysis

For biologists and students in life sciences, acquiring basic bioinformatics skills is greatly advantageous, as biology has become more and more data rich and data driven. Understanding the basics of bioinformatics also facilitates communication with bioinformaticians on the conduct of more advanced tasks. In general, these skills include use of common computing environments, bioinformatic algorithms, and software packages. Below is a short list of bioinformatics skills required of biologists for NGS data handling:

- Familiarity with Unix/Linux, and the most commonly used commands in the Unix/Linux computing environment. This is essential to run jobs on an HPC cluster, a local machine running a Unix/Linux operating system such as Ubuntu or macOS, or a Windows Subsystem for Linux (or WSL) that provides a compatibility layer for running Linux tools natively on Windows.
- Basic knowledge of programming languages that are commonly used for NGS data analysis. These languages include R, Perl, and Python, all of which are open-source, easy to learn, and have a large user base for help and support. While programming is not required of biologists, understanding how an algorithm is executed step-by-step can be helpful, especially when a pre-existing tool does not work ideally for a special case and needs modification.
- Knowledge of key concepts in computational biology and biostatistics. Some computational methodologies developed in the field of computer science, especially machine learning and data mining, have been widely applied to high-throughput biological data processing. Artificial neural networks (ANNs), hidden Markov models (HMMs), and support vector machines (SVMs) serve as good examples in this domain. Statistical approaches such as linear and nonlinear regression are integrated into many genomics data analysis tools and should also be integrated into a biologist's knowledge base.
- Basic understanding of relational database. Most of the information currently available for the annotation and interpretation of NGS data is captured in various databases. Knowledge of database design and

structure is the basis to extract, manipulate, and process the information stored in these databases for generation of new biological knowledge. Knowing how to interact with the databases via SQL (Standard Query Language) or APIs (Application Programmer Interfaces) is also beneficial. The knowledge on relational databases and their operation also helps biologists to curate, organize, and disseminate the tremendous amount of information generated from NGS projects.

- Basic understanding and handling of computer hardware such as CPU, RAM, and storage. Although strictly speaking computer hardware is not in the realm of bioinformatics, it is nevertheless advantageous and economical to know how to put together a workstation and put it to use. It is also beneficial to understand how an HPC cluster, or a heterogeneous computing system, works through parallel processing, as NGS tools that are designed to take advantage of these computing systems usually function better and this knowledge can help evaluate and select those that maximize performance built in a computing system.

For bioinformaticians who deal with NGS data, on the other hand, the following is a list of skills that are needed:

- Proficiency with Unix-based operating systems;
- Familiarity with a programming language such as Python, Perl, Java, or Ruby;
- Familiarity with statistical software such as R, MATLAB, or Mathematica;
- Understanding of supercomputing, HPC (including parallel computing), and enterprise data storage systems;
- Knowledge of database management languages such as MySQL or Oracle;
- Familiarity with web authoring and web-based user interface implementation technologies;
- Understanding of molecular biology & genetics, cell biology, and biochemistry.

References

1. Li R, Zhu H, Ruan J, Qian W, Fang X, Shi Z, Li Y, Li S, Shan G, Kristiansen K *et al.* De novo assembly of human genomes with massively parallel short read sequencing. *Genome Res* 2010, 20(2):265–272.
2. Lampa S, Dahlo M, Olason PI, Hagberg J, Spjuth O. Lessons learned from implementing a national infrastructure in Sweden for storage and analysis of next-generation sequencing data. *GigaScience* 2013, 2(1):9.

3. Supernat A, Vidarsson OV, Steen VM, Stokowy T. Comparison of three variant callers for human whole genome sequencing. *Sci Rep* 2018, 8(1):17851.
4. Rasche A, Lienhard M, Yaspo ML, Lehrach H, Herwig R. ARH-seq: identification of differential splicing in RNA-seq data. *Nucleic Acids Res* 2014, 42(14):e110.
5. Farazi TA, Brown M, Morozov P, Ten Hoeve JJ, Ben-Dov IZ, Hovestadt V, Hafner M, Renwick N, Mihailovic A, Wessels LF *et al.* Bioinformatic analysis of barcoded cDNA libraries for small RNA profiling by next-generation sequencing. *Methods* 2012, 58(2):171–187.
6. Schadt EE, Linderman MD, Sorenson J, Lee L, Nolan GP. Computational solutions to large-scale data management and analysis. *Nat Rev Genet* 2010, 11(9):647–657.
7. Galaxy C. The Galaxy platform for accessible, reproducible and collaborative biomedical analyses: 2022 update. *Nucleic Acids Res* 2022, 50(W1):W345–W351.
8. Schatz MC, Philippakis AA, Afgan E, Banks E, Carey VJ, Carroll RJ, Culotti A, Ellrott K, Goecks J, Grossman RL *et al.* Inverting the model of genomics data sharing with the NHGRI Genomic Data Science Analysis, Visualization, and Informatics Lab-space. *Cell Genom* 2022, 2(1):100085.

Part III

Application-Specific NGS Data Analysis

7

Transcriptomics by Bulk RNA-Seq

7.1 Principle of RNA-Seq

Transcriptomic analysis deals with the questions of which parts of the genome are transcribed, and how active they are transcribed. In the past, these questions were mostly answered with microarray, which is based on hybridization of RNA samples to DNA probes that are specific to individual gene-coding regions. With this hybridization-based approach, the repertoire of hybridization probes, which are designed based on the current annotation of the genome, determines what genes in the genome or which parts of the genome are analyzed, and genomic regions that have no probe coverage are invisible. An NGS-based approach, on the other hand, does not depend on the current annotation of the genome. Because it relies on sequencing of the entire RNA population, hence the term RNA-seq, this approach makes no assumption as to which parts of the genome are transcribed. After sequencing, the generated reads are mapped to the reference genome in order to search for their origin in the genome. The total number of reads mapped to a particular genomic region represents the level of transcriptional activity at the region. The more transcriptionally active a genomic region is, the more copies of RNA transcripts it produces, and the more reads it will generate. RNA-seq data analysis is essentially based on counting of reads generated from different regions of the genome.

By counting the number of reads from transcripts and therefore being digital in nature, RNA-seq does not suffer from the problem of signal saturation that is observed with microarray at very high values. RNA-seq also offers a native capability to differentiate alternative splicing variants, which is basically achieved by detecting reads that fall on different splice junctions. While some specially designed microarrays, like the Affymetrix Exon Arrays, can be used to analyze alternative splicing events, standard microarrays cannot usually make distinctions between different splicing isoforms. Also different from microarray signals, which are continuous, raw RNA-seq signals (i.e., read counts) are discrete. Because of this difference, distribution model and methods of differential expression analysis designed for microarray data cannot be directly applied to RNA-seq data without modification.

DOI: 10.1201/9780429329180-10

7.2 Experimental Design

7.2.1 Factorial Design

Before carrying out an RNA-seq experiment, the biological question to be answered must be clear and well defined. This will guide experimental design and subsequent experimental workflow from sample preparation to data analysis. For experimental design, factorial design is usually used. Many experiments compare transcriptomic profile between two conditions, e.g., cancer vs. normal cells. This is a straightforward design, involving only one biological factor (i.e., cell type). Experiments involving a single factor may also have more than two conditions, e.g., comparison of samples collected from multiple tissues in the body in order to detect tissue-specific gene expression.

If a second biological factor (e.g., treatment of a drug) is added to the example of cancer vs. normal cell comparison, the experiment will have a total of four (2×2) groups of samples (Table 7.1). In this two-factor design, besides detecting the effects of each individual factor, i.e., cell type and drug treatment, respectively, the interacting effects between the two factors are also detected, e.g., drug treatment may have a larger effect on cancer cells than normal cells. If the factors contain more conditions, there will be a total of m×n groups of samples, with m and n representing the total number of conditions for each factor. Experiments involving more than two factors, such as adding a time factor to the above example to detect time-dependent drug effects on the two cell types, are inherently more complex and therefore more challenging to interpret, because in this circumstance it is not easy to attribute a particular gene expression change to a certain factor, or especially, to the interaction of these factors due to the existence of multiple interactions (three factors involve four different types of interactions).

7.2.2 Replication and Randomization

As with any experiment that requires proper statistical analysis, replication and randomization is an essential component of RNA-seq experimental design. Randomization refers to the random assignment of experimental

TABLE 7.1

Experimental Design Involving Two Biological Factors

	Cancer Cells	Normal Cells
Drug Treated	Cancer + Drug	Normal + Drug
Vehicle Treated	Cancer + Vehicle	Normal + Vehicle

subjects or targets into each group. This is to avoid introducing unwanted biases to the sample collection process. To generalize the gene expression differences observed from groups of samples to the respective populations, within-group variability in the expression of each gene has to be estimated, which requires replication. To meet this requirement at least three replicates need to be included within each group. The more replicates each group has, the more accuracy there is in within-group biological variability estimation, and therefore the more confidence to call a gene differentially expressed. While differential gene expression can be detected from unreplicated data, the results are limited to the tested samples and not easily generalizable. Due to the lack of knowledge on biological variation within each group, it is unrealistic to draw conclusions on the population from an unreplicated experiment.

7.2.3 Sample Preparation and Sequencing Library Preparation

Since gene expression is highly plastic and varies greatly with internal (such as tissue and cell type, developmental stage, circadian rhythm, etc.) and external (such as environmental stress) conditions, samples should be collected in a way that minimizes the effects of irrelevant factors. If the influence of such factors cannot be totally avoided, they should be balanced across groups. As many biological samples contain different cell types, this heterogeneity in cell composition is another factor that may confound data interpretation (potential heterogeneity in cell composition can be revealed by single-cell RNA-seq, to be covered in the next chapter). Use of homogeneous target cells is preferred whenever possible as this will greatly improve data quality and experimental reproducibility.

To prepare samples for RNA-seq, total RNA is first extracted from samples of contrasting conditions. High-quality RNA extraction is essential to obtain quality RNA-seq data. The leading cause of low-quality RNA is degradation. To detect the intactness of RNA molecules in samples, quality metrics such as RNA Integrity Number (or RIN) are often used. RIN is generated using a neural network-based algorithm from Bioanalyzer fragment size profile of an RNA sample, having values from 1 to 10 with 10 being the best possible sample quality. One prerequisite to extracting high-quality RNA is to snap-freeze tissue samples whenever possible to avoid potential RNA degradation. Under circumstances where this is not possible (e.g., sample collection in the field), RNA stabilizing reagent (such as RNAlater) can be used. For RNA samples prepared under certain circumstances such as those from historical samples or formalin-fixed paraffin-embedded (FFPE) clinical tissues, however, RNA degradation can be unavoidable. Even from highly degraded RNA samples such as these, useful data may still be generated with the use of RNA-seq library construction strategies that are more tolerant of RNA degradation [1]. Among other issues that might affect RNA-seq data quality is genomic DNA contamination. To remove DNA contaminants, a DNase treatment step during RNA extraction is recommended. It should also be noted that many RNA extraction protocols do not retain small RNA species including miRNAs. If these species are also of interest (more on small RNA sequencing in the Chapter 9), alternative protocols (such as the TRIzol method) need to be used.

Besides quality, the quantity of RNA sample available also determines RNA-seq library prep strategy. With enzyme engineering and as a result improvement in library construction chemistry, preparing sequencing libraries from increasingly small quantities of RNA is no longer a barrier. This usually involves signal amplification in order to produce enough library molecules for sequencing. While the needed amplification can introduce bias to the process, its impact on the detection of differential gene expression has been found to be limited [2]. The greatly increased sensitivity in RNA-seq library making has also made sequencing of transcripts from a single cell a reality (see next chapter for single-cell RNA-seq).

There are two general approaches to constructing RNA-seq sequencing libraries. One approach is based on direct enrichment of mRNA molecules, the major detection targets for the majority of RNA-seq work. Because most eukaryotic mRNAs have a poly-A tail (Chapter 3, Section 3.3.4), this approach is carried out through the use of poly-T capture probes to enrich for mRNA molecules carrying such a tail. The other approach is based on depletion of ribosomal RNAs (rRNAs), since rRNAs are usually the predominant but uninformative component in total RNA extractions. Depletion of rRNAs is typically based on hybridization using rRNA-specific probes, followed by their capture and subsequent removal. Other rRNA depletion strategies include degradation by duplex-specific nuclease (DSN), which relies on denaturation-reassociation kinetics to remove extremely abundant RNA species including rRNAs [3], and RNase H selective depletion, on the basis of binding rRNAs with rRNA-specific DNA probes and then using RNase H to digest bound rRNAs. Library prep based on the rRNA depletion approach is more tolerant of RNA degradation issues.

After mRNA enrichment or rRNA depletion, subsequent RNA sequencing library preparatory process typically involves reverse transcription to cDNA using random primers, followed by fragmentation (for short-read platforms) and attachment of sequencing adapters. This sequencing library construction process may also introduce bias to the subsequent sequencing and data generation. For example, the use of poly-T based mRNA enrichment introduces 3′ end bias, as this procedure precludes analysis of those mRNAs and other non-coding RNAs that do not have the poly-A tail structure [4]. If these RNA species are of interest, a library prep process based on the use of rRNA depletion can be employed.

Compared to short-read sequencing, long-read sequencing offers new RNA-seq capabilities and options that are impossible or difficult to perform using short-read sequencing. For example, both ONT and PacBio platforms provide full-length transcript sequencing, and thereby enable characterization and quantification of gene expression at the splicing isoform level. Besides sequencing cDNA, ONT sequencing can also sequence RNA molecules directly that removes some of the biases introduced during cDNA-based library construction.

7.2.4 Sequencing Strategy

Sequencing depth and read length are two major factors to consider when sequencing bulk RNA-seq libraries especially with short-read sequencing. The factor of sequencing depth, that is, how many reads to obtain, is based on a number of factors, mainly the size of the organism's genome, the purpose of the study, and ultimately statistical rigor (effect size and statistical power). Small genomes, such as those of bacteria, require less reads to analyze than large genomes, such as those of mammalian species. If the purpose of a study is to identify differentially expressed genes among those expressed at intermediate to high abundance levels, it requires fewer reads than studies that aim to encompass low-abundance genes, study alternative splicing events, or discover new transcripts. As a general guideline, for a gene expression profiling experiment targeting intermediate- to high-abundance transcripts, a sequencing depth of 5–25 million reads, depending on the size the genome, is suggested. To cover transcripts of lower abundance or common alternative splicing variants, 20–50 million reads are suggested. For more thorough coverage of the transcriptome and/or discovery of new transcripts, 100–300 million reads are often needed. As for sequencing read length, for gene expression profiling, single end 50–75 bp reads are typically long enough to map to their originating genes in the genome. For assembly of new transcripts and/or identification of alternative splicing isoforms, longer and often paired-end sequencing reads, such as paired-end 150 bp reads, are often acquired.

The number of sample replicates also largely affects the detection power of an RNA-seq study, as sample replication provides estimation on gene expression dispersion across biological subjects within a group. While a minimum of three biological replicates is commonly used, specially designed RNA-seq power analysis tools can be used to calculate sample size to achieve a detection power. These tools, including Scotty [5], ssizeRNA [6], PROPER [7], and RnaSeqSampleSize [8], are designed based on different statistical models. For example, while PROPER and RnaSeqSampleSize are based on negative binomial model, Scotty and ssizeRNA use Poisson-lognormal and linear models, respectively. As to be covered more later in this chapter (Section 7.3.3 "Identification of Differentially Expressed Genes"), these models provide different approximations to RNA-seq data distribution. Sample size calculations using these tools require as input a number of parameters, including the total number of genes expressed, the percentage of genes expected to be differentially expressed, the minimal fold change needed to call differential expression, false discovery rate, average read count (related to sequencing depth), and the desired statistical power. Because most of the parameters are not known *a priori*, recently published data collected from similar conditions may be used to provide some guidance. To start on a species or cell type that has not yet been studied, it might be useful to try out a small number of samples first to get a general idea on the composition of the target transcriptome and the variability between biological replicates. Besides detection power, experimental and sequencing costs are also key factors in deciding sample size and sequencing depth. For projects on a budget, it has been reported that increasing the number of biological replicates is more effective in boosting detection power than increasing sequencing depth [9].

Besides sequencing depth and read length, other considerations when planning for sequencing samples include how to arrange samples on a sequencer in terms of flow cell or lane assignment. Here a balanced block design [10] should be used to minimize technical variation due to flow cell-to-flow cell or lane-to-lane difference. In such a design, samples from different conditions are multiplexed on the same flow cell(s) or lanes, instead of running different samples or conditions on separate flow cell(s) or lanes.

7.3 RNA-Seq Data Analysis

7.3.1 Read Mapping

The first step after an RNA sequencing run is to examine the run summary with regard to the total number of reads generated, quality score distribution, GC content, and other indices of the sequencing run as detailed in Chapter 5. Based on such QC results, reads filtering and base trimming can be conducted to remove low-quality reads or basecalls. The distribution of sequencing depth across samples should also be checked to ensure all samples receiving expected numbers of reads. In addition, there are tools such as subSeq [11] that can be used to determine whether the desired detection power is reached and whether additional sequencing would lead to significant increase in detection sensitivity.

The subsequent mapping of RNA-seq reads to a reference genome is in general more complex than the general read mapping procedure described in Chapter 5. Because eukaryotic mRNAs are generated from splicing out of introns and joining of exons, many RNA-seq reads may not map contiguously to the reference genomic sequence. As prokaryotic mRNA generation does not involve splicing, the aligners introduced in Chapter 5 for contiguous (ungapped) mapping, such as Bowtie and BWA, can still be used. Mapping of eukaryotic RNA-seq reads, however, creates a challenge to these aligners. To address this challenge, two approaches have been developed. One is to use the current gene exonic annotation in the reference genome to build a database of reference transcript sequences that join currently annotated exons. RNA-seq reads are then searched against this reference transcripts database using the ungapped read aligners. Examples of annotation-guided mappers are RUM [12] and SpliceSeq [13]. These mappers may produce better outcomes when high accuracy and reliability are emphasized. This approach, however, does not provide the capability to discover novel transcripts. In addition, it leads to high rate of multi-mapping, as a read that maps to common exon(s) shared by multiple splicing isoforms of a gene is counted multiple times.

The other approach conducts *ab initio* splice junction detection, and therefore does not depend on genome annotation. Depending on their methodology, *ab initio* spliced mappers can be classified into two categories: methods using "exon-first" and those using "seed-and-extend." The exon-first methods include TopHat/TopHat2 [14, 15], MapSplice [16], SpliceMap [17], and GEM [18]. They first align reads to a reference genome to identify unspliced continuous reads (i.e., exonic reads first), and then predict splice junctions out of the initially unmapped reads based on the initial mapping results. Taking TopHat/TopHat2 as an example, they first use Bowtie/Bowtie2 to align reads to the reference genome. Reads that map to the reference continuously without interruption are then clustered based on their mapping position. The clusters, supposedly representing exonic regions, are used to search for splicing junctions from the remaining reads. The seed-and-extend methods, on the other hand, use part of reads as substrings (or *k*-mers) to initiate the mapping process, followed by extension of candidate hits to locate splicing sites. Examples of methods in this category include STAR [19], HISAT/HISAT2 [20, 21], and GMAP/GSNAP [22]. Among these methods, STAR employs a two-step process for gapped alignment. The first is a seed searching step, aiming to sequentially locate substrings of maximum length from a read that each matches exactly to one or more substrings in the reference genome. If this step does not reach the end of the read due to the presence of mismatches, it will use the located seed region(s) as anchors to extend the alignment. In the second step, alignment of the entire read sequence is built by joining all the seed regions located in the first step. HISAT applies an algorithm called hierarchical indexing to achieve splice-aware alignment. It starts with a global search using FM indexing of the whole reference genome to identify the genomic location(s) of a read using part of its sequence as seed. Such a location is then used as an anchor to extend the alignment. Once the alignment cannot be extended further, e.g., reaching a splicing junction, a local search is then performed using FM indexing of the local region to map the rest of the read (Figure 7.1). A hybrid strategy combining the two is also used sometimes, with the exon-first approach employed for mapping unspliced reads and the seed-and-extend approach for spliced reads. As they do not rely on current genomic annotations, these *ab initio* methods are suitable to identify new splicing events and variants.

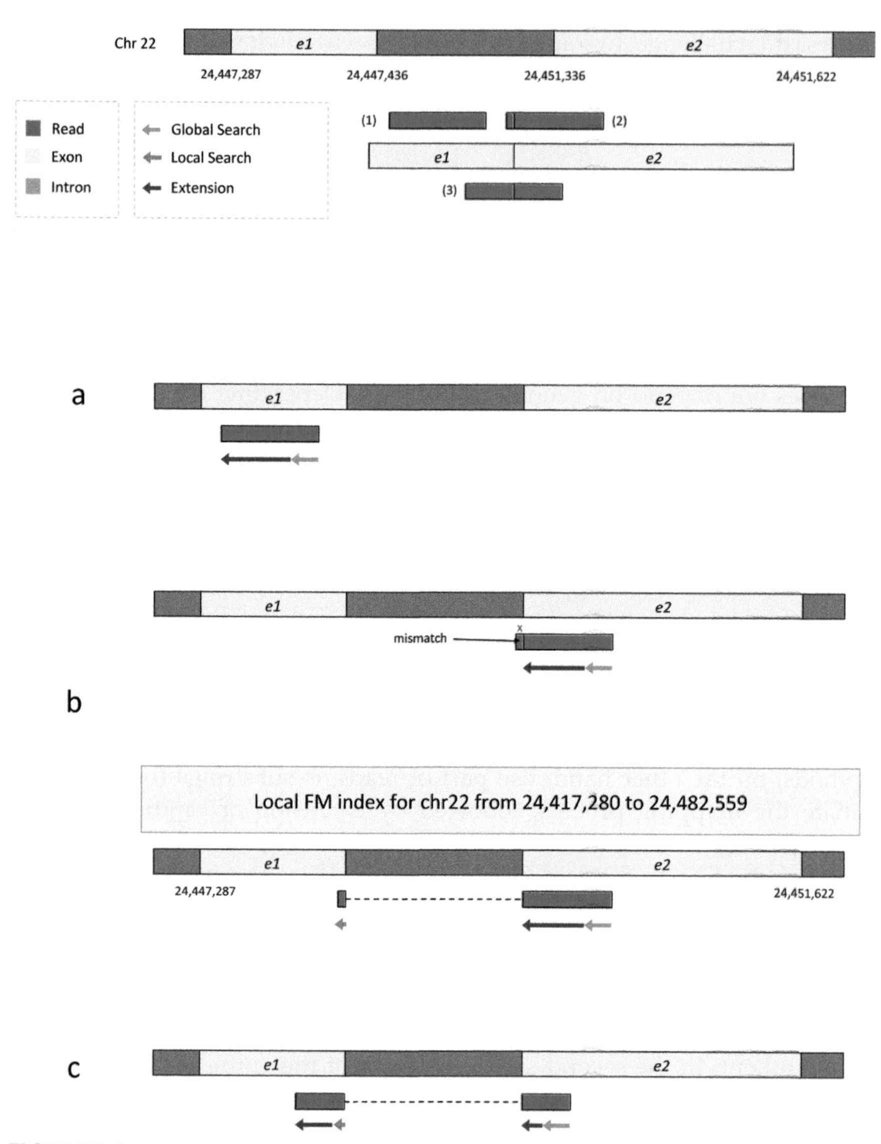

FIGURE 7.1
Mapping of RNA-seq reads with HISAT. Three representative reads are shown on the top, i.e., one exonic read (1), one read spanning a splice junction with a short anchor in one exon (2), and another junction-spanning read with long anchor in each exon (3). Panel (a) shows alignment of Read1 with a global FM genome index search using partial read sequence, followed by an extension step to align the rest of the read. Panel (b) show that when the global search and extension are halted at the junction, a local search using the region's FM index is performed to align the remaining short sequence. Panel (c) shows that to align Read 3, a second extension step is conducted after the local FM index search. The shown exemplary reads are error-free and 100 bases in length. (Adapted by permission from Springer Nature Customer Service Centre GmbH: Springer Nature, *Nature Methods*, HISAT: a fast spliced aligner with low memory requirements, Kim, D., Langmead, B. & Salzberg, S., Copyright 2015.)

To map long RNA-seq reads generated on the PacBio and ONT platforms, some of the short-read mappers introduced above can still be used, such as GMAP [23]. More commonly, however, this task is performed using tools specially designed for long reads. For example, minimap2 (introduced in Chapter 5) has a splice-aware option for mapping long RNA sequencing reads. Other currently available tools include deSALT [24], GraphMap2 [25], and uLTRA [26].

The percentage of reads that are mapped to the genome is an important QC parameter. While it is variable depending on a number of factors such as aligning method and species, this number usually falls within the range of 70–90%. The percentage of reads that map to rRNA regions is dependent on and a measure of the efficiency of the rRNA depletion step. Due to technical and biological reasons, it is usually impossible to remove all rRNA molecules. The percentage of rRNA reads can vary greatly, from 1–2% to 35% or more. For downstream analysis rRNA reads are filtered out so they do not usually affect subsequent normalization. Duplicate reads, a common occurrence in an RNA-seq experiment, can be caused by biological factors, such as over-presentation of a small number of highly expressed genes, and/or technical reasons, such as PCR over-amplification. It is possible to have a high percentage of duplicate reads, e.g., 40–60%, in a run. While it is still debatable as to how to treat duplicate reads, because of the biological factors involved in their formation they should not be simply removed. Some experimental approaches, such as removing some of the highly expressed genes prior to library construction, or using paired-end reads, can help reduce the number of duplicate reads. With regard to genomic coverage, RNA-seq QC tools, including RNA-SeQC 2 [27], RSeQC [28], and QoRTs [29], report on the percentage of reads that are intragenic, that is, those that map within genes (including exons or introns), or intergenic, for those that map to genomic space between genes. These tools also report other data quality metrics, including percentage of total aligned reads, percentage of rRNA reads, as well as rate of duplicate reads.

If the species under study does not have a sequenced reference genome against which to map RNA-seq reads, two approaches exist. One is to map the reads to a related species that has a reference genome, while the alternative is to assemble the target transcriptome *de novo*. The *de novo* assembly approach is more computationally intensive, but it does not rely on reference genomic sequence. Currently available *de novo* transcriptome assemblers include rnaSPAdes [30], Trinity [31], Bridger [32], Trans-ABySS [33], SOAPdenovo-Trans [34], Oases [35], and StringTie/StringTie2 [36, 37]. Among these assemblers rnaSPAdes and StringTie2 can be used with long RNA reads, or a hybrid of long and short reads [38, 39]. These *de novo* assemblers are suited when no related species or only very distantly related species with a reference genome exists, or the target genome, despite with available reference sequence, is heavily fragmented or altered (such as in tumor cells). It should also be noted that if a related reference genome exists with 85%

or higher sequence similarity with the species under study, mapping to the related genome may work equally well, or even better, compared to the *de novo* assembly approach. This is especially true when studying alternative splicing variants. These *de novo* approaches are also applicable to cases where aberrant transcripts are expected, or novel transcripts are the detection targets. For these *de novo* assembly approaches, paired-end sequencing or long-read sequencing are more advantageous compared to single-end short reads.

7.3.2 Quantification of Reads

After mapping of reads, the number of reads mapped to each gene/transcript needs to be counted to generate a table with rows representing genes/transcripts and columns different samples. Such an expression matrix is the basis for subsequent differential expression determination. This read quantification process can be performed using tools such as featureCounts as part of Subread [40], htseq-count as distributed with the HT-Seq Python framework [41], RSEM [42], eXpress [43], or Cufflinks [44]. Among these tools, featureCounts and htseq-count are read count-based, requiring as input SAM/BAM alignment files and a genomic feature annotation GFF/GTF file. They generally discard reads that map to multiple regions in the genome or overlap multiple genomic features. RSEM, eXpress, and Cufflinks, on the other hand, are model-based, requiring as input SAM/BAM alignments as well as a transcriptome reference file containing transcript sequences. They assign multi-mapped or ambiguous reads to transcripts based on probability from the use of the expectation-maximization algorithm. Because of the differences in how they quantify genes/transcripts, the selection of counting methods has been shown to have an effect on quantification results [45].

Gene/transcript expression can also be quantified without mapping reads to a reference genome or transcriptome. Examples of such mapping-independent algorithms include kallisto [46], Salmon [47], and Sailfish [48]. These methods rely on pseudo- or quasi-alignment of *k*-mers extracted from reads, instead of the entire reads, to achieve transcript quantification at vastly faster speed. For example, kallisto is a pseudoaligner that aligns *k*-mers in reads to a hash table built from *k*-mers that represent different transcripts in the reference transcriptome. Although not relying on mapping of entire reads, this method enables rapid determination of the compatibility of reads with target transcripts, as it preserves the key information needed for transcript quantification. Since only *k*-mers need to be aligned to the hash table, the speed is greatly increased with similar quantification performance to the mapping-based methods above. They perform well on highly expressed protein-coding genes but less so on rare or short transcripts [49].

7.3.3 Normalization

As mentioned previously, the basic principle of determining gene expression levels through RNA-seq is that the more active a gene is transcribed, the more reads we should be able to observe from it. To apply this basic principle to gene expression quantification and cross-condition comparison, at least two factors must be taken into consideration. The first is sequencing depth. If a sample is split into two halves, and one half is sequenced to a depth that is twice of that of the other, for the same gene the former will generate twice as many reads as the latter although both are from the same sample. The other factor is the length of gene transcript. If one gene transcript is twice the length of another gene transcript, the longer transcript will also produce twice as many reads as the shorter one. Because of these confounding factors, prior to comparing abundance of reads from different genes across samples in different conditions, the number of reads for each gene needs to be normalized against both factors using the following formula to ensure different samples and genes can be directly compared,

$$e_{i,j} = \frac{g_{i,j} \times SF}{a_i \times l_j}$$

where $e_{i,j}$ is the normalized expression level of gene j in sample i, $g_{i,j}$ is the number of reads mapped to the gene in the same sample, a_i is the total number of mapped reads (depth) in sample i, and l_j is the length of gene j. SF is a scaling factor and equals to 10^9 when $e_{i,j}$ is presented as RPKM or FPKM (Reads, or Fragments [for paired-end reads], per Kilobase of transcript per Million mapped reads).

The calculation of RPKM or FPKM is the simplest form of RNA-seq data normalization. In a nutshell, normalization deals with non-intended factors and/or technical bias, such as those that lead to unwanted variation in total read counts in different samples. By correcting for the unwanted effects of these factors or bias, the normalization process puts the focus on the biological difference of interest, and makes samples comparable. Since the introduction of RKPM or FKPM as an early normalization approach for RNA-seq data, other methods of normalization have also been developed. Some of these methods employ a similar strategy to adjust for sequencing depth. This group of methods normalize RNA-seq data through dividing gene read counts by either (1) the total number of mapped reads (i.e., the total count approach); (2) the total read count in the upper quartile (the upper quartile approach) [50]; and (3) the median read count (the median approach). Another method called quantile normalization sorts gene read count levels and adjusts quantile means to be the equal across all samples, so that all samples have the same empirical distribution [51]. These sequencing depth-based methods do not normalize against gene length, as it is not needed if the goal is to detect relative expression changes of same genes between groups, rather than compare relative abundance levels of different genes in the same samples.

Subsequently, more sophisticated normalization approaches are developed based on the assumption that the majority of genes are not differentially expressed, and for those that show differential expression, the proportion of up- and down-regulation is about equal. These approaches include those that are employed by two commonly used RNA-seq analysis tools, DESeq2 [52] and edgeR [53]. DESeq2 employs a method called relative log expression, or RLE, which is essentially carried out through dividing the read count of each gene in each sample by a scaling factor. To compute the scaling factor for each sample, the ratio of each gene's read count in a sample over its geometric mean across all samples is first calculated. After calculating this ratio for all genes in the sample, the median of this ratio is used as the scaling factor. The edgeR package employs a approach called Trimmed Means of M-values, or TMM [54]. In this approach, one sample is used as the reference and others as test samples. TMM is computed as the weighted mean of gene count log ratios between a test sample and the reference, excluding genes of highest expression and those with the highest expression log ratios. Based on the assumption of no differential expression in the majority of genes, the TMMs should be 1 (or very close to 1). If not, a scaling factor should be applied to each sample to adjust their TMMs to the target value of 1. Multiplying the scaling factor with the total number of mapped reads generates effective library size. The normalization is then carried out through dividing raw reads count by the effective library size, i.e., normalized read count = raw read count/(scaling factor × total number of mapped reads).

Among other approaches are those that use iterative processes to achieve normalization, as exemplified by TbT [55], DEGES [56], and PoissonSeq [57]. Based on the same assumption that there is no overall differential expression, these methods use a multi-iteration process. For example, DEGES, or Differentially Expression Genes Elimination Strategy, uses a process to repeatedly remove potential differential genes until their elimination, prior to calculating the final normalization factor. It starts with using any of the normalization methods introduced above, e.g., TMM, followed by a test for differential expression using a differential detection method (to be introduced next). After removal of the DE genes, the same process is repeated until convergence.

There are also normalization methods that use a list of housekeeping genes or spike-in controls as normalization standard. The use of housekeeping genes or spike-in controls is for conditions in which the assumption that the majority of genes are not differentially expressed might be violated. In this approach, a set of constitutively expressed housekeeping genes that are known to stay unchanged in expression under the study conditions, or a panel of artificial spike-in controls that mimic natural mRNA and are added to biological samples at known concentrations, is used as the basis against which other genes are normalized.

7.3.4 Batch Effect Removal

Normalization cannot fully address signal variation caused by collection of samples in batches. While it is ideal to collect all samples at once using a balanced experimental design, due to logistical or realistic considerations samples are sometimes collected in different batches that may involve different personnel, instrument, or protocol. Data collected in batches may introduce unwanted, non-biological variation in the RNA-seq signal, which will confound downstream analysis if not removed. Principal components analysis (PCA) can be used as an intuitive tool to detect batch effects, as it provides a convenient way to visualize relative relationships among samples, and samples are expected to cluster together based on experimental design (Figure 7.2). Tools such as BatchQC [58] can also be used to detect potential batch effects. Once batch effects are detected, specially designed tools, such as ComBat [59], RUVSeq [60], and svaseq [61], can be used to remove them. Among these tools, ComBat uses an empirical Bayes approach to adjust for batch effects, making it robust for data collected from batches with small sample sizes. ComBat-seq is an adaptation of the original ComBat for RNA-seq count data, including the use of a negative binomial regression model [62]. Besides removing batch effects from a single experiment, these tools can also be employed for integrated analysis of large datasets collected from multiple experiments or groups to boost detection power.

7.3.5 Identification of Differentially Expressed Genes

To compare gene expression in different groups and identify DE genes, the distribution model of the data needs to be established first in order to decide on the appropriate statistic tests to be used. While microarray data can be treated as normally distributed variables after log transformation, the RNA-seq read count values, being discrete in nature, cannot be approximated by continuous distributions even after transformation. In general, count data, including the RNA-seq data, follows the Poisson distribution, which is characterized by the mean of the distribution being equal to the variance. While this distribution can be and has been used to model RNA-seq data [50, 63], it has also been observed that in RNA-seq data genes with larger mean counts tend to have greater variance, causing the over-dispersion problem [64] (see Figure 7.3). To deal with this problem, an over-dispersed Poisson process, or as an approximation the negative binomial distribution, is often applied. Other distribution models that have been used in RNA-seq data analysis tools, including the Poisson log-linear model used by PoissonSeq [57] and the log-normal model used by limma [65] and Ballgown [66], have also been found to perform well under many circumstances.

On the identification of DE genes based on these models, there is a list of methods (Table 7.2) to choose from, among which the commonly used ones are baySeq [67], Cuffdiff 2 [68], DEGseq [69], DESeq2, EBSeq [70], and

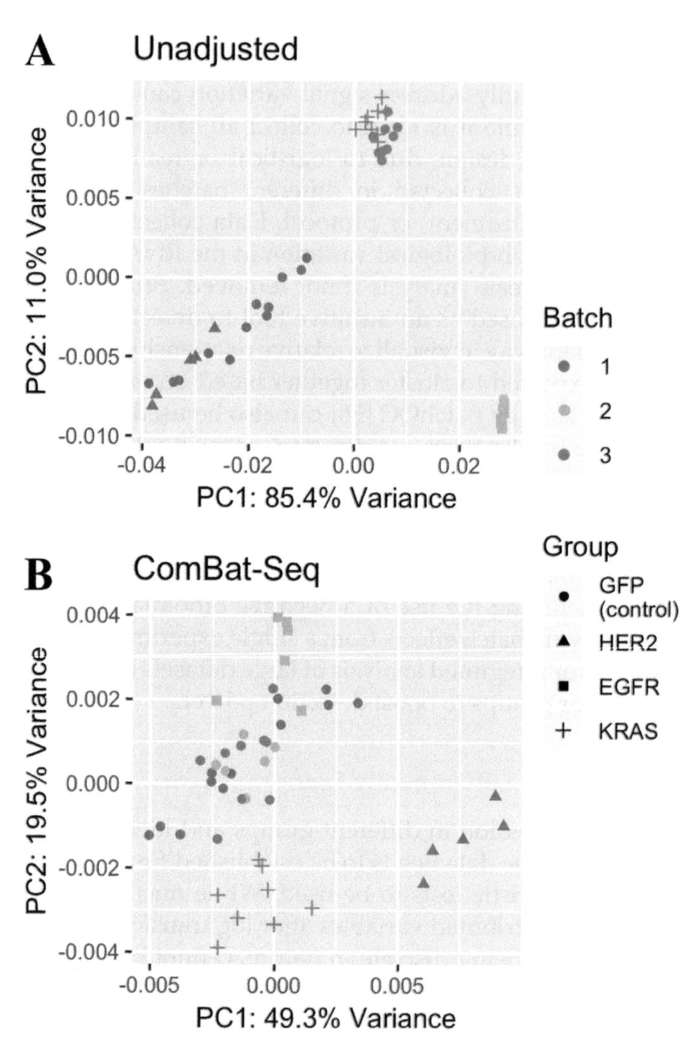

FIGURE 7.2
Removal of batch effects using ComBat-seq. PCA plots are shown before (A) and after (B) batch effects correction. In this example, three batches of breast cancer tissue samples that overexpress three genes (HER2, EGFR, and KRAS) separately with their controls expressing GFP are shown. The correction effectively removes batch effects on the control samples from the three batches, while maintains the effects of the transgenes. (Adapted from Y Zhang, G Parmigiani, *ComBat-seq*: batch effect adjustment for RNA-seq count data, *NAR Genomics and Bioinformatics* 2020, 2(3):lqaa078.)

edgeR [53]. While DEGseq has been developed based on the Poisson distribution, baySeq, Cuffdiff 2, DESeq2, EBSeq, and edgeR have been designed on the negative binomial distribution. To detect DE genes, these packages use different approaches. For example, baySeq and EBSeq employ empirical

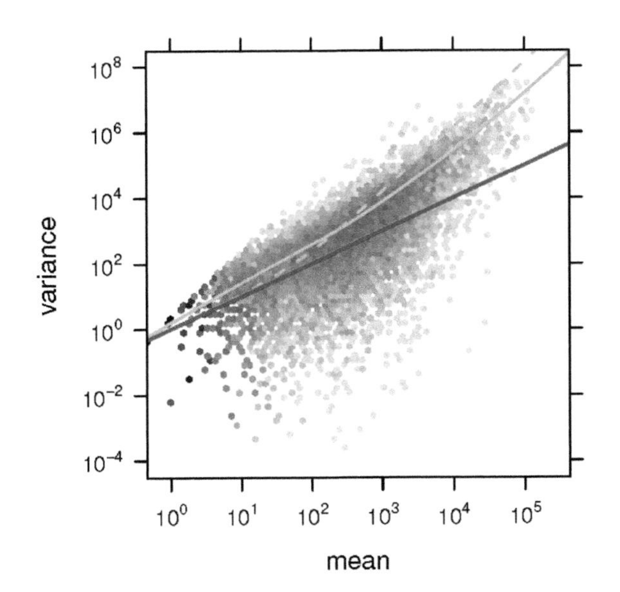

FIGURE 7.3
The overdispersion problem in RNA-seq data. Poisson distribution is often used to model RNA-seq data, but instead of the variance/dispersion being approximately equal to the mean as assumed by the distribution, the variance in RNA-seq data is often dependent on the mean. The purple line represents the relationship between variance and mean based on the Poisson distribution, while the solid and dashed orange lines represent local regressions used by DESeq and edgeR, respectively, based on negative binomial distribution. (Modified from Anders S. and Huber W. (2010) Differential expression analysis for sequence count data. Genome Biology, 11, R106. Used under the terms of the Creative Commons Attribution License [http://creativecommons.org/licenses/by/2.0] © 2010 Anders et al.)

TABLE 7.2
Tools for Detection of DE Genes

Name	Description	Reference
DESeq2	Employs negative binomial generalized linear modeling and Wald test to detect DE genes. Uses empirical Bayes estimation to shrink dispersions and fold changes to increase detection stability	[52]
edgeR	Detects DE genes based on negative binomial distribution, using techniques including exact test, generalized linear modeling, quasi-likelihood F-test, and empirical Bayes	[53]
limma	Fits a gene-based linear model for DE analysis. Originally developed for microarray gene expression data, it uses the voom function for RNA-seq data to apply empirical Bayes to estimate gene-wise variability	[65]
NOISeq	A nonparametric and data-adaptive method that detects DE genes based on simultaneous comparison of fold change and absolute expression difference	[72]

(continued)

TABLE 7.2 (Continued)

Tools for Detection of DE Genes

Name	Description	Reference
Cuffdiff 2	Combines estimates of count uncertainty and overdispersion to model fragment count variability, which is then used to test for significant DE changes	[68]
EBSeq	Applies empirical Bayesian methods to identify DE genes and splicing isoforms	[70]
baySeq	Builds on negative binomial distribution, and uses an empirical Bayesian approach to estimate posterior probabilities for each of a set of models to determine DE	[67]
SAMseq	Another nonparametric method originally designed for microarray data adapted for RNA-seq data. Uses resampling to account for differences in sequencing depth when performing DE analysis	[71]

Bayesian based methods, in which two alternative models are proposed for each gene, with one assuming differential expression and another assuming null. Given the observed counts, the posterior likelihood for the differential expression model is used to identify DE genes. Cuffdiff 2 uses the T statistic, which equals the ratio of mean(log[y]) over variance(log[y]), with y representing the expression ratio of a gene between two groups. Since this statistic approximately follows a normal distribution, t-test is used for DE analysis. DEGseq employs several methods to identify DE genes, including methods based on the MA-plot, Fisher's exact test, likelihood ratio test, and samrWrapper (a wrapper of functions in SAM, which is originally designed for identifying differential gene expression from microarray data). DESeq2 applies a generalized linear model (GLM) approach to identify DE genes with a Wald test. Another commonly used method, edgeR, tests for differential gene expression using an exact test that is highly parallel to Fisher's exact test for experiments with one factor, and GLM likelihood ratio test or quasi-likelihood F-test for general multifactorial experiments.

Packages that are not based on the Poisson distribution or negative binomial distribution are also used for differential expression analysis. For example, limma uses linear modeling after logarithmic transformation of read counts, and a moderated *t*-statistic to find DE genes. PoissonSeq conducts DE analysis based on a test of a correlation term between gene and experimental conditions, which follows a chi-squared distribution model. The adaptation of SAM for RNA-seq data analysis has led to the development of SAMseq, which, different from the original SAM, is based on a non-parametric approach [71]. NOISeq provides another example of using the non-parametric approach for cases where a probability distribution model of the data may not hold [72]. Despite the differences in the statistical models used, benchmarking studies have shown that many of the tools introduced above generate similar results for well-powered studies, and those based on the non-parametric approach are often equally effective [73, 74].

Most of the currently available methods are designed to handle samples with biological replicates. Under non-ideal circumstances when RNA-seq is performed without replication, it becomes impossible to estimate biological variability for a satisfactory statistical analysis. For such cases, the only indicator of differential expression is fold change. Some of the above tools can still take such data, such as edgeR, which offers an option to manually input a dispersion value estimated from similar studies, and NOISeq, which provides technical replicates through simulation of the data assuming a multinomial distribution, acting as alternative means to estimate biological variability for a significance analysis. There are also tools especially designed for RNA-seq experiments without replication, including GFOLD [75] and ACDtool [76]. ACDtool is an implementation of the method originally proposed by Audic and Claverie developed on the basis of Poisson distribution [77]. Although originally designed for analyzing relatively small data sets (<10 K reads), the A-C statistic and its implementation through ACDtool is equally sensitive and applicable to the much larger NGS data sets that contain millions of reads without replicates.

7.3.6 Multiple Testing Correction

A typical RNA-seq experiment involves a small number of samples, but a large number of genes, to test for differential expression between conditions. If only one gene is compared, a p of 0.05 is usually used as a threshold for type I error (false positive), i.e., accepting that there is a 5% chance of calling a gene differentially expressed while it is not. When a large number of genes are compared simultaneously, however, using the same p of 0.05 for each gene will lead to a significant number of false discoveries. For example, if 10,000 genes are compared, at $p = 0.05$ there will be 500 genes to be called differentially expressed just by random chance. Therefore, an adjustment is needed to correct for the high rate of false discoveries caused by multiple testing. One adjustment approach is to divide the p by the total number of genes/comparisons being conducted, e.g., using an adjusted p of 5×10^{-6} when comparing 10,000 genes. This approach, called Bonferroni correction, controls the family wise error rate (FWER), i.e., the probability of detecting one or more false-positive genes. While it is straightforward to implement, it is very conservative and often too stringent, leading to false negatives (also called type II errors).

A more practical and powerful approach was proposed by Benjamini and Hochberg (1995) [78]. Instead of controlling FWER, this approach controls the proportion of false positives among identified genes, i.e., the false discovery rate (FDR). To illustrate, when using an FDR of 0.05, we accept the probability that 5% of the genes detected to be differentially expressed are false positives. As an example, if 1,000 genes are identified to be differentially expressed, 50 genes are expected to be false positives by chance. FDR-adjusted p values are often called q values. If a gene has a q value of 0.05, it means that 5% of the genes with q values less than 0.05 are expected to be false positives. This approach, through relaxing the expectation and allowing a small proportion of false positives among all discoveries, leads to a gain in detection power. All DE detection tools introduced above offer options for multiple testing correction.

7.3.7 Gene Clustering

Genes showing differential expression pattern between conditions can be grouped into different clusters based on their overall expression pattern. This unsupervised process helps uncover different patterns of overall gene expression changes, and thereby serves as an important exploratory step to find key target genes for further investigation and hypothesis generation. Among the most widely used clustering algorithms are hierarchical and k-means clustering. Hierarchical clustering aims to build a dendrogram based on similarity of expression between genes. This clustering method can take either a "bottom-up," also called agglomerative, approach in which each gene is in their own cluster initially and then recursively merged until only one cluster remains, or a "top-down" (divisive) process that employs a reverse process. For RNA-seq data, the agglomerative approach is used more commonly. Besides clustering genes, samples are often clustered at the same time to uncover relationships among them (Figure 7.4). With *k*-means clustering, the number of clusters, i.e., the *k* value, needs to be defined *a priori*. Performing hierarchical clustering first can help assess what *k* value to use. The objective of *k*-means clustering is to assign each gene to the nearest cluster mean. For both hierarchical and *k*-means clustering, an often-used similarity measure is Pearson correlation coefficient. Besides these two commonly used clustering algorithms, other clustering methods include Self-Organizing Map (SOM) and Partitioning Around Medoids (PAM). These clustering processes can be performed using functions in R such as "hclust" for hierarchical clustering and "kmeans" for *k*-means clustering.

7.3.8 Functional Analysis of Identified Genes

Besides clustering, functional analysis of identified DE genes is also necessary to connect the genes, usually in large numbers, to the biological question under study. This analysis can be conducted at multiple levels, including Gene Ontology (GO), biological pathway, and gene network. GO provides hierarchically structured, standardized terms to describe biological processes, molecular functions, and cellular components associated with a gene [79]. A biological pathway, on the other hand, refers to a series of molecular reactions that leads to a cellular event or product. Commonly used pathway

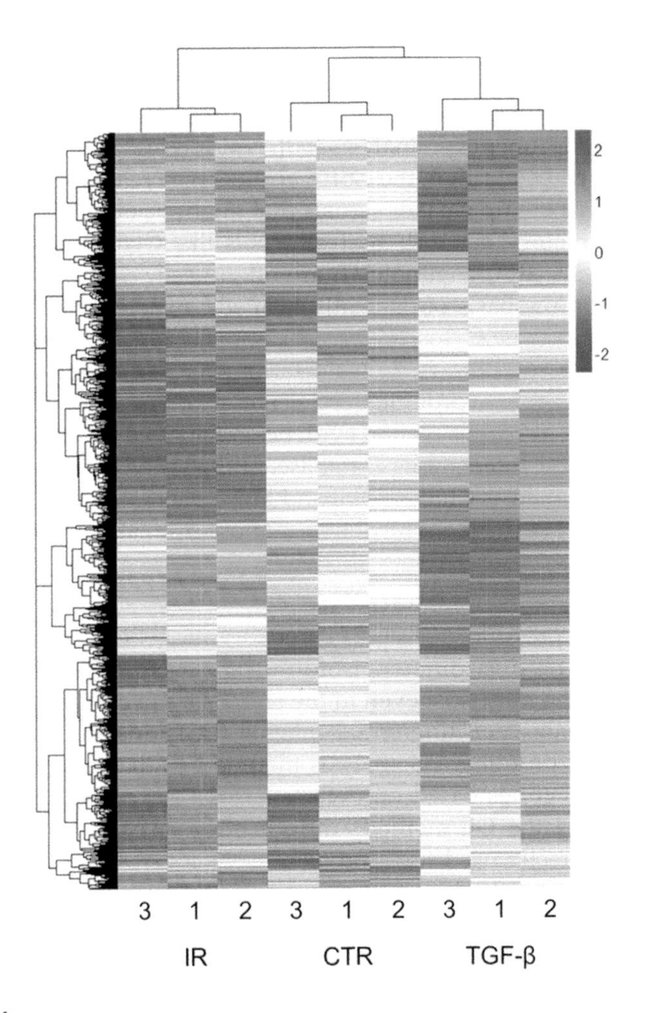

FIGURE 7.4
Hierarchical clustering of DE genes as well as experimental samples. RNA-seq data shown here is collected from cultured fibroblasts that were subjected to two treatment conditions (irradiation [IR] and TGF-β1 [TGF-β], in comparison with control [CTR]). (From Mellone M, Hanley CJ, Thirdborough S, Mellows T, Garcia E, Woo J, Tod J, Frampton S, Jenei V, Moutasim KA, Kabir TD, Brennan PA, Venturi G, et al. (2016) Induction of fibroblast senescence generates a non-fibrogenic myofibroblast phenotype that differentially impacts on cancer prognosis. *Aging (Albany NY)*. 159:114–132. Used under the terms of the Creative Commons Attribution License (CC BY 3.0) [https://creativecommons.org/licenses/by/3.0/] © 2016 Mellone et al.)

databases include KEGG [80], Pathway Commons [81], WikiPathways [82], and Reactome [83]. Tools such as Enrichr [84], GOseq [85], GOrilla [86], g:Profiler [87], DAVID (or Database for Annotation, Visualization and Integrated Discovery) [88], and ToppGene [89] detect enrichment of GO terms and biological pathways in identified genes. The statistical significance of this enrichment is usually calculated using the hypergeometric distribution, or Fisher's exact test. An alternative approach is the Gene Set Enrichment Analysis (GSEA), which instead of using a filtered list of genes, uses the entire gene set for functional analysis [90]. Not relying on a somewhat arbitrary cutoff for gene selection, the GSEA approach increases sensitivity of the analysis and can pick up weaker signals that might be otherwise missed.

For gene regulatory network analysis, tools like Cytoscape or Ingenuity Pathway Analysis (IPA, commercial) are often used. Gene network can be reconstructed on the basis of currently available experimental evidence, or co-expression patterns. Cytoscape, for example, provides a range of apps for gene network analysis and visualization, such as stringApp [91] and GeneMANIA [92], which provide gene-gene (or protein-protein) interaction information based on experimental evidence catalogued in databases such as BioGrid, Integrated Interactions Database, and STRING. In addition, some Cytoscape Apps, such as ClueGO [93] and EnrichmentMap [94], offer visualization of gene set enrichment results (e.g., those from GSEA) as a network.

7.3.9 Differential Splicing Analysis

Besides overall expression level changes, eukaryotic genes also undergo alternative splicing to produce different isoforms of transcripts (see Chapter 3). As differential splicing may exist even in the absence of overall gene expression level changes, analysis of differential splicing (DS) adds another dimension to transcriptomic profiling. DS analysis methods can be generally classified into different categories based on detection target (exon, splicing event, or isoform). Some of the DE methods introduced above, such as edgeR and limma, can be directly used for exon-level DS analysis to detect differential usage of exons. Specially designed exon-level DS tools include DEXSeq [95] and JunctionSeq [96]. DEXSeq, as an example, implements a test for differential exon usage between experimental conditions based on relative exon usage, which is defined as the number of reads that map to an exon divided by the total number of reads that map to the entire gene.

Event-based methods focus on alternative splicing events, such as skipped exons, alternative 3′/5′ splicing sites, or retained introns. These methods quantify splicing events using "percent-spliced-in," or Psi (ψ), as a measure of the frequency of an alternative splicing event into final transcripts. Examples of event-based methods include MISO [97], rMATS [98], MAJIQ [99], SpliceSeq, SUPPA2 [100], SplAdder [101], Leafcutter [102], and Whippet [103].

Isoform-centric methods attempt to reconstruct different splicing isoforms first, and then quantify their abundance levels in samples, prior to applying statistical tests to detect those that show differential expression between conditions. Previously introduced DE methods Cuffdiff 2 and EBSeq are in this category when performing DS analysis. DiffSplice, as another example, first maps reads to the reference genome using tools such as MapSplice, and then constructs splice graph to detect alternatively spliced regions in the form of alternative splicing modules. After estimation of their abundance, DiffSplice uses nonparametric permutation to test for DS between experimental groups.

7.4 Visualization of RNA-Seq Data

RNA-seq data visualization is required to explore trends and patterns embedded in a dataset, verify results from an established analytic workflow, and furthermore, appreciate the complexity in gene transcription, including alternative splicing. There are a variety of data visualization tools that can be used for data exploration and verification of findings, including those offered in R through packages such as ggplot2, plotly, reshape, and viridis. Histogram, boxplot, density, violin, and volcano plots can be created using these tools to explore a dataset, while PCA plot, heatmap, and dendrogram can be used to provide a summarized view of the data. Verification of findings using the same tools helps confirm that a proper workflow including data model is used. Among the tools used to map RNA-seq reads to the reference genome to develop a better understanding of the complexity of gene transcription, IGV and Integrated Genome Browser (IGB) are often used. In addition, RNA-seq data can be exported as custom tracks for display in a genome browser such as the UCSC Genome Browser. For visualization of alternative splicing, tools like SpliceSeq and JunctionSeq have their own built-in visualization capabilities. DiffSplice generates GFF-style files that can be visualized in the genomic browser GBrowse.

7.5 RNA-Seq as a Discovery Tool

RNA-seq is an important exploratory tool to study the molecular mechanisms of biological processes. Based on the DE genes identified, new hypotheses can be formulated for testing. Prior to carrying out functional studies to establish causational relationship between change of gene expression pattern to the biological process of interest, the first step is typically to validate RNA-seq findings with other experimental techniques, most commonly Western blotting and quantitative PCR. Beyond DE gene detection for basic research, the rich information in RNA-seq data can also be used clinically to develop new ways to classify diseases and improve diagnostic outcomes. For example, with combined deployment of informative gene feature selection and machine learning techniques RNA-seq profiles are used to stratify breast cancer into different subtypes [104]. As another example, RNA sequencing has significantly improved diagnostic outcome for hereditary cancer by resolving uncertainties from DNA genetic sequencing alone [105].

Besides interrogating currently catalogued genes, RNA-seq, being an unbiased approach, is also a powerful technology for discovering novel transcripts, splicing events, and other transcription-related phenomena. RNA-seq studies of the transcriptional landscape of the genome have found that besides protein-coding regions, the majority of the genome produces RNA transcripts. The finding that 75% of the human genome is transcribed (see Chapter 3), made with extensive use of RNA-seq, shows the power of this technology in discovering currently unknown transcripts. RNA-seq has also been used to discover novel alternative splicing isoforms. For example, the discovery of circular RNAs (also see Chapter 3), which are formed as a result of non-canonical RNA splicing, is also due to the application of RNA-seq [106]. RNA-seq has also been applied to uncover other transcription-related phenomena, such as gene fusion. Gene fusion is caused by genomic rearrangement and is a common occurrence under certain conditions such as cancer. Because RNA-seq has the capability to locate transcripts generated from a fused gene, detection of gene fusion events has been greatly facilitated by this powerful technology [107, 108]. With rapid technological developments in advancing RNA sequencing as an even more powerful discovery tool, RNA-seq has now entered into a new era – single-cell RNA-seq, which is covered next.

References

1. Li J, Fu C, Speed TP, Wang W, Symmans WF. Accurate RNA sequencing from formalin-fixed cancer tissue To represent high-quality transcriptome from frozen tissue. *JCO Precis Oncol* 2018, 2:PO.17.00091.
2. Parekh S, Ziegenhain C, Vieth B, Enard W, Hellmann I. The impact of amplification on differential expression analyses by RNA-seq. *Sci Rep* 2016, 6:25533.
3. Zhulidov PA, Bogdanova EA, Shcheglov AS, Vagner LL, Khaspekov GL, Kozhemyako VB, Matz MV, Meleshkevitch E, Moroz LL, Lukyanov SA *et al.* Simple cDNA normalization using kamchatka crab duplex-specific nuclease. *Nucleic Acids Res* 2004, 32(3):e37.
4. Yang L, Duff MO, Graveley BR, Carmichael GG, Chen LL. Genomewide characterization of non-polyadenylated RNAs. *Genome Biol* 2011, 12(2):R16.
5. Busby MA, Stewart C, Miller CA, Grzeda KR, Marth GT. Scotty: a web tool for designing RNA-Seq experiments to measure differential gene expression. *Bioinformatics* 2013, 29(5):656–657.
6. Bi R, Liu P. Sample size calculation while controlling false discovery rate for differential expression analysis with RNA-sequencing experiments. *BMC Bioinformatics* 2016, 17:146.

7. Wu H, Wang C, Wu Z. PROPER: comprehensive power evaluation for differential expression using RNA-seq. *Bioinformatics* 2015, 31(2):233–241.
8. Zhao S, Li CI, Guo Y, Sheng Q, Shyr Y. RnaSeqSampleSize: real data based sample size estimation for RNA sequencing. *BMC Bioinformatics* 2018, 19(1):191.
9. Ching T, Huang S, Garmire LX. Power analysis and sample size estimation for RNA-seq differential expression. *RNA* 2014, 20(11):1684–1696.
10. Auer PL, Doerge RW. Statistical design and analysis of RNA sequencing data. *Genetics* 2010, 185(2):405–416.
11. Robinson DG, Storey JD. subSeq: determining appropriate sequencing depth through efficient read subsampling. *Bioinformatics* 2014, 30(23):3424–3426.
12. Grant GR, Farkas MH, Pizarro AD, Lahens NF, Schug J, Brunk BP, Stoeckert CJ, Hogenesch JB, Pierce EA. Comparative analysis of RNA-Seq alignment algorithms and the RNA-Seq unified mapper (RUM). *Bioinformatics* 2011, 27(18):2518–2528.
13. Ryan MC, Cleland J, Kim R, Wong WC, Weinstein JN. SpliceSeq: a resource for analysis and visualization of RNA-Seq data on alternative splicing and its functional impacts. *Bioinformatics* 2012, 28(18):2385–2387.
14. Trapnell C, Pachter L, Salzberg SL. TopHat: discovering splice junctions with RNA-Seq. *Bioinformatics* 2009, 25(9):1105–1111.
15. Kim D, Pertea G, Trapnell C, Pimentel H, Kelley R, Salzberg SL. TopHat2: accurate alignment of transcriptomes in the presence of insertions, deletions and gene fusions. *Genome Biol* 2013, 14(4):R36.
16. Wang K, Singh D, Zeng Z, Coleman SJ, Huang Y, Savich GL, He X, Mieczkowski P, Grimm SA, Perou CM *et al*. MapSplice: accurate mapping of RNA-seq reads for splice junction discovery. *Nucleic Acids Res* 2010, 38(18):e178.
17. Au KF, Jiang H, Lin L, Xing Y, Wong WH. Detection of splice junctions from paired-end RNA-seq data by SpliceMap. *Nucleic Acids Res* 2010, 38(14):4570–4578.
18. Marco-Sola S, Sammeth M, Guigo R, Ribeca P. The GEM mapper: fast, accurate and versatile alignment by filtration. *Nat Methods* 2012, 9(12):1185–1188.
19. Dobin A, Davis CA, Schlesinger F, Drenkow J, Zaleski C, Jha S, Batut P, Chaisson M, Gingeras TR. STAR: ultrafast universal RNA-seq aligner. *Bioinformatics* 2013, 29(1):15–21.
20. Kim D, Langmead B, Salzberg SL. HISAT: a fast spliced aligner with low memory requirements. *Nat Methods* 2015, 12(4):357–360.
21. Kim D, Paggi JM, Park C, Bennett C, Salzberg SL. Graph-based genome alignment and genotyping with HISAT2 and HISAT-genotype. *Nat Biotechnol* 2019, 37(8):907–915.
22. Wu TD, Reeder J, Lawrence M, Becker G, Brauer MJ. GMAP and GSNAP for Genomic Sequence Alignment: Enhancements to Speed, Accuracy, and Functionality. *Methods Mol Biol* 2016, 1418:283–334.
23. Krizanovic K, Echchiki A, Roux J, Sikic M. Evaluation of tools for long read RNA-seq splice-aware alignment. *Bioinformatics* 2018, 34(5):748–754.
24. Liu B, Liu Y, Li J, Guo H, Zang T, Wang Y. deSALT: fast and accurate long transcriptomic read alignment with de Bruijn graph-based index. *Genome Biol* 2019, 20(1):274.
25. Marić J, Sović I, Križanović K, Nagarajan N, Šikić M. Graphmap2-splice-aware RNA-seq mapper for long reads. *bioRxiv* 2019, doi: https://doi.org/10.1101/720458

26. Sahlin K, Makinen V. Accurate spliced alignment of long RNA sequencing reads. *Bioinformatics* 2021, 37(24):4643–4651.
27. Graubert A, Aguet F, Ravi A, Ardlie KG, Getz G. RNA-SeQC 2: Efficient RNA-seq quality control and quantification for large cohorts. *Bioinformatics* 2021, 37(18):3048–3050.
28. Wang L, Wang S, Li W. RSeQC: quality control of RNA-seq experiments. *Bioinformatics* 2012, 28(16):2184–2185.
29. Hartley SW, Mullikin JC. QoRTs: a comprehensive toolset for quality control and data processing of RNA-Seq experiments. *BMC Bioinformatics* 2015, 16:224.
30. Bushmanova E, Antipov D, Lapidus A, Prjibelski AD. rnaSPAdes: a de novo transcriptome assembler and its application to RNA-Seq data. *GigaScience* 2019, 8(9):giz100.
31. Grabherr MG, Haas BJ, Yassour M, Levin JZ, Thompson DA, Amit I, Adiconis X, Fan L, Raychowdhury R, Zeng Q *et al.* Full-length transcriptome assembly from RNA-Seq data without a reference genome. *Nat Biotechnol* 2011, 29(7):644–652.
32. Chang Z, Li G, Liu J, Zhang Y, Ashby C, Liu D, Cramer CL, Huang X. Bridger: a new framework for de novo transcriptome assembly using RNA-seq data. *Genome Biol* 2015, 16:30.
33. Robertson G, Schein J, Chiu R, Corbett R, Field M, Jackman SD, Mungall K, Lee S, Okada HM, Qian JQ *et al.* De novo assembly and analysis of RNA-seq data. *Nat Methods* 2010, 7(11):909–912.
34. Xie Y, Wu G, Tang J, Luo R, Patterson J, Liu S, Huang W, He G, Gu S, Li S *et al.* SOAPdenovo-Trans: de novo transcriptome assembly with short RNA-Seq reads. *Bioinformatics* 2014, 30(12):1660–1666.
35. Schulz MH, Zerbino DR, Vingron M, Birney E. Oases: robust de novo RNA-seq assembly across the dynamic range of expression levels. *Bioinformatics* 2012, 28(8):1086–1092.
36. Pertea M, Pertea GM, Antonescu CM, Chang TC, Mendell JT, Salzberg SL. StringTie enables improved reconstruction of a transcriptome from RNA-seq reads. *Nat Biotechnol* 2015, 33(3):290–295.
37. Kovaka S, Zimin AV, Pertea GM, Razaghi R, Salzberg SL, Pertea M. Transcriptome assembly from long-read RNA-seq alignments with StringTie2. *Genome Biol* 2019, 20(1):278.
38. Shumate A, Wong B, Pertea G, Pertea M. Improved transcriptome assembly using a hybrid of long and short reads with StringTie. *PLoS Comput Biol* 2022, 18(6):e1009730.
39. Prjibelski AD, Puglia GD, Antipov D, Bushmanova E, Giordano D, Mikheenko A, Vitale D, Lapidus A. Extending rnaSPAdes functionality for hybrid transcriptome assembly. *BMC Bioinformatics* 2020, 21(Suppl 12):302.
40. Liao Y, Smyth GK, Shi W. featureCounts: an efficient general purpose program for assigning sequence reads to genomic features. *Bioinformatics* 2014, 30(7):923–930.
41. Anders S, Pyl PT, Huber W. HTSeq—a Python framework to work with high-throughput sequencing data. *Bioinformatics* 2015, 31(2):166–169.
42. Li B, Dewey CN. RSEM: accurate transcript quantification from RNA-Seq data with or without a reference genome. *BMC Bioinformatics* 2011, 12:323.

43. Roberts A, Pachter L. Streaming fragment assignment for real-time analysis of sequencing experiments. *Nat Methods* 2013, 10(1):71–73.
44. Trapnell C, Roberts A, Goff L, Pertea G, Kim D, Kelley DR, Pimentel H, Salzberg SL, Rinn JL, Pachter L. Differential gene and transcript expression analysis of RNA-seq experiments with TopHat and Cufflinks. *Nat Protoc* 2012, 7(3):562–578.
45. Robert C, Watson M. Errors in RNA-Seq quantification affect genes of relevance to human disease. *Genome Biol* 2015, 16:177.
46. Bray NL, Pimentel H, Melsted P, Pachter L. Near-optimal probabilistic RNA-seq quantification. *Nat Biotechnol* 2016, 34(5):525–527.
47. Patro R, Duggal G, Love MI, Irizarry RA, Kingsford C. Salmon provides fast and bias-aware quantification of transcript expression. *Nat Methods* 2017, 14(4):417–419.
48. Patro R, Mount SM, Kingsford C. Sailfish enables alignment-free isoform quantification from RNA-seq reads using lightweight algorithms. *Nat Biotechnol* 2014, 32(5):462–464.
49. Wu DC, Yao J, Ho KS, Lambowitz AM, Wilke CO. Limitations of alignment-free tools in total RNA-seq quantification. *BMC Genomics* 2018, 19(1):510.
50. Bullard JH, Purdom E, Hansen KD, Dudoit S. Evaluation of statistical methods for normalization and differential expression in mRNA-Seq experiments. *BMC Bioinformatics* 2010, 11:94.
51. Hansen KD, Irizarry RA, Wu Z. Removing technical variability in RNA-seq data using conditional quantile normalization. *Biostatistics* 2012, 13(2):204–216.
52. Love MI, Huber W, Anders S. Moderated estimation of fold change and dispersion for RNA-seq data with DESeq2. *Genome Biol* 2014, 15(12):550.
53. Robinson MD, McCarthy DJ, Smyth GK. edgeR: a Bioconductor package for differential expression analysis of digital gene expression data. *Bioinformatics* 2010, 26(1):139–140.
54. Robinson MD, Oshlack A. A scaling normalization method for differential expression analysis of RNA-seq data. *Genome Biol* 2010, 11(3):R25.
55. Kadota K, Nishiyama T, Shimizu K. A normalization strategy for comparing tag count data. *Algorithms Mol Biol* 2012, 7(1):5.
56. Sun J, Nishiyama T, Shimizu K, Kadota K. TCC: an R package for comparing tag count data with robust normalization strategies. *BMC Bioinformatics* 2013, 14 :219.
57. Li J, Witten DM, Johnstone IM, Tibshirani R. Normalization, testing, and false discovery rate estimation for RNA-sequencing data. *Biostatistics* 2012, 13(3):523–538.
58. Manimaran S, Selby HM, Okrah K, Ruberman C, Leek JT, Quackenbush J, Haibe-Kains B, Bravo HC, Johnson WE. BatchQC: interactive software for evaluating sample and batch effects in genomic data. *Bioinformatics* 2016, 32(24):3836–3838.
59. Johnson WE, Li C, Rabinovic A. Adjusting batch effects in microarray expression data using empirical Bayes methods. *Biostatistics* 2007, 8(1):118–127.
60. Risso D, Ngai J, Speed TP, Dudoit S. Normalization of RNA-seq data using factor analysis of control genes or samples. *Nat Biotechnol* 2014, 32(9):896–902.
61. Leek JT. svaseq: removing batch effects and other unwanted noise from sequencing data. *Nucleic Acids Res* 2014, 42(21):e161.

62. Zhang Y, Parmigiani G, Johnson WE. ComBat-seq: batch effect adjustment for RNA-seq count data. *NAR Genom Bioinform* 2020, 2(3):lqaa078.
63. Marioni JC, Mason CE, Mane SM, Stephens M, Gilad Y. RNA-seq: an assessment of technical reproducibility and comparison with gene expression arrays. *Genome Res* 2008, 18(9):1509–1517.
64. Anders S, Huber W. Differential expression analysis for sequence count data. *Genome Biol* 2010, 11(10):R106.
65. Ritchie ME, Phipson B, Wu D, Hu Y, Law CW, Shi W, Smyth GK. limma powers differential expression analyses for RNA-sequencing and microarray studies. *Nucleic Acids Res* 2015, 43(7):e47.
66. Frazee AC, Pertea G, Jaffe AE, Langmead B, Salzberg SL, Leek JT. Ballgown bridges the gap between transcriptome assembly and expression analysis. *Nat Biotechnol* 2015, 33(3):243–246.
67. Hardcastle TJ, Kelly KA. baySeq: empirical Bayesian methods for identifying differential expression in sequence count data. *BMC Bioinformatics* 2010, 11:422.
68. Trapnell C, Hendrickson DG, Sauvageau M, Goff L, Rinn JL, Pachter L. Differential analysis of gene regulation at transcript resolution with RNA-seq. *Nat Biotechnol* 2013, 31(1):46–53.
69. Wang L, Feng Z, Wang X, Wang X, Zhang X. DEGseq: an R package for identifying differentially expressed genes from RNA-seq data. *Bioinformatics* 2010, 26(1):136–138.
70. Leng N, Dawson JA, Thomson JA, Ruotti V, Rissman AI, Smits BM, Haag JD, Gould MN, Stewart RM, Kendziorski C. EBSeq: an empirical Bayes hierarchical model for inference in RNA-seq experiments. *Bioinformatics* 2013, 29(8):1035–1043.
71. Li J, Tibshirani R. Finding consistent patterns: a nonparametric approach for identifying differential expression in RNA-Seq data. *Stat Methods Med Res* 2013, 22(5):519–536.
72. Tarazona S, Furio-Tari P, Turra D, Pietro AD, Nueda MJ, Ferrer A, Conesa A. Data quality aware analysis of differential expression in RNA-seq with NOISeq R/Bioc package. *Nucleic Acids Res* 2015, 43(21):e140.
73. Corchete LA, Rojas EA, Alonso-Lopez D, De Las Rivas J, Gutierrez NC, Burguillo FJ. Systematic comparison and assessment of RNA-seq procedures for gene expression quantitative analysis. *Sci Rep* 2020, 10(1):19737.
74. Stupnikov A, McInerney CE, Savage KI, McIntosh SA, Emmert-Streib F, Kennedy R, Salto-Tellez M, Prise KM, McArt DG. Robustness of differential gene expression analysis of RNA-seq. *Comput Struct Biotechnol J* 2021, 19:3470–3481.
75. Feng J, Meyer CA, Wang Q, Liu JS, Shirley Liu X, Zhang Y. GFOLD: a generalized fold change for ranking differentially expressed genes from RNA-seq data. *Bioinformatics* 2012, 28(21):2782–2788.
76. Claverie JM, Ta TN. ACDtool: a web-server for the generic analysis of large data sets of counts. *Bioinformatics* 2019, 35(1):170–171.
77. Audic S, Claverie JM. The significance of digital gene expression profiles. *Genome Res* 1997, 7(10):986–995.
78. Benjamini Y, Hochberg Y. Controlling the false discovery rate – a practical and powerful approach to multiple testing. *J R Stat Soc Ser B Methodol* 1995, 57(1):289–300.

79. Gene Ontology C. The Gene Ontology resource: enriching a GOld mine. *Nucleic Acids Res* 2021, 49(D1):D325–D334.
80. Kanehisa M, Furumichi M, Tanabe M, Sato Y, Morishima K. KEGG: new perspectives on genomes, pathways, diseases and drugs. *Nucleic Acids Res* 2017, 45(D1):D353–D361.
81. Rodchenkov I, Babur O, Luna A, Aksoy BA, Wong JV, Fong D, Franz M, Siper MC, Cheung M, Wrana M *et al*. Pathway Commons 2019 update: integration, analysis and exploration of pathway data. *Nucleic Acids Res* 2020, 48(D1):D489–D497.
82. Martens M, Ammar A, Riutta A, Waagmeester A, Slenter DN, Hanspers K, R AM, Digles D, Lopes EN, Ehrhart F *et al*. WikiPathways: connecting communities. *Nucleic Acids Res* 2021, 49(D1):D613–D621.
83. Gillespie M, Jassal B, Stephan R, Milacic M, Rothfels K, Senff-Ribeiro A, Griss J, Sevilla C, Matthews L, Gong C *et al*. The reactome pathway knowledgebase 2022. *Nucleic Acids Res* 2022, 50(D1):D687–D692.
84. Xie Z, Bailey A, Kuleshov MV, Clarke DJB, Evangelista JE, Jenkins SL, Lachmann A, Wojciechowicz ML, Kropiwnicki E, Jagodnik KM *et al*. Gene set knowledge discovery with Enrichr. *Curr Protoc* 2021, 1(3):e90.
85. Young MD, Wakefield MJ, Smyth GK, Oshlack A. Gene ontology analysis for RNA-seq: accounting for selection bias. *Genome Biol* 2010, 11(2):R14.
86. Eden E, Navon R, Steinfeld I, Lipson D, Yakhini Z. GOrilla: a tool for discovery and visualization of enriched GO terms in ranked gene lists. *BMC bioinformatics* 2009, 10:48.
87. Raudvere U, Kolberg L, Kuzmin I, Arak T, Adler P, Peterson H, Vilo J. g:Profiler: a web server for functional enrichment analysis and conversions of gene lists (2019 update). *Nucleic Acids Res* 2019, 47(W1):W191–W198.
88. Huang da W, Sherman BT, Lempicki RA. Systematic and integrative analysis of large gene lists using DAVID bioinformatics resources. *Nat Protoc* 2009, 4(1):44–57.
89. Chen J, Bardes EE, Aronow BJ, Jegga AG. ToppGene Suite for gene list enrichment analysis and candidate gene prioritization. *Nucleic Acids Res* 2009, 37(Web Server issue):W305–311.
90. Subramanian A, Tamayo P, Mootha VK, Mukherjee S, Ebert BL, Gillette MA, Paulovich A, Pomeroy SL, Golub TR, Lander ES *et al*. Gene set enrichment analysis: a knowledge-based approach for interpreting genome-wide expression profiles. *Proc Natl Acad Sci U S A* 2005, 102(43):15545–15550.
91. Doncheva NT, Morris JH, Gorodkin J, Jensen LJ. Cytoscape stringapp: network analysis and visualization of proteomics data. *J Proteome Res* 2019, 18(2):623–632.
92. Montojo J, Zuberi K, Rodriguez H, Kazi F, Wright G, Donaldson SL, Morris Q, Bader GD. GeneMANIA Cytoscape plugin: fast gene function predictions on the desktop. *Bioinformatics* 2010, 26(22):2927–2928.
93. Bindea G, Mlecnik B, Hackl H, Charoentong P, Tosolini M, Kirilovsky A, Fridman WH, Pages F, Trajanoski Z, Galon J. ClueGO: a Cytoscape plug-in to decipher functionally grouped gene ontology and pathway annotation networks. *Bioinformatics* 2009, 25(8):1091–1093.
94. Merico D, Isserlin R, Stueker O, Emili A, Bader GD. Enrichment map: a network-based method for gene-set enrichment visualization and interpretation. *PLoS One* 2010, 5(11):e13984.

95. Anders S, Reyes A, Huber W. Detecting differential usage of exons from RNA-seq data. *Genome Res* 2012, 22(10):2008–2017.
96. Hartley SW, Mullikin JC. Detection and visualization of differential splicing in RNA-Seq data with JunctionSeq. *Nucleic Acids Res* 2016, 44(15):e127.
97. Katz Y, Wang ET, Airoldi EM, Burge CB. Analysis and design of RNA sequencing experiments for identifying isoform regulation. *Nat Methods* 2010, 7(12):1009–1015.
98. Shen S, Park JW, Lu ZX, Lin L, Henry MD, Wu YN, Zhou Q, Xing Y. rMATS: Robust and flexible detection of differential alternative splicing from replicate RNA-Seq data. *Proc Natl Acad Sci U S A* 2014, 111(51):E5593–E5601.
99. Vaquero-Garcia J, Barrera A, Gazzara MR, Gonzalez-Vallinas J, Lahens NF, Hogenesch JB, Lynch KW, Barash Y. A new view of transcriptome complexity and regulation through the lens of local splicing variations. *Elife* 2016, 5:e11752.
100. Trincado JL, Entizne JC, Hysenaj G, Singh B, Skalic M, Elliott DJ, Eyras E. SUPPA2: fast, accurate, and uncertainty-aware differential splicing analysis across multiple conditions. *Genome Biol* 2018, 19(1):40.
101. Kahles A, Ong CS, Zhong Y, Ratsch G. SplAdder: identification, quantification and testing of alternative splicing events from RNA-Seq data. *Bioinformatics* 2016, 32(12):1840–1847.
102. Li YI, Knowles DA, Humphrey J, Barbeira AN, Dickinson SP, Im HK, Pritchard JK. Annotation-free quantification of RNA splicing using LeafCutter. *Nat Genet* 2018, 50(1):151–158.
103. Sterne-Weiler T, Weatheritt RJ, Best AJ, Ha KCH, Blencowe BJ. Efficient and Accurate Quantitative Profiling of Alternative Splicing Patterns of Any Complexity on a Laptop. *Mol Cell* 2018, 72(1):187–200 e186.
104. Cascianelli S, Molineris I, Isella C, Masseroli M, Medico E. Machine learning for RNA sequencing-based intrinsic subtyping of breast cancer. *Sci Rep* 2020, 10(1):14071.
105. Karam R, Conner B, LaDuca H, McGoldrick K, Krempely K, Richardson ME, Zimmermann H, Gutierrez S, Reineke P, Hoang L *et al*. Assessment of Diagnostic Outcomes of RNA Genetic Testing for Hereditary Cancer. *JAMA Netw Open* 2019, 2(10):e1913900.
106. Salzman J, Gawad C, Wang PL, Lacayo N, Brown PO. Circular RNAs are the predominant transcript isoform from hundreds of human genes in diverse cell types. *PLoS One* 2012, 7(2):e30733.
107. Davare MA, Tognon CE. Detecting and targeting oncogenic fusion proteins in the genomic era. *Biol Cell* 2015, 107(5):111–129.
108. Haas BJ, Dobin A, Li B, Stransky N, Pochet N, Regev A. Accuracy assessment of fusion transcript detection via read-mapping and de novo fusion transcript assembly-based methods. *Genome Biol* 2019, 20(1):213.

8

Transcriptomics by Single-Cell RNA-Seq

Bulk RNA-seq, as covered in Chapter 7, reveals the transcriptome at the level of an organ, a tissue, or a population of primary or cultured cells. Bulk RNA-seq is very effective to reveal overall gene expression change under contrasting conditions, but the detected change is a result of comparing average gene expression of all cells between conditions. It was highly challenging, if not impossible, to analyze the transcriptome of each cell individually not long ago. But with the recent rapid technological developments in the single-cell sequencing field, interrogating the transcriptome of thousands to tens of thousands of single cells simultaneously has become a reality. For the first time single-cell RNA-seq, or scRNA-seq, enables researchers to study cell-to-cell variation, even among those that appear to be similar morphologically and reside in close proximity. Single-cell RNA-seq has offered unprecedented opportunities for the biomedical community to identify distinct subpopulations of cells in a heterogenous population, characterize their functional states, and infer cellular trajectories during differentiation.

While creating unprecedented opportunities, scRNA-seq also presents new challenges that do not exist in bulk RNA-seq. Some of the challenges come from the tremendous amount of cellular heterogeneity in a typical tissue sample, and therefore the richness of bio-information embedded in the scRNA-seq dataset from the much-increased resolution and change of scale. Other challenges, however, are due to the inherent characteristics of scRNA-seq data. Compared to bulk RNA-seq data, scRNA-seq data is sparser, i.e., containing a lot of zeros (often called dropouts), and has higher technical noise. This signal sparsity is partially biological as a result of normal gene transcription oscillation at the single-cell level. In the meantime, this is also caused by technical challenges. Based on estimation, a typical mammalian cell has on average 360,000 mRNA molecules representing 12,000 different transcripts, with wide variation from cell type to cell type [1]. By weight, a cell typically contains ~10 pg RNA, of which mRNA molecules are only a small fraction (1–5%) [2]. Generating scRNA-seq signals from such low quantities of cellular mRNA is technically challenging. For example, the rate of mRNA molecular capture is still quite low in the range of 10–20% [3–5]. Such technical challenges, along with non-exhaustive sequencing of captured molecules, lead to signal stochasticity, false-negative detection, and

DOI: 10.1201/9780429329180-11

high measurement noise in the scRNA-seq data. To effectively extract the rich information embedded in the scRNA-seq data, these characteristics demand development and deployment of algorithms and tools that are different from those designed for bulk RNA-seq.

Furthermore, the questions that can be answered from scRNA-seq data are beyond those from bulk RNA-seq. Detection of cell-to-cell variation, identification and visualization of different cell types/identities in a population, and inference of cellular developmental trajectories are all new realms that have emerged from the development of scRNA-seq. Some other data analytical topics such as data preprocessing, normalization, batch effect correction, and clustering are also significantly different from those covered in Chapter 7, and are presented in detail in this chapter. On topics that are similar to and/or have significant overlap with those covered in Chapter 7, such as identification and functional analysis of differentially expressed genes, only aspects that are specific to or need adjustment for scRNA-seq data are presented to avoid redundancy.

8.1 Experimental Design

8.1.1 Single-Cell RNA-Seq General Approaches

The landscape of single-cell sequencing, including RNA-seq, has been rapidly evolving. There have been a multitude of strategies or platforms in existence today, while new ones are continuously being developed and existing ones improved upon. In general, existing scRNA-seq strategies can be divided into two broad approaches, i.e., low- and high-throughput, based on the number of cells that can be analyzed simultaneously at a time. Low-throughput methods can process up to a few hundred cells at a time, while those of the high-throughput approach allow simultaneous analysis of thousands to tens of thousands of cells, or even more with the technologies continuing to evolve. A good example of the low-throughput approach is Smart-seq3 [6]. The low-throughput approach suits situations where the focus is on a small number of cells that need detailed molecular characterization. This approach has higher sensitivity, which leads to the identification of more transcripts and genes from each cell. It also provides full-length coverage of the transcriptome, thus enabling recognition of different splicing isoforms and allelic gene expression.

There are currently a multitude of platforms for the high-throughput approach. These platforms include Drop-seq [7], inDrop [8], sci-RNA-seq for single-cell combinatorial indexing RNA sequencing [9], and 10× Chromium [10]. Among these various platforms, the commercially available 10×

Chromium platform has been more widely adopted so far in the biomedical research community. Benchmark studies have shown that 10× generally has more consistent performance, better sensitivity and precision, and lower technical noise [11, 12]. In general, the high-throughput approach is more suitable to study cellular heterogeneity in a large population of cells. The transcript detection sensitivity is typically lower with this approach, as a result the data is sparser. The detection and counting of transcripts are usually based on sequencing of either the 3′ or 5′ end and not full length. Because of such differences between the two approaches, it is advisable to evaluate the particular needs of a project in order to decide which approach is more appropriate.

Despite the differences, the two general approaches share basically the same workflow in wet lab operational procedures. As detailed in Section 8.2, for both approaches single cells need to be prepared from dissociation of input materials, such as an organ or a tissue biopsy, followed by partitioning of the cells. The mechanism of cell partitioning varies with the approach and particular platform chosen, e.g., cells can be partitioned into 96- or 384-well plates (such as for the low-throughput approach), microfluidic droplets (used by many of the high-throughput platforms), etc. On the 10× Chromium platform, single cells are partitioned into nanoliter-scale GEMs (or Gel beads-in-EMulsion), with each GEM carrying a unique barcode. After partitioning cells are lysed to release cellular RNA, which is then reverse transcribed into cDNA. This is followed by cDNA amplification and subsequently library construction. Because of the shared commonalities in lab operation as well as data analysis, the two general approaches are not covered separately in the following sections unless otherwise noted.

8.1.2 Cell Number and Sequencing Depth

As many scRNA-seq projects aim to investigate cellular heterogeneity in a large cell population, to adopt the high-throughput approach some of the major technical questions to address first include how many cells to sequence, how deep to sequence, and how to balance cell number and sequencing depth in cases of budgetary constraints to maximize the amount of information to be obtained. On the question of how many cells to sequence, this depends on (1) the number of cell types present in the target population, (2) the minimum fraction of any of the cell types, and (3) the minimum number of cells desired to be observed in any of the cell types. Based on this information, negative binomial (NB) distribution can be used to estimate the number of cells to sample. For example, if 10 cell types are expected in a population, with each type being present at a minimum fraction of 3% in the total population, in order to observe at least 20 cells in any of the cell types, we will need to sample at least 1,104 cells in total at the 95% confidence level [13]. If the needed information on the heterogeneity of the population is not available, it is advisable to first conduct a pilot study using the maximum number of cells possible.

The maximum allowable cell number varies with platform, for example, on the current configuration of 10× Chromium 10,000 cells is the upper limit. For any droplet-based platforms including the Chromium, the limit on maximum cell number is affected by the law of Poisson distribution, as the loading of single cells into droplets (or GEMs for Chromium) is a Poisson process. This process leads to the formation of doublets (or multiplets), i.e., two (or more) cells being loaded into the same droplet/GEM and treated as one cell, and the rate of their formation follows Poisson statistics. To ensure scRNA-seq data quality, the rate of doublets/multiplets needs to be controlled at a manageable level (usually <5%).

On the question of sequencing depth, this depends on how transcriptionally active cells are in the sample, and the diversity of their transcriptomes. Without this knowledge a priori, the generally suggested depth for 3′ (or 5′) end gene expression profiling on the 10× Chromium platform is a minimum of 20,000 reads per cell using the current v3 chemistry. The number of detected genes at this depth varies with cell type, e.g., 1,000–2,000 for peripheral blood mononuclear cells and 2,000–3,000 for neurons. Sequencing depth may be fine-tuned based on specific project needs, and increasing sequencing depth from the suggested minimum depth generally leads to identification of more genes. But beyond a certain point (varies with cell type), further increase in sequencing depth leads to diminished return.

On the question of how to balance cell number and sequencing depth to maximize the amount of information obtained, the options are either sequencing fewer cells at greater depth, or more cells with fewer reads per cell. The former option allows identification of more transcripts and genes, and as a result generates a more accurate picture of each cell's transcription status. However, the fewer cells used may not offer a sufficient representation of the cellular population under study. The latter option, on the other hand, allows analysis of more cells to increase cell representation, but at the expense of identifying less transcripts and genes per cell. To quantify the tradeoff between sequencing depth and cell number, computational simulation using a multivariate generative model showed that increasing sequencing depth is better than increasing cell number before reaching the depth of 15,000 reads per cell, beyond which point there is a diminished return [14]. Another modeling demonstrated that under a fixed budget the strategy is to sequence as many cells as possible at the depth of one read per gene per cell [15]. Commonly used scRNA-seq pipeline tools like the open source 10× Cell Ranger [10] and zUMIs [16] have downsampling function to help determine whether the library was sequenced to saturation, or whether additional sequencing would increase the number of detected genes cost effectively.

To further compare the two options, another factor to consider is that many genes are often co-regulated in a cell, forming functional modules. The modularity of the cellular gene transcriptional system is manifested in the extensive gene-gene covariance embedded in the sequencing data, which suggests that scRNA-seq data is inherently low dimensional [17]. Because of this nature, a number of downstream data analysis steps are quite robust and can tolerate technical noise caused by shallow sequencing. Because of this, for many studies especially those that involve highly heterogeneous cell populations, limited resources are better spent on sequencing more cells instead of increasing sequencing depth. For studies in which cells are more homogeneous with more nuanced differences, sequencing fewer cells at greater depth might offer better distinguishing power. In this situation, the depth of 1–2 million reads per cell has been shown to be sufficient, beyond which detection sensitivity barely increases [18, 19].

8.1.3 Batch Effects Minimization and Sample Replication

The general principles of factorial design, randomization, and replication as covered in the last chapter for bulk RNA-seq equally apply to single-cell studies. In an ideal design, samples are prepared and sequenced simultaneously, with different groups or experimental conditions randomly distributed across all samples. For sequencing, all libraries should be mixed and sequenced together across all lanes. Compared to bulk RNA-seq, however, scRNA-seq samples are more often divided into batches. This is usually due to practical considerations, such as technical (as it is more challenging to prepare and process a large number of single-cell samples at once), or budgetary (a small-scale pilot run is often run first due to the associated high cost of running scRNA-seq). To minimize batch effects, the same principles should still be adhered to as much as possible within the same batch. For example, to study lung cell compositional change before and after an infection, when samples are divided into batches, within each batch dissociated cells should be prepared from lung tissue dissected from infected and control animals of the same age, with the procedure performed by a single operator following the same procedure. For subsequent cell partitioning, the cells should be partitioned using the same microfluidic chip if using 10× Chromium. If multiple chips are needed, loading of samples to chips and wells should be randomized and balanced. For sequencing, libraries prepared from all samples need to be barcoded differently, mixed together, and loaded onto multiple lanes. An unbalanced design, such as processing and sequencing of infected and control samples on two separate chips and different sequencer lanes, can make it challenging to distinguish cell compositional change caused by the infection from the variation resulted from the technical differences in processing and sequencing.

While it is a common practice to use biological replicates in bulk RNA-seq to measure within-group variation, there is still a lack of consensus on the use of replicates in scRNA-seq experiments. This is partly because scRNA-seq data contains more information than bulk RNA-seq, and therefore there are more diverse analyses that can be performed on scRNA-seq data beyond the identification of differentially expressed genes. How biological replicates can be incorporated into and help these analyses is still not very clear. As a general trend, use of biological replicates has become more common, allowing examination of reproducibility on these analyses. The number of biological replicates to run can vary based on the specific questions asked, as well as the computational power available to the researcher as inclusion of biological replicates can significantly increase the demand on computing resources in some of the analytic steps. It is not necessary to run *technical* replicates if using a standardized platform such as 10× Chromium.

8.2 Single-Cell Preparation, Library Construction, and Sequencing

8.2.1 Single-Cell Preparation

Generating high-quality scRNA-seq data is a multi-step process (see Figure 8.1), and preparing high-quality single cells is an important first step. To prepare the large number of cells needed for high-throughput scRNA-seq, cell preparatory procedure needs to be optimized for each target tissue in order to generate cells that are of high viability, with minimal cell death/debris and no cell aggregation. If present, dead and dying cells need to be removed first before proceeding. If they are not removed, the RNA they release to the environment from their breakdown leads to high levels of ambient RNA, which in turn results in increased background noise [20] that is difficult, if not impossible, to remove bioinformatically during data analysis. Preparing cells from liquid tissues such as blood and bone marrow is relatively straightforward. When starting from solid tissues, enzymatic and mechanical dissociation is required first to break down extracellular matrix and cleave cell-cell junctions. If certain types of cells need to be enriched, approaches such as fluorescence-activated cell sorting (FACS), microchip-, or magnetic-activated cell sorting are commonly used. If using FACS, gentle settings such as large nozzle sizes and/or slower flow velocities are recommended to minimize cell damage and death. If using a low-throughput platform, besides the above methods, manual microscope-guided precision cell picking techniques, such as micro-pipetting used in micromanipulation or laser capture microdissection, are also often used. If using these techniques, target cells can be directly deposited in lysis buffer that contains RNase inhibitor.

Cell preparatory procedures may also lead to alteration of cellular gene expression profiles. For example, early stress response genes such as *Fos* and *Jun* have been found to be expressed after cell dissociation and FACS [21]. To avoid such unintended changes, once prepared single cells should be processed as quickly as possible without delay to subsequent steps, including

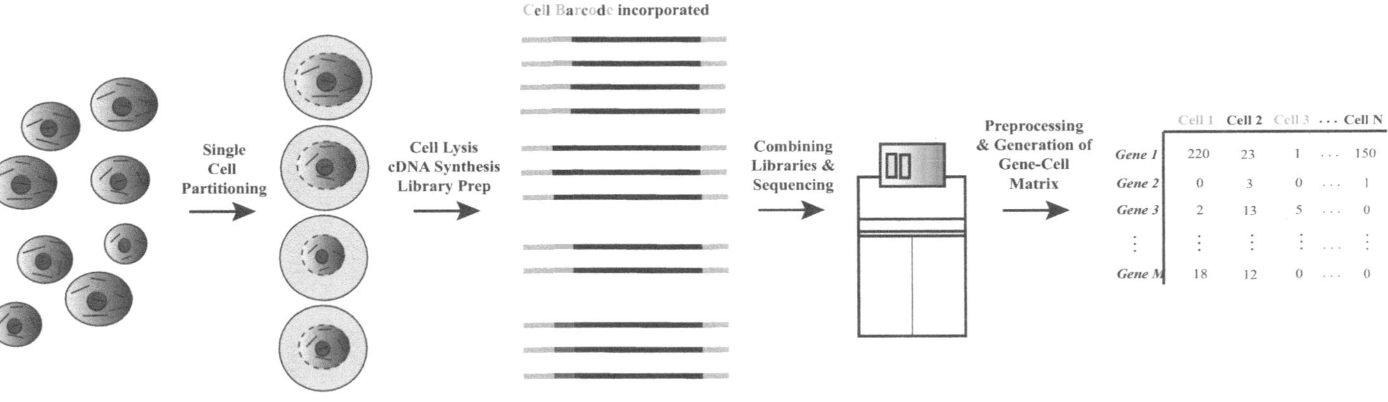

FIGURE 8.1
Single-cell RNA-seq general lab process. Single cells are first partitioned into individual droplets, wells, tubes, etc. Once partitioned, cells are lysed to release RNA for reverse transcription into cDNA. During cDNA synthesis and subsequent sequencing library preparation, cell-specific barcodes are incorporated to track transcripts from each cell. After sequencing of scRNA-seq libraries, reads are demultiplexed as part of preprocessing to reveal transcripts from each cell using cell barcodes. Counting of transcripts from each cell generates a gene-cell matrix for further analyses.

cell partitioning, lysis, and cDNA conversion. If immediate processing is not possible due to practical limitations (e.g., distance, time, and processing capacity), alternative strategies including cryopreservation and fixation can be used. For cryopreservation, dimethyl sulfoxide (DMSO) is often added to prevent the formation of ice crystals that result in cell damage. If fixation is preferred, methanol can be used. Among such methods it has been found that DMSO cryopreservation offers a better approach than fixation, or other specially developed preservatives [22].

8.2.2 Single Nuclei Preparation

Variations in cell size and shape may prevent conduct of scRNA-seq on certain platforms. For example, the width of the microfluidic channel used in the 10× Chromium controller is 50–60 μm. Most cells in the body are below this size in the range of 10–15 μm in diameter. Some cells, however, are larger (e.g., human egg cells, 100 μm in diameter), or have highly sophisticated shape (e.g., neurons, with dendrites and axons that can reach over one meter in length), preventing their passing through the channel. One solution for such cells is to use single nucleus RNA-seq (or snRNA-seq). Side-by-side comparisons of scRNA-seq and snRNA-seq data have shown that they reveal similar transcriptional profiles [23–25], lending support to the use of snRNA-seq as a surrogate for scRNA-seq. These comparative studies have also shown that snRNA-seq data contain more reads that map to intronic regions, and the number of genes detected from snRNA-seq is in general lower than that from scRNA-seq. These observations are consistent with the fact that the population of transcripts assayed by snRNA-seq is the nuclear, still being processed portion of the cell's entire transcriptome (nuclear and cytoplasmic) assayed by scRNA-seq. As detailed in Chapter 3, to generate mature mRNA, the spliceosome needs to remove introns from pre-mRNAs before mature mRNAs are exported out of the nucleus to the cytoplasm.

Besides accommodating cells of unusual sizes and shapes, snRNA-seq may also offer advantages over scRNA-seq in sample prep and availability because of procedural differences between the two approaches. For example, cells in certain tissues (such as the retina) are more prone to damage and cell death during tissue dissociation, which can lead to their underrepresentation in the final scRNA-seq data. In processing such tissues snRNA-seq can offer an advantage because single nuclei prep is usually faster, and it is usually easier to maintain the intactness of prepared nuclei than entire cells [26]. With snRNA-seq, archival frozen tissues may also be used [25], which as indicated in the last section is an issue with scRNA-seq if appropriate cryopreservation processes were not followed for freezing the tissues. In addition, for cells that have a high tendency to aggregate after dissociation, such as monocytes and granulocytes, snRNA-seq also offers a good alternative since prepared single nuclei have a low propensity to aggregate.

8.2.3 Library Construction and Sequencing

To construct scRNA/snRNA-seq libraries, the released RNA from lysed cells (or nuclei) is first reverse transcribed into cDNA. This step is most commonly carried out using oligo-dT primers that target polyadenylated transcripts. To generate full-length cDNA, the so-called SMART technology, or Switching Mechanism at 5′ End of RNA Template [27], is often used. Because of the low amounts of cDNA generated from each cell, PCR amplification is needed to boost cDNA quantity. To make the amplified cDNA into sequenceable libraries, different scRNA-seq platforms use different strategies. For the 10× Chromium, the amplified cDNA is enzymatically fragmented into smaller pieces to accommodate Illumina shorts read sequencing. The cDNA fragments are then end repaired, followed by A-tailing and ligation of Illumina adapters. A subsequent PCR is the last step of library construction that incorporates sample barcodes, as well as the rest of Illumina adapter sequences required for sequencing. For the low-throughput Smart-seq3, a quicker, Nextera-based method is used, in which the amplified cDNA is fragmented using tagmentation, followed by a library amplification process prior to sequencing.

As a result of cDNA amplification, one original transcript molecule generates multiple copies and therefore is represented multiple times in the final sequencing library. Without correcting for this amplification effect, reads derived from PCR duplicates of the same original transcript cannot be differentiated from those from different mRNA molecules transcribed from the same gene (the true detection target). To correct for this effect and thereby reduce experimental noise, unique molecular identifiers (or UMIs) are used, which are essentially a large number of randomly synthesized, unique nucleotide combinations. UMIs are introduced into each cDNA before the amplification step, and after amplification all PCR products from the same original cDNA will carry the same UMI. After sequencing, reads carrying the same UMI are bioinformatically collapsed into one, thereby removing the amplification effect. The use of UMI has been adopted by both high- and low-throughput scRNA-seq approaches.

The specifics of sequencing scRNA-seq libraries may vary depending on the particular platform chosen. To sequence 10× libraries, for example, the specific read length required of Illumina sequencing for 3′ v3.1 dual indexed libraries (Figure 8.2) is: Read 1–28 bp, covering cell barcode (16 bp) and UMI (12 bp); Index 1 (i7) – 10 bp; Index 2 (i5) – 10 bp; Read 2–90 bp, covering cDNA insert. For Smart-seq3 libraries, regular single end 50 or 75 bp sequencing can be used for general gene expression profiling, and paired end 150 bp sequencing can be used for splicing isoform and/or allelic expression analyses. With regard to sequencing depth, the reader should refer to above, Section 8.1.2.

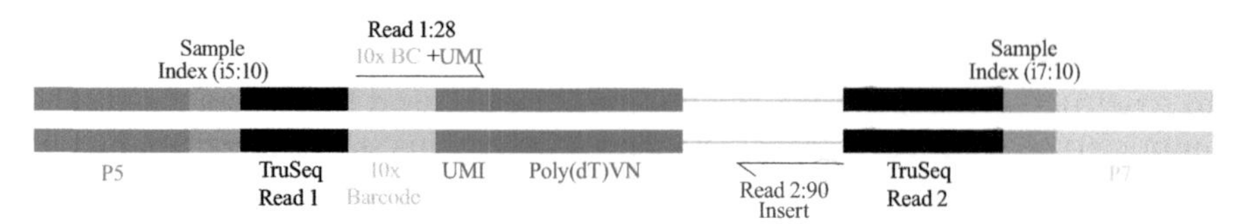

FIGURE 8.2
Structure of a 10× scRNA-seq 3′ library. To sequence such a library using Illumina sequencers, Read 1 covers 10× barcodes (16 bp) that track individual cells and UMI (12 bp) for removal of PCR duplicates, while Read 2 is used to sequence actual cDNA fragments for gene detection. The dual sample indices (i5 and i7, 10 bp each) are used to separate reads into different samples, each of which may contain thousands of cells. (This illustration is based on 10× scRNA-seq v3.1 chemistry. Image provided by 10× Genomics.)

8.3 Preprocessing of scRNA-Seq Data

After sequencing, raw scRNA-seq reads need to be preprocessed before further analysis can be conducted (Figure 8.3 provides an overview of the basic scRNA-seq data analysis workflow). The tasks of the preprocessing step include assigning reads to their cells of origin, collapsing reads according to their UMIs, quantifying the abundance of transcripts in each cell, and generating a cell-transcript count matrix, among others to be detailed next. It should be noted that these tasks put high demands on computational resources, as the substantial increase in resolution to the single-cell level is associated with significant increase in raw data volume in comparison to bulk RNA-seq. For preprocessing of 10× scRNA-seq data using the Cell Ranger software, for example, the recommended requirements are a Linux system with 16 CPU cores, 128 GB RAM, and 1 TB free disk space.

8.3.1 Initial Data Preprocessing and Quality Control

As performed for other NGS applications, the original sequencing data in the BCL format needs to be first demultiplexed (if two or more samples are sequenced together), and converted to FASTQ files. While this is typically achieved using tools such as Illumina's bcl2fastq, if using the 10× Chromium platform, the 10× Cell Ranger, which offers a series of pipelines for analyzing 10× scRNA-seq data, has a built-in bcl2fastq wrapper in its "cellranger mkfastq" pipeline. For quality control (QC) of the FASTQ data, while general NGS data QC tools such as FastQC can be used, specialized scRNA-seq software such as Cell Ranger provides specific scRNA-seq QC metrics, including total number of reads from each sample, % of Read 1 and 2 bases ≥Q30, % of cell barcode bases ≥Q30, % of GEM barcodes from the 10× GEM barcode "whitelist," etc. If using the low-throughput scRNA-seq approach, tools such as zUMIs can take FASTQ files as input to perform data QC. As part of initial

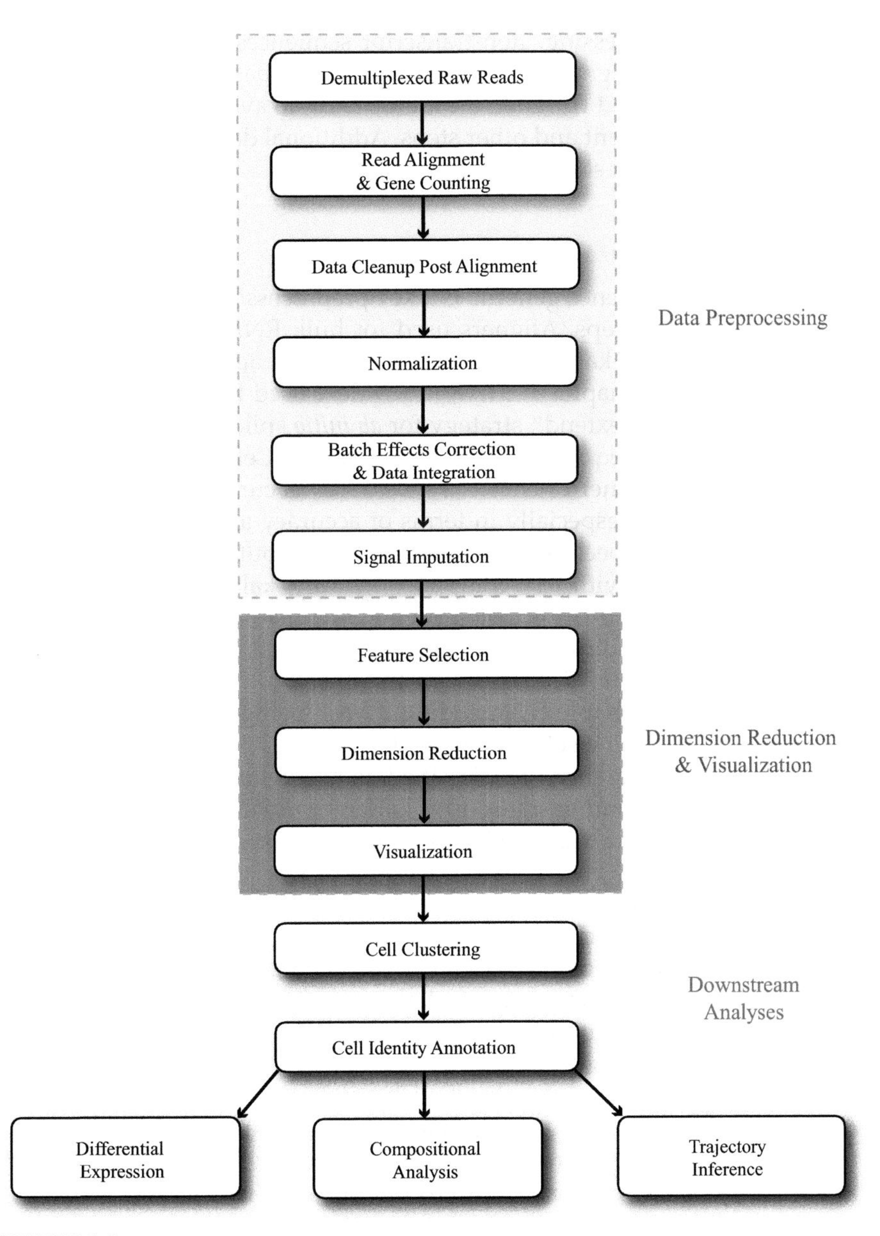

FIGURE 8.3
Basic scRNA-seq data analysis workflow.

data QC and preprocessing, non-transcript sequences such as polyA and template switching oligo (required for the SMART-based cDNA synthesis) need to be checked, and trimmed off if they exist, to avoid their interference on subsequent alignment and other steps. Additional data cleanup is carried out after the alignment step.

8.3.2 Alignment and Transcript Counting

Alignment to the reference genome is a key preprocessing step prior to most other preprocessing steps. Aligners used for bulk RNA-seq, such as STAR and the later aligner kallisto [28], can be directly applied to scRNA-seq data. As covered in Chapter 7, STAR is a widely used RNA-seq aligner that employs a "seed-and-extend" strategy for *ab initio* splice junction detection. For aligning scRNA-seq reads, the aforementioned Cell Ranger and zUMIs use STAR as their aligner of choice. While STAR is currently one of the top performing methods, especially in terms of accuracy and alignment rate, it is relatively slow in speed and demanding on computational resources. To achieve faster speed with less computer memory, kallisto is developed to use a de Bruijn-based process called pseudoalignment. Instead of mapping sequencing reads, it is designed to map *k*-mers extracted from reads to speed up the process (Chapter 7, Section 7.3.2). Other bulk RNA-seq alignment methods covered in Chapter 7, including BWA, Bowtie2, and TopHat2, may also be used for mapping scRNA-seq reads.

After alignment, reads can be classified as exonic, intronic, or intergenic for the purpose of counting transcripts. Different pipelines may use slightly different definitions for this classification process. For example, Cell Ranger considers a read to be exonic if at least 50% of the sequence intersects an exon. Traditionally, only exonic reads are counted. When snRNA-seq is used, however, intronic reads derived from nascent transcripts, or transcripts that were in the middle of being spliced, can also be informative and counted. To perform transcript counting, for platforms that incorporate cell-specific barcodes, such as 10× Chromium, sequencing reads need to be assigned to their cells of origin using the cell barcodes they carry. The 10× platform has a "whitelist" of all GEM barcodes to help track cells. All reads associated with a particular barcode are grouped together and considered to come from one cell. As sequencing errors may exist in actual cell barcode reads, this grouping process can be specified to tolerate such errors (e.g., one base deviation from the whitelisted barcodes). For platforms that incorporate UMIs, reads carrying the same UMI within the same cell are expected to be PCR duplicates, and therefore need to be collapsed. Similarly, sequencing errors in the UMIs may also be tolerated during collapsing of reads. Depending on the method used, the collapsing of UMIs may be based on the UMI sequence alone, or a combination of reads' alignment location and the UMI sequence. After the assignment of reads to cell barcodes and collapsing of reads using

	Cell 1	Cell 2	Cell 3	...	Cell N
Gene 1	1	220	23	...	150
Gene 2	0	0	3	...	1
Gene 3	5	2	13	...	0
⋮	⋮	⋮	⋮	...	⋮
Gene M	0	18	12	...	0

FIGURE 8.4
Gene-cell count matrix.

the UMIs, the number of UMIs is counted for each gene and each cell barcode, thereby generating a gene-cell matrix (Figure 8.4). This matrix becomes the basis of nearly all downstream analyses.

The above processes of genome alignment, reads classification, barcode assignment, UMI collapsing, and transcript counting can be accomplished together in a single workflow using various pipeline tools. For instance, for 10× data the "cellranger count" pipeline is typically used. However, it requires significant computational resources and is not very fast. To increase processing efficiency and speed up these preprocessing steps, newer tools such as STARsolo [29], kallisto/bustools [30], and Alevin [31] have been developed. Besides 10× data, these later tools can also be used to process scRNA-seq data generated from other platforms. As suggested by its name, STARsolo is built around the high-performing STAR aligner for read mapping, as well as cell barcode assigning, UMI collapsing, and gene-cell matrix creation. Kallisto, which aims to achieve a balance between computing efficiency and accuracy, can generate pseudoaligned scRNA-seq data in a new format called BUS, which provides a binary representation of the data in the form of barcode, UMI, and sets of equivalence classes. Once generated, BUS files can be manipulated with bustools [30] to produce a data matrix consisting of gene count and barcode. To improve the accuracy of gene abundance estimates, Alevin includes gene-ambiguous reads in its quantification, i.e., those that multimap between genes and are usually discarded by other tools.

8.3.3 Data Cleanup Post Alignment

For a high-throughput platform such as 10× Chromium, additional data cleanup is needed post alignment to remove signal associated with doublets (or multiplets), dead cells, or ambient RNA. On the 10× platform, not all reads associated with GEM barcodes from the "whitelist" are derived from bona fide single cells. As mentioned earlier, doublets (or multiplets) form through partitioning of two (or more) cells into the same GEM. Dead cells and ambient RNA molecules can also be partitioned into GEMs like authentic cells. Detection and removal of reads associated with such unwanted GEMs are based on three indicators: the number of uniquely mapped genes per GEM barcode, the total count of UMIs per GEM barcode, and the fraction of reads per barcode that map to the mitochondrial genome. If a barcode has unusually high gene and UMI counts, it is possibly associated with a doublet (or multiplet). Conversely, if a barcode is associated with few mapped genes, a low UMI count, and/or a high fraction of mitochondrial genes, it is an indication of ambient RNA or a dead cell in the original sample. A dead cell may have most of its cytoplasmic mRNA leaked out due to compromised cell membrane, with only mitochondrial RNA preserved because of the organelle's double membrane system (Chapter 1, Section 1.4.9). To detect such unwanted reads, these three quality indicators should be used in combination rather than alone. It should also be noted that under some cellular conditions these commonly used quality indicators may be violated. For example, cells in high metabolic state may have unusually higher mitochondrial RNA fraction, or very large cells may appear to be doublets.

Doublets/multiplets represent hybrid- or super-transcriptomes and compound downstream data analyses if not removed. Besides the basic detection method using the total count of genes and UMIs as mentioned above, a number of specially developed doublet/multiplet detection tools are available, including DoubletFinder [32], Scrublet [33], DoubletDetection [34], cxds and/or bcds [35], and solo [36]. In general, these methods work by first building artificial doublets from combination of randomly selected droplets (or GEMs), then generating a "doublet score" for each droplet based on their similarity to the artificial doublets, and finally calling doublets if the score surpasses a threshold. The major difference between these methods is on how the doublet score is generated. For example, DoubletFinder, one of the top performing tools based on a benchmark study [37], uses the *k*-nearest neighbors (kNN) method to calculate the proportion of nearest artificial doublet neighbors as the score for each droplet. It should be noted that none of these tools works well for every case. In addition, to avoid removing large sized cells that appear to be doublets, the rate of doublets/multiplets identified by these tools for removal should not exceed that expected from Poisson statistics for the experimental condition.

Besides using software tools alone, experimental techniques can also be used to improve their detection, such as cell hashing [38], or mixing of cells from different species (e.g., human and mouse cells). With cell hashing, antibodies against ubiquitous cell surface proteins are tagged with oligonucleotide barcodes to distinguish cells from different samples for robust identification of cross-sample doublets/multiplets. Along the same line, genotypic differences between cells, such as those collected from unrelated individuals, can also be used to for detection of doublets/multiplets. For example, tools like demuxlet [39], scSplit [40], souporcell [41], and Vireo [42] can separate mixed cells into individual samples based on each sample's unique SNP profile, and thereby identify doublets/multiplets if mutually exclusive SNP profiles are associated with the same droplet. Among these tools, demuxlet requires knowledge of genotypic differences between individual samples *a priori*, while the later tools (scSplit, souporcell, and Vireo) infer genotypic differences from the observed scRNA-seq reads.

To keep the rate of doublets/multiplets low, the majority of droplets are empty without any cells. However, ambient RNA molecules in the cell sample buffer, if exist, can still be captured by empty droplets as well as those containing cells. Reads generated from droplets associated with only ambient RNA do not represent real cells and need to be removed. Since only a small number of genes and UMIs are expected to be detected in these droplets, they can be identified based on their low gene and UMI counts and the associated reads filtered out. This filtering approach, however, may also remove small cells that have low RNA content. To overcome this problem, tools such as EmptyDrops [43], dropkick [44], and CB2 [45] can be used. These tools rely on modeling of ambient RNA profile, against which the RNA profile of each droplet is compared to determine whether it is associated with a genuine cell. For example, EmptyDrops first constructs an ambient RNA profile using droplets with low UMI counts that most likely represent empty droplets. For droplets that are not associated with high UMI counts (those with high UMI counts are automatically considered to represent cell-containing droplets), their RNA profiles are tested for deviations from the ambient RNA profile using a Dirichlet-multinomial distribution model [43]. Droplets with RNA profiles that significantly deviate from the ambient RNA model are identified as genuine cells, even if they have low UMI counts (Figure 8.5). This method is incorporated into Cell Ranger from version 3.0, which leads to improvement in identifying cells of low RNA content. Besides these tools, others such as DecontX [46] and SoupX [47] have been developed to remove ambient mRNA contamination in cell-containing droplets to reveal their true gene/UMI counts.

The fraction of UMIs from the mitochondrial transcriptome is another indicator of cell preparation and data quality. High-quality cell preparations should have minimal cell death and typically generate only low percentages of reads of mitochondrial origin (<5% for cells of low energy demands, such as adrenal, lung, and white blood cells) [20, 48]. Mitochondria have their own small genome and transcriptome, but in a typical cell the amount of mitochondrial RNA is relatively low compared to the large amount of RNA transcribed from the nuclear genome. If the fraction of mitochondrial UMI counts is significantly higher in a droplet, however, it is a strong indication that it captured a dead cell. If a significant percentage of droplets is associated with high mitochondrial reads, it is an indication of low cell sample quality. It should also be noted that some cells, due to biological factors such as stress or high metabolic activity (such as cardiomyocytes, hepatocytes, kidney cells), may display higher mitochondrial gene/UMI counts because of the presence

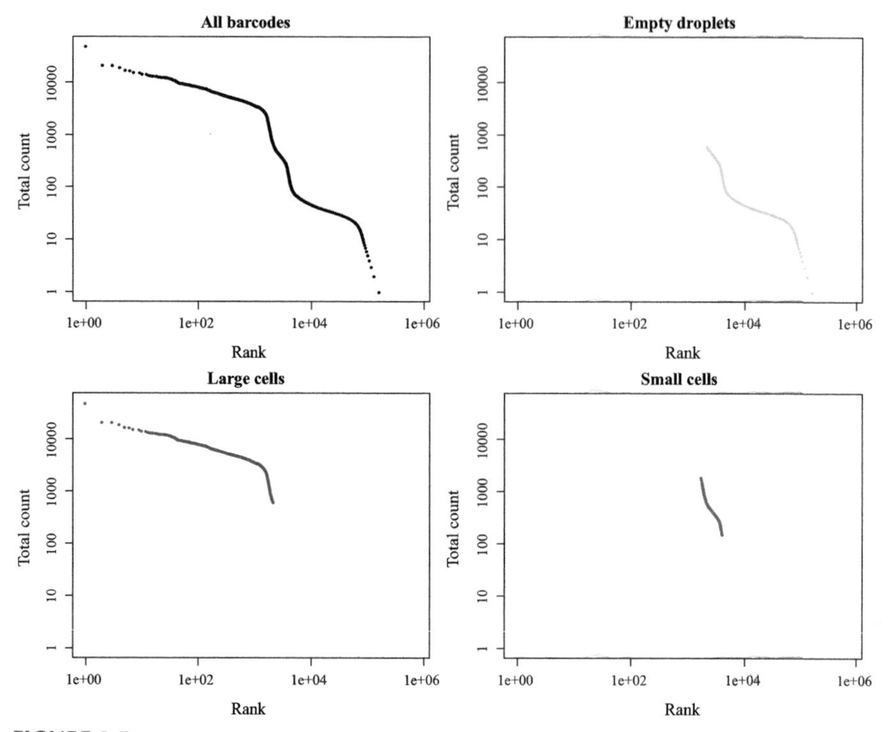

FIGURE 8.5
Detection of empty droplets vs. those containing cells. Total UMI count is plotted against the rank for each barcode in a scRNA-seq dataset from Lun et al. (2019). Plots are shown for all barcodes, and those associated with empty droplets, large and small cells, respectively. Using this method (EmptyDrops) barcodes associated with small cells can be distinguished from those with empty drops. (Adapted from: ATL Lun, S Riesenfeld, T Andrews, TP Dao, T Gomes, et al., EmptyDrops: distinguishing cells from empty droplets in droplet-based single-cell RNA sequencing data, *Genome Biology* 2019, 20(1):63. Used under the terms of the Creative Commons Attribution 4.0 International License, https://creativecommons.org/licenses/by/4.0, ©2019 Lun et al.)

of larger numbers of mitochondria and/or increased mitochondrial gene expression [48].

8.3.4 Normalization

While the use of UMIs removes the effects of PCR duplicates, other factors during single-cell sample processing can still introduce undesirable variations among cells or samples. Such factors include RNA capture rate, reverse transcription efficiency, random sampling of molecules during sequencing, and sequencing depth. If uncorrected, such variations may lead to inaccurate results in downstream analytic steps. The goal of normalization is to correct for such variations in order to make cells and samples directly comparable. Normalization approaches developed for bulk RNA-seq (Chapter 7, Section 7.3.3) may be used for scRNA-seq data, especially those generated on low-throughput platforms with full-length transcript coverage. In general, however, normalization of scRNA-seq data faces unique challenges mostly due to the issue of signal sparsity. To address the challenges, a number of methods have been developed specifically for normalizing scRNA-seq data. Examples of these methods are SCnorm [49], Linnorm [50], BASiCS [51], Census [52], ZINB-WaVE [53], and sctransform [54]. Besides these dedicated normalization methods, commonly used scRNA-seq pipeline toolkits such as Cell Ranger, Seurat [55], Scanpy [56], scran [57], and scVI [58] also contain their own normalization methods.

These normalization methods can be generally classified into two general categories: global scaling-based and modeling-based. The former approach, represented by BASiCS and those employed by Cell Ranger, Seurat, Scanpy, and scran, is based on the use of a global "size factor." As an example, Cell Ranger performs normalization in its *aggr* pipeline, through subsampling of libraries of higher sequencing depth until all libraries have on average the same number of confidently mapped reads per cell. Seurat, as another example, includes a similarly simple normalization process performed as follows: first divide the UMI count of each gene by the total number of UMIs in each cell, then multiply the resultant ratios with a scaling factor (typically in the range of 10^4 to 10^6), and lastly perform a log transformation of the scaled values (actually log(x+1) to accommodate zero count genes) to generate normalized data. This global scaling approach assumes that RNA content is constant across all cells, and that one size factor fits all genes/cells. This assumption may not be true at times, especially with highly heterogeneous population containing cells of different sizes and RNA content. To address this concern, methods in the other category, as represented by SCnorm, Linnorm, Census, ZINB-WaVE, sctransform, and scVI, are based on modeling of cellular molecule counts using probabilistic approaches. SCnorm, for example, employs quantile regression modeling, which is used to group genes based on the relationship between their UMI count and sequencing depth. Genes in different groups are normalized using group-specific size factors. With sctransform, which is included in Seurat from ver3 as a normalization option, a regularized negative binomial (NB) regression model is used. This method first constructs a generalized linear model (GLM) for each gene to estimate the relationship between its UMI count and sequencing depth. The estimated parameters are then regularized based on gene expression level. To generate normalized values for each gene, the regularized parameters are applied to an NB regression model. Overall, these methods are based on models that make different assumptions about the sparsity and underlying distribution of gene expression values in cells. Comparative studies on these methods have reported that the performances of different methods vary from dataset to dataset, and therefore it is advisable to use more than one method on the dataset at hand and then select the one that has the best performance [59, 60].

Variance stabilization is often an inherent goal of normalization. The log transformation used in many methods toward the end achieves this goal while also makes the normalized gene expression approximate a normal distribution to facilitate downstream analyses. Without this stabilization, the magnitude of a gene's average gene expression correlates with the magnitude of its variance, i.e., the so-called mean-variance relationship. Based on this relationship, highly expressed genes also tend to have high levels of variance, even if they do not contribute to cellular heterogeneity, such as housekeeping genes. Conversely, genes that are expressed at low levels have relatively low variance, even if they are biologically significant including those coding for transcription factors. The goal of variance stabilization is to remove the unwanted effects of this relationship, so that genes with greater-than-expected variance between cells, regardless of the magnitude of their expression, can be revealed.

8.3.5 Batch Effects Correction

If samples must be divided into batches, batch effects correction needs to be performed. If left uncorrected, technical variations between batches caused by factors such as different reagent lots, operators, protocols, sequencers, or flow cells/lanes will compound data analysis and affect results. Batch effects can be detected through exploratory data analysis, such as visual inspection of cells in a low-dimensional space (Figure 8.6; more details on data visualization can be found in Section 8.4.3). To correct for batch effects between highly similar biological replicates in an experiment, where the technical variations between batches are mostly due to the use of different flow cells and sequencing depth, a simple approach is to regress them out. Implemented in pipeline tools such as Seurat and Scanpy, this approach uses linear regression to regress out the unwanted effects of experimental batches and other technical covariates, as well as biological covariates (such as cell cycle). Tools that are specially designed for batch effects correction, such as ComBat, have also been shown to work well [61]. ComBat, originally developed for correcting batch effects in bulk gene expression analysis [62], uses a linear regression model to model normalized gene expression with experimental batch as a covariate. It first generates batch-specific mean and variance for each gene, and then an empirical Bayes-based adjustment is applied to all genes to remove batch effects, thereby generating a batch corrected expression matrix as output.

For data integration that involves datasets generated from different experiments or laboratories, where technical variations can be due to more diverse factors including different cell preparatory procedures, sequencing setups, or even overall approaches and platforms, the above methods for

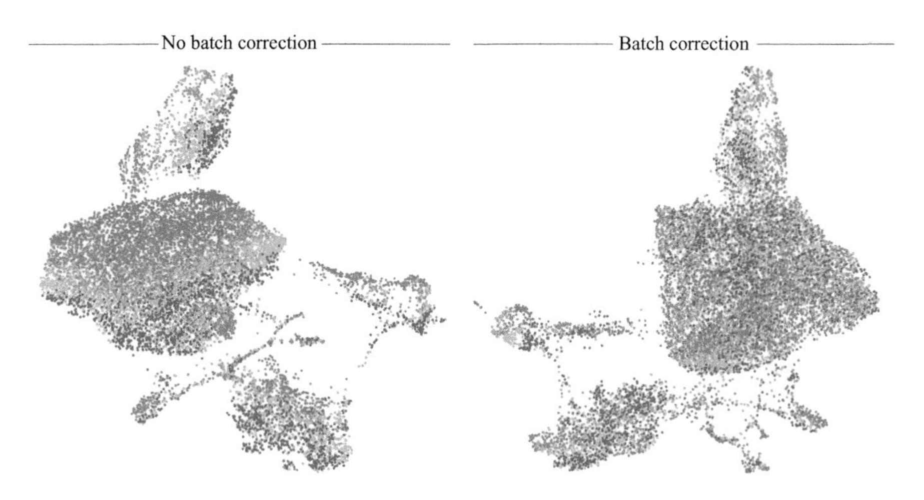

FIGURE 8.6
Correction of batch effects. Without batch correction (left) batch effects are evident with cells being colored by their original batches. With batch correction (right) batch effects are removed. (Adapted from: MD Luecken, FJ Theis, Current best practices in single-cell RNA-seq analysis: a tutorial, *Molecular Systems Biology* 2019, 15(6):e8746. Used under the terms of the Creative Commons Attribution 4.0 License, https://creativecommons.org/licenses/by/4.0, ©2019 Luecken et al.)

intra-experiment batch effects correction may underperform. Here methods that can be used include fastMNN [63], Seurat Integration [64], LIGER [65], Scanorama [66], BBKNN [67], Harmony [68], and Conos [69]. The fastMNN method is one implementation of the mutual nearest neighbors (MNN) approach, which is based on generalized non-linear modeling. MNN detects pairs of cells in two datasets that are most similar to each other (mutual neighbors) based on their gene expression profiles and therefore assumed to be equivalent cell types. Assuming the batch effects are orthogonal to the biological subspace, the systematic differences between the cells in all the MNN pairs are used to estimate the direction and magnitude of the batch effects, which is then applied to transform the original data of all cells to remove the batch effects. Seurat Integration, as implemented in Seurat 3, also uses the MNN approach to detect equivalent cell types, but this is performed in a subspace created by another algorithm called Canonical Correlation Analysis (CCA) [55], which aims to find shared sources of variation between datasets. The CCA process starts with identifying linear combinations of genes that have maximum correlation between datasets. This is followed by aligning the resultant vectors across the datasets using a nonlinear transformation process called dynamic time warping, thereby creating a low-dimensional subspace containing all datasets prior to applying the MNN procedure. Scanorama is based on a similar approach performing searches for MNNs in a low-dimensional subspace created by transformation of the original gene expression using a method called singular value decomposition (SVD). The Chemistry Batch Correction algorithm used by Cell Ranger since v3 for correcting batch effects between chemistries is also based on the MNN approach.

Besides the MNN-based methods above, BBKNN applies the kNN algorithm to remove batch effects from diverse datasets and create batch balanced data. The kNN procedure first constructs a cell neighborhood graph to identify the k nearest neighbors for each individual cell within each batch of data. This is followed by merging of the nearest neighborhood sets across different batches to connect equivalent cell types, while still keeping dissimilar cell types separated. Through creating connections between equivalent cell types from different datasets batch effects are quantified and corrected. LIGER, or Linked Inference of Genomic Experimental Relationships, offers another approach that has been shown to perform well on integrating different scRNA-seq datasets [70]. It employs a method called integrative non-negative matrix factorization (iNMF) to identify shared and dataset-specific factors. The low dimensional shared factor space is then used to identify similar cell types across datasets, and joint cell clusters for batch correction. All of the dataset *integration* methods introduced above aim mostly to correct batch effects between technically more variable datasets collected from different experiments or labs. If they are used for intra-experiment batch effects correction, they might lead to over-correction. To determine how a batch effects correction method has performed, visualization of the data before and after the correction in a low-dimensional space should offer a quick qualitative assessment (Figure 8.6). For a quantitative assessment, metrics such as kBET [61] and average silhouette width (ASW) [71] can be used.

8.3.6 Signal Imputation

As mentioned at the beginning of this chapter, technical factors such as low mRNA molecular capture rate and non-exhaustive sequencing cause signal dropout, i.e., the inability to detect a transcript that is present in a cell. This leads to high signal stochasticity, signal sparsity, and zero inflation, all of which can affect downstream data analyses. Signal imputation, also called denoising or expression recovery, aims at inferring missing transcript values to help alleviate this problem. Some of the commonly used scRNA-seq signal imputation methods include MAGIC [72], kNN-smoothing [73], SAVER and SAVER-X [74, 75], ALRA [76], CIDR [77], DCA [78], scImpute [79], mcImpute [80], and DrImpute [81]. Some of the pipeline tools introduced earlier have built-in imputation function (such as scVI), or use wrappers to run externally developed imputation tools (such as Seurat running imputation using ALRA). These methods can be separated into three groups based on the general approach they employ [82]: (1) modeling-based: SAVER, CIDR, and scImpute, for example, use probabilistic models to quantify signal uncertainty from sparse count data to differentiate technical zeroes (for which imputation will be provided) from biological zeroes (no imputation needed); (2) data-smoothing-based: MAGIC, kNN-smoothing, and DrImpute adjust all expression values, not just zeros, based on gene counts in similar cells; and (3) data-reconstruction-based: ALRA, DCA, mcImpute, and scVI use machine learning or low-rank matrix methods to create a latent space representation of the original data to reconstruct the cell-gene count matrix.

While signal imputation can lead to increased performance in downstream analyses, it should also be noted that these tools have the inherent limitation of being "circular," i.e., inferring missing transcript values using information from within the original data. As a result, they run the risk of generating false-positive results [83]. Benchmark studies also show that the performance of these methods varies with the type of downstream analyses performed [81, 84]. There is currently no consensus on whether signal imputation is an essential preprocessing step, largely due to the difficulty in assessing imputed data. The reader should be cautious on applying signal imputation techniques on normalized or raw data. If performed it is prudent to visually inspect and, better yet, experimentally validate results produced from the imputed data. Like in the case of genotype imputation [85], use of a reference database of single-cell transcriptomic profiles, such as the Human Cell Atlas, may help improve imputation performance. In this direction there are some emerging methods, such as SAVER-X, that use information from reference database or external dataset for transfer learning.

8.4 Feature Selection, Dimension Reduction, and Visualization

8.4.1 Feature Selection

Among the many genes contained in a dataset, many of them may not be very informative because, for example, their expression remains constant across cells (such as housekeeping genes). While it is not mandatory to preclude these genes from further analysis, including them leads to high memory usage and slow processing speed. To increase computational performance, feature selection approaches can be employed to identify the most biologically informative features (i.e., genes, with their numbers usually in the range of 500 to 2,000), and remove uninformative features for downstream analyses such as dimension reduction and visualization. Uninformative genes can be selected based on their expression levels, with those showing zero or low expression often being the target. This approach only needs to set a threshold on the level of mean gene expression across all cells. The most informative genes are often those that display high expression variability between different cell groups or identities, and therefore most likely contribute to cell-to-cell variation. This relies on the premise that genes showing high variability across cells are a result of biological effects, not experimental noise. Methods for selecting highly variable genes (HVGs) include the squared coefficient of variation method proposed by Brennecke et al. (2013) [86], the FindVariableGenes method used by Seurat, or those incorporated in other pipeline tools scran and scLVM [87]. Because of the heteroscedasticity of cellular gene expression, i.e., the aforementioned mean-variance relationship, the selection of HVGs should not be based on variance alone. Instead, these methods fit the relationship between variance and the mean into their respective models, based on which HVGs are selected using different statistic tests. For example, scran performs LOESS fit on the variance-mean relationship, and then uses the fit as the model to infer biological variation across cells for HVG selection. Evaluation of commonly used HVG selection methods reported large differences among the methods, and that different tools perform optimally for different datasets [88]. Besides using the variability of a gene's expression across cells, alternative feature selection strategies include using average gene expression level to select genes with the highest average expression [89], deviance to identify genes that deviate from the null model of constant expression [90], or dropout rate to retain genes with higher number of dropouts than expected [91].

8.4.2 Dimension Reduction

Collected from a large number of cells (e.g., tens of thousands) for a large number of genes (potentially all genes in the genome), scRNA-seq data has high dimensionality. This, in data science terms, causes the curse of dimensionality, i.e., the amount of data needed for accurate generalization grows exponentially with the increase in dimensions. Computationally, high dimensionality leads to mathematical intractability for many modeling and statistical calculations. Feature selection is one preliminary step toward dimensionality reduction. However, even after this step the dimensionality of the data is still very high. For example, there are still 2,000 gene dimensions if the top 2,000 HVGs are retained. For many downstream analytical steps, such as visualization and clustering to be detailed next, the number of dimensions must first be significantly reduced to only a very small number of dimensions (e.g., 10). To achieve this, specialized dimensionality reduction methods are required. The goal of these methods is not to select a small number of original features, but to transform the data to create new features so that the information contained in the original data can be preserved in the low-dimensional space.

Among the most commonly used dimensionality reduction methods are those based on linear transformation. Principal components analysis (PCA), independent component analysis (ICA) [92], non-negative matrix factorization

(NMF) [93], and factor analysis [94] are all examples that have been applied to scRNA-seq data. Among these methods, PCA is perhaps the best known. It projects the cell-gene count matrix onto a subspace that is defined by a few principal components, which are linear combinations of the original genes. The first principal component, or PC1, is one axis in the new subspace along which the maximal amounts of variation in the original data are captured. The second axis, corresponding to PC2, is orthogonal to the first axis catching the second most variation in the data. Although straightforward, PCA does not take signal dropout into consideration. ZIFA (zero-inflated factor analysis), often considered to be a variation of PCA, is developed to address this issue [94]. Although effective, such PCA-based approaches have one downside, i.e., the principal components that catch the majority of variance in the original data are sometimes difficult to interpret biologically. Other methods, such as f-scLVM (or factorial single-cell latent variable model) and NMF, address this difficulty through generation of reduced dimensions that are more biologically relevant. For example, reduced dimensions from f-scLVM are based on explicit modeling of bio-pathway annotations of gene sets [95].

The linear dimensionality reduction methods introduced above are based on the assumption that the underlying data structure is linear in nature. Methods that are based on non-linear transformation do not make this assumption, instead these methods operate under the premise that in a high-dimensional space most relevant information concentrates in a small number of low-dimensional manifolds. Currently available non-linear methods include t-SNE (t-distributed Stochastic Neighbor Embedding) [96], MDS (Multi-Dimensional Scaling) [97], Isomap (Isometric Feature Mapping) [98], LLE (Locally Linear Embedding) [99], diffusion maps [100], spectral embedding [101], and UMAP (Uniform Manifold Approximation and Projection) [102, 103]. The t-SNE algorithm, for example, works by modeling transcriptionally similar cells around each cell based on probability distribution, with Gaussian and t-distributions being used in the original and dimension-reduced space, respectively. The process first computes cell-cell similarity in the original space using a Gaussian kernel and then maps the cells to a dimension-reduced space that best preserves that similarity. The strength of t-SNE is to reveal local data structure, but this is often achieved at the expense of global data structure. An important parameter in the t-SNE algorithm is perplexity, which effectively controls the number of transcriptionally similar cells that each cell has. Proper adjustment of this parameter can help regulate the balance between local and global data structure [104]. UMAP is a more recent algorithm that is designed to provide better preservation of global data structure without losing performance on local data structure, better scalability, and faster computation speed. While it has a similar procedure to t-SNE in that it first constructs a high-dimensional representation of the original dataset and then projects it to a low-dimensional space, UMAP differs from t-SNE in how cell-cell similarity and the topology of cellular relations in the original high-dimensional space are computed and represented. UMAP achieves these tasks through building the so-called Cech complex, which is essentially a mathematically efficient way of representing the topological structure of cell-cell similarities embedded in the original data structure. UMAP has two key tunable parameters, i.e., the number of nearest neighbors a cell can have in the high-dimensional space, and the minimum distance between cells in the output low-dimensional space. By adjusting their values, UMAP can reach a balance between local and global structure based on project needs.

Other classes of dimensionality reduction methods include those based on the use of autoencoders and those that use an ensemble approach. An autoencoder, a type of unsupervised artificial neural network (ANN), aims to compress a dataset to a lower dimensional space and then reconstruct it. It achieves this by searching for representation of the dataset in the dimension-reduced space by focusing on key distinguishing features and removing noise and redundant information. Methods using autoencoders include VAE (Variational Autoencoder) [105], DCA (Deep Count Autoencoder) [78], scvis [106], scScope [107], and scVI. SIMLR (single-cell interpretation via multikernel learning) is a method that uses the ensemble approach. It uses t-SNE for dimension reduction, but instead of using the Gaussian-based cell similarity measure as input to t-SNE, it uses a cell similarity measure learned from the multikernel learning framework to improve scalability and performance of downstream steps [108].

Among the various dimensionality reduction methods introduced above, benchmark comparisons [98, 109] show that no method excels for all datasets. Non-linear and autoencoder-based methods can have better performance on stability and accuracy especially for highly heterogeneous datasets, but to achieve the performance users need to adjust parameters as parameter settings can have significant impacts on results. Linear methods (such as PCA) and UMAP tend to perform better in terms of computing speed, and scalability especially when there are a large number of cells and genes in the dataset. In practice, because of their efficiency, linear methods especially PCA are often used to reduce dimensionality first, followed by visualization of the resultant data using some of the non-linear methods such as t-SNE and UMAP.

8.4.3 Visualization

One goal of dimensionality reduction is to enable graphical depiction of the underlying data so that the researcher can visualize the major cell types (or conditions) in the dataset, and thereby intuitively understand the inherent heterogeneity of the represented cellular population. In such a visual (Figure 8.7), scatter plots are often used, in which each point represents a cell projected into a two- or three-dimensional space, with each dimension corresponding

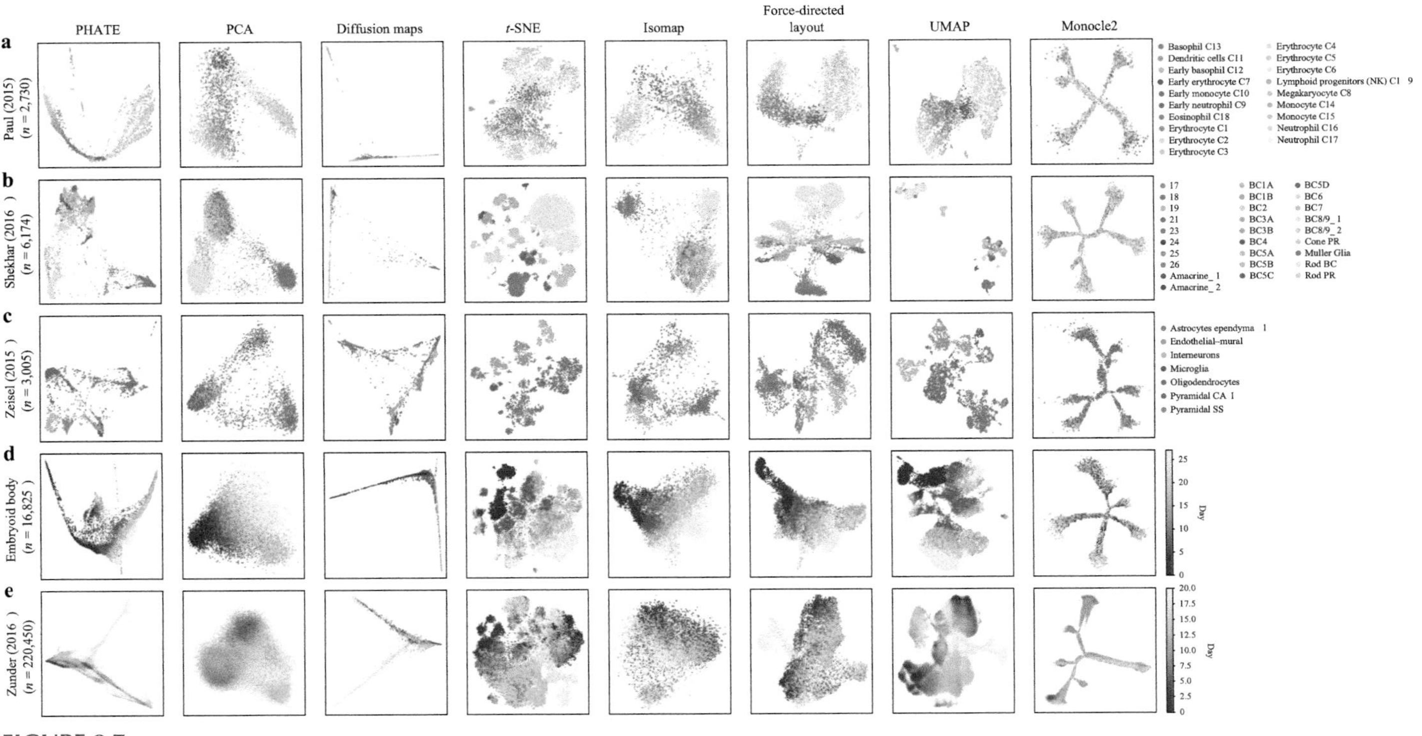

FIGURE 8.7
Visualization of single-cell RNA-seq data in 2-D space. Different visualization methods are applied to different datasets to show how these methods differ from each other in generating visualizations for datasets of different characteristics. (Adapted by permission from Springer Nature Customer Service Centre GmbH: Springer Nature, *Nature Biotechnology*, Visualizing structure and transitions in high-dimensional biological data, Kevin R. Moon et al., Copyright 2019.)

to the top reduced dimensions, or principal components if PCA was used for dimension reduction. The actual rendering of the visualization from the dimension reduced data is mostly achieved by non-linear methods t-SNE and UMAP, which have better performance than other non-linear methods. Although PCA as the primary dimensionality reduction method may also be used to visualize global data structure, and as a linear approach the distances between cells or clusters in a PCA plot provide consistent depiction of their relationships, its disadvantage as a visualization tool is its relative lack of power to provide sufficient separation between distinct cell types.

As a non-linear approach, t-SNE typically offers a better presentation of the same dataset, by placing similar cells close to each other and dissimilar cells distant. As indicated earlier, using this method the similarities between cells are measured locally based on their transcriptomic profiles. Because of its emphasis on local similarity, t-SNE tends to have limited preservation of the dataset's global structure. Although the perplexity parameter can be increased to reveal more global structure, this usually leads to dramatically increased computational time. It should be noted that even at low perplexity values t-SNE is not a fast method. Because t-SNE emphasizes local structure and may overestimate the differences between cells or cell clusters, the researcher should be cautioned not to over-interpret the distance between cells or clusters, the size of clusters, as well as the relative position of cells or clusters in a t-SNE visualization. As t-SNE is a stochastic algorithm, the final output may not be fully reproducible between runs. While revised implementations of the t-SNE approach aim to overcome such shortcomings [104], new visualization methods, with UMAP being a good example and gaining popularity, have been established as robust alternatives [102]. Compared to t-SNE, UMAP is faster, more scalable to very large datasets in terms of both data size and dimensionality, and able to preserve more global structure while retaining local data structure. While it is also stochastic, UMAP's run-to-run performance is generally more reproducible than that of t-SNE. It should be noted that as in the case of t-SNE the distance between cells or clusters in a UMAP visualization may not have biological bearing.

Besides t-SNE and UMAP, other visualization tools are also available. Example of these tools are PAGA (or partition-based graph abstraction) [110], PHATE (or potential of heat diffusion for affinity-based transition embedding) [111], SPADE (or spanning-tree progression analysis of density-normalized events [112], and SPRING [113]. Compared to t-SNE and UMAP, which are better for visualizing distinct groups of cells in a mixed population, these other tools are more suited to display cells along a continuous trajectory, such as those that are undergoing differentiation. Among these tools PAGA provides a manifold learning algorithm that can faithfully preserve both continuous and distinct cell identifies/groups at multiple resolutions. It is robust in terms of computational efficiency, scalability, and interpretability of results for continuous cell trajectories (Section 8.7 covers "Trajectory Inference").

8.5 Cell Clustering, Cell Identity Annotation, and Compositional Analysis

8.5.1 Cell Clustering

Based on their gene expression profiles, cells in a dataset can be grouped into clusters. By placing transcriptionally similar cells into clusters, the process can uncover different cell groups or identities in a sample, achieving a common goal of scRNA-seq analysis. This process often starts with dimension-reduced data, usually from PCA. To capture most of the gene expression variation across cells in the dataset, enough principal components should be used. In the dimension-reduced space, different metrics can be used to quantify the similarity (or distance) between cells based on their transcriptomic profiles. Such similarity metrics can be mostly classified as distance- or correlation-based. Two examples of distance-based metrics are Euclidean distance, which represents the shortest or straight-line distance between two points (cells), and Manhattan distance (also called city-block distance), measuring the distance between two points in a grid like path. Correlation-based metrics, such as Pearson and Spearman's correlation coefficients, measure the similarity between the general shapes of gene expression in two cells. For comparison, distance-based metrics capture gene expression levels and measure how far two cells are from each other in the PCA space, while correlation-based metrics track relative expression trends between two cells and are insensitive to the scale of gene expression levels. A comparison of their clustering performance showed that correlation-based metrics are more robust than distance-based metrics [114]. Other similarity metrics have also been used, e.g., cosine similarity, a measure of angle instead of magnitude between two vectors, that is used for spherical K-means clustering [115].

There are different algorithms to cluster cells, and among the widely used are hierarchical, K-means, and graph-based clustering. These algorithms identify major cell identities (encompassing different cell types, or different states in the same cell type) present in a population as clusters, and assign each cell to one of these clusters without relying on any *a priori* information. Hierarchical clustering constructs a dendrogram in which different cell clusters merge or split at different branches, creating a hierarchy. One process to create such a hierarchy is agglomerative, or "bottom-up," in which at the start each cell is assigned to its own cluster, and at each subsequent iteration the most similar nodes are merged. This process recurs hierarchically, until all cells are merged. The opposite process takes a divisive or "top-down" approach, which starts with all cells being in a single cluster, and then splits it into two branches at each subsequent step, until each cell is assigned to its own cluster. From the dendrogram, clusters can be obtained through "cutting the tree" to generate the desired number of clusters, or at desired height.

The K-means clustering algorithm first requires the user to specify *k*, the number of clusters. The process starts with all cells randomly assigned to one of *k* clusters. Then the centroid of each cluster is determined, and each cell re-clustered based on their distance to each of the *k* centroids. This process is reiterated until each cell's cluster assignment no longer changes. The third commonly used approach, graph-based clustering, is based on graph construction that connects cells to their nearest neighbors. In a kNN graph (different from K-means), for example, two nodes (cells) are connected by an edge, if the distance from cell A to B is within the *k*-th lowest distances from A to other cells. The edge may have a weight assigned based on the similarity between the cells. In a Shared Nearest Neighbor (SNN) graph, an edge is weighted based on their proximity to each other, or similarity in terms of the number of mutual neighbors the two cells share. After such a graph is constructed, dense regions that contain a large number of highly connected nodes can be detected as the so-called communities, representing distinct cell identities. Within each community (or cluster), cells are more highly connected with each other indicative of high similarity, than those in other communities. To partition cells into distinct communities, community detection techniques, such as the Louvain and Leiden methods [116, 117], can be used.

To carry out these clustering approaches, either general-purpose clustering methods or tools specifically designed for scRNA-seq data can be used. As examples of general-purpose methods, the hclust() and kmeans() functions in R can be directly used on PCA dimension reduced data to perform hierarchical and K-means clustering, respectively. Examples of tools specially developed for clustering single cells include SC3 [118], pcaReduce [119], CIDR, RaceID2 [120], SIMLR [108], and SNN-Cliq [121] (Table 8.1). Many pipeline tools, such as Seurat, Cell Ranger, Pagoda2 [122], and Scanpy, scran, ascend [97], and SINCERA [123], also provide built-in clustering functionalities. Seurat, for example, provides graph-based clustering. This process first constructs a kNN graph using Euclidean distance in the PCA space, with the edges weighted by Jaccard similarity that measures the number of neighbors they share. To find clusters in the graph, the Louvain community detection method is then applied. Seurat clustering uses a user adjustable parameter called resolution to control the number of clusters generated, with a higher resolution (e.g., >1.0) leading to a larger number of clusters. Cell Ranger provides a similar nearest neighbor graph-based clustering approach, with an additional step to merge clusters that show no differential gene expression. In addition, it also offers K-means clustering as another option. Benchmarking studies on most of these clustering methods show that there is wide variation in actual performance and poor concordance among them [89, 124]. Some methods, such as Seurat, SC3, and Cell Ranger, showed better overall performance than others. In terms of running times, Seurat also showed consistently faster speed than most other methods.

TABLE 8.1
Single-Cell Clustering Methods

Name	Description	Reference
SC3 (Single-cell consensus clustering)	PCA + K-means + hierarchical clustering. Uses multiple parallel clustering parameter tests to reach a consensus matrix by hierarchical clustering	[118]
pcaReduce	PCA + K-means + hierarchical clustering. An agglomerative clustering process that integrates PCA and hierarchical clustering	[119]
SIMLR (Single-cell interpretation via multi-kernel learning)	Dimensionality reduction + K-means. Learns proper weights for multiple kernels, and constructs a symmetric similarity matrix for dimension reduction and clustering	[108]
SNN-Cliq	Graph-based clustering. Constructs an SNN graph and then clusters cells using quasi-clique finding techniques	[121]
RaceID2	Pearson's correlation distance matrix for all pairs of cells + K-means clustering	[120]
CIDR (Clustering through Imputation and Dimensionality Reduction)	PCA + hierarchical clustering. Uses implicit imputation to help alleviate the effects of dropouts, then performs hierarchical clustering on principal coordinates	[77]
Seurat Pagoda2 Scanpy	PCA + kNN graph-based clustering	Seurat [55], Pagoda2 [122], Scanpy [56],
Cell Ranger	PCA + graph-based clustering, or K-means clustering	[10]
scran	PCA + hierarchical clustering, or SNN graph-based clustering	[57]
ascend	PCA + hierarchical clustering	[97]
SINCERA	Z-score/trimmed mean normalization + hierarchical (default)/consensus/tight clustering	[123]

8.5.2 Cell Identity Annotation

After placing cells into different clusters, the next question is usually what cell type, or what cellular state, is represented by each cluster. To help answer this question, comprehensive pipeline toolkits such as Seurat, Cell Ranger, Scanpy, and scran have built-in functions to identify marker genes that are uniquely highly expressed in a cluster compared to other clusters. For example, in Seurat the FindMarkers() function identifies candidate marker genes for each cluster by differential gene expression analysis, which is conducted using the default non-parametric Wilcoxon rank sum test among other available tests.

Cell Ranger finds differential genes for each cluster using edgeR (covered in Chapter 7), or sSeq, which is a modified NB exact test [125]. From the identified differentially expressed, candidate marker genes, manual annotation can be used to identify cell type or state in each cluster, based on previously characterized, canonical cell identity specific marker genes. For example, *GFAP* (glial fibrillary acidic protein) gene expression is a marker of astroglial cells in the brain, and *CD79a* and *CD79b* are markers for B cells. The identification of cell type or state based on such classic gene markers is an extension of the traditional method often used in the lab for cell identity recognition. This process, however, is typically labor-intensive and time-consuming, as it often needs extensive review of currently available literature, or deep domain knowledge about the cell system under study, which requires close interactions between bench scientists and informaticians. In addition, some cell types may not have well characterized gene markers, or the expression of marker genes in host cells may be undetectable due to signal dropout.

To address some of the issues and speed up the cell identity recognition process, a relatively easy and semi-automatic approach is through Gene Ontology (GO) and/or bio-pathway enrichment analysis of the candidate marker genes (detailed in Chapter 7, Section 7.3.8), since the GO terms or pathways significantly enriched in the genes can produce insights into cell identity. For example, if "hepatocyte homeostasis" is identified as a significantly enriched GO term, it is indicative that at least some of cells in the cluster are hepatocytes. Further, the cell identity detection process on the basis of canonical marker genes can be automated, with the use of specialized tools such as Garnett [126], Digital Cell Sorter [127], SCINA [128], CellAssign [129], and scANVI [130]. The list of marker genes required by these automated tools to classify different cell identities can be provided by databases such as PanglaoDB [131], CellMarker [132], the Mouse Brain Atlas [133], the BRAIN Initiative Cell Census Network (or BICCN) [134], and DropViz.org [135]. Garnett, as an example, first uses as input a list of gene markers to train a regression-based classifier, which is then applied to classify cells in a new dataset. As another example, scANVI, a semi-supervised variant of scVI, also classifies cells based on their expression of canonical marker genes. This tool goes even further to use these cells as "seeds" to classify other cells in the same dataset with unobserved expression of the marker genes, on the basis of how close these other cells are to the "seeds."

Instead of expression of marker genes, overall gene expression pattern may also be used for automated inference of cellular identities. This approach uses information embedded in the overall gene expression of annotated cells in a reference dataset, to predict cell identities in a query dataset. Without reliance on prior knowledge in the form of a pre-defined list of marker genes, this approach directly harnesses the power of rapidly accumulating single-cell data that can be used as reference, such as those from the Human Cell Atlas [136] and the Tabula Muris Atlas for mouse [137]. Table 8.2 lists some of the

TABLE 8.2

Cell Identity Annotation Tools

Name	Description	Reference
Classifiers that do not require marker genes		
scmap	Mapping cells in a scRNA-seq dataset onto the cell-types or individual cells annotated in a reference dataset	[138]
SingleCellNet	Uses a random forest classifier trained on annotated scRNA-seq data to classify cells in a query dataset	[139]
scPred	Employs a combination of feature selection from a reduced-dimension space, and a support vector machine model to classify single cells	[136]
ACTINN	Uses a 4-layer neural network for automated cell type identification	[140]
SingleR	Assigns cell identify based on comparison of their transcriptomes to reference bulk transcriptomic datasets of pure cell types	[148]
CHETAH	Characterizes cell types by using correlation and confidence score in comparison to reference dataset using a hierarchical classification tree	[149]
Cell BLAST	Cell querying and annotation based on a neural network-based generative model, and a posterior-based cell-cell similarity metric	[150]
ItClust	Iterative transfer learning with neural network to improve cell type classification	[151]
CaSTLe	Classifies single cells based on feature selection and XGBoost classification	[152]
OnClass	Uses Cell Ontology to embed different cell types into a low-dimensional space, and then maps cells to this partitioned space based on gene expression for cell type annotation	[144]
CellO	Performs hierarchical cell classification against the directed acyclic graph structure of the CO system	[145]
Classifiers that require marker genes		
Garnett	First trains a regression-based classifier using the gene marker definition file, and then applies to classify cells from similar tissues or sample types	[126]
SCINA	Leverages established gene signatures to perform cell type classification	[128]
CellAssign	Automates assignment of cells to known cell types using a probabilistic model with known marker genes as input	[129]
Digital Cell Sorter (DCS)	Assigns cell types automatically using a voting process based on known molecular markers	[127]
scANVI	A semi-supervised classifier that uses Gaussian mixture model to provide posterior probability of assigning a cell to different classes	[130]

methods that use this strategy. Among them, scmap [138] projects cells in a query dataset onto a reference dataset (or combined references), to identify matching individual cells (the scmap-cell mode) or specific cell types (the scmap-cluster mode). In this method, the similarity between individual cells in the query sample and cells or cell types in the reference is measured using three distance metrics (Pearson, Spearman, and cosine). Some of the other methods are machine learning based, i.e., they first construct classifier models from a reference dataset and then use the classifiers to annotate individual cells or clusters from a query dataset. SingleCellNet (SCN), for example, uses a random forest (RF) classifier trained on annotated reference scRNA-seq data to transfer to query cell classification [139]. ACTINN (or Automated Cell Type Identification using Neural Networks) [140] is based on neural network for such transfer learning. Another generalizable method, scPred uses support vector machines (SVMs), combined with singular value decomposition for unbiased feature selection, to perform probability-based cell type prediction [141]. Besides these machine learning-based tools specially designed for single-cell data, general-purpose classifiers including SVM and RF may also be used directly [142]. Among other cell classification methods that use overall gene expression pattern instead of marker genes are those developed for integrated analysis of multiple scRNA-seq datasets, such as Seurat Integration, Scanorama, and Conos as introduced in Section 8.3.5 (therefore not listed on Table 8.2). These methods achieve automated annotation of cells from a query dataset through detection of equivalent cells in a reference dataset. As an example, Seurat performs annotation of query cells by mapping the query dataset onto a reference, which is accomplished through projecting the query cells onto the reference UMAP structure.

Because the diverse array of methods introduced above may not always use the same term to label the same cell type, for consistency cell annotation can benefit from the use of standardized cell type terms from the Cell Ontology (CO). CO is a community effort to organize cell types anatomically and hierarchically through a structured and controlled vocabulary [143]. To further leverage the inherent hierarchical relationships built into the CO terms, specialized CO-based cell annotation tools, such as OnClass [144] and CellO [145], have also been developed. Cello, for example, performs hierarchical classification to annotate cells based on the graph structure of the CO system. Compared to methods that do not use CO terms, such tools provide more consistent and standardized annotation of individual cells or clusters, through the use of CO terms.

To evaluate the performance of these methods, comprehensive benchmark studies have been performed [146, 147]. Using 27 scRNA-seq datasets of varying cell numbers, platforms, species, and cellular heterogeneity, one of the studies [146] compared 22 automated cell identification methods and showed that the general-purpose SVM classifier had the best overall performance. Other top performers included SingleCellNet, scmap-cell, scPred, and ACTINN. After evaluating 10 methods on 6 datasets, the other study [147] found Seurat to be the top performer (this method was not included in the other study by Abdelaal et al.). SingleR and SingleCellNet also performed reasonably well. Both studies found that methods that require marker genes did not outperform those that use the overall gene expression pattern. It should be noted that many of the methods evaluated by both studies leave unknown cell types (i.e., those that display low or no expression of all marker genes, or do not exist in the reference dataset) as unclassified, which is desired. Such cells need to be further characterized using additional canonical marker genes or biochemical techniques. In addition, many of the evaluated methods, besides labeling distinct cell types/identities, can also be used to identify transitional or intermediate cell types as that they may represent part of a developmental trajectory.

8.5.3 Compositional Analysis

The composition of different cell identities in a population varies with internal and external conditions. For example, upon bacterial pathogen infection, in the small intestinal epithelium there is a change in the proportions of different cell types as part of antimicrobial response (Figure 8.8) [153]. Compositional analysis involves examination of the proportions of different cell identities in different samples. For this analysis, different statistical approaches have been used to assess the significance associated with changes of cellular composition. For example, to detect the change of cell composition in the intestinal epithelium upon infection, Haber et al. (2017) applied a Poisson process to model

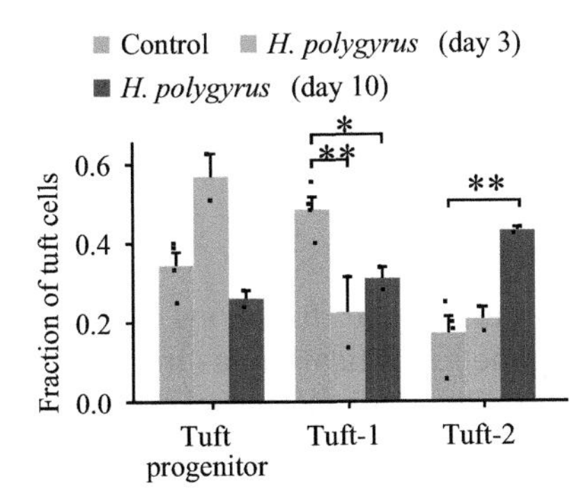

FIGURE 8.8
Changes in cellular composition in intestinal epithelium caused by pathogen infection. Shown here are changes in the fraction of three different types of tuft cells (having chemosensory function in the gut lining) after infection with the parasitic helminth *Heligmosomoides polygyrus*. Significant changes in frequency are marked (* FDR < 0.25, ** FDR < 0.05; Wald test). (Adapted by permission from Springer Nature Customer Service Centre GmbH: Springer Nature, *Nature*, A single-cell survey of the small intestinal epithelium, Adam L. Haber et al., Copyright 2017.)

the detected proportions of different cell types as a random count variable, which used the grouping of each sample (treatment or control) as a covariate and the total number of detected cells in each sample as an offset variable. As another example, Hashimoto et al. (2019) used a Wilcoxon rank-sum test to compare the fractions of different classes of circulating lymphocytes in supercentenarians vs. younger controls to uncover the mechanism of their exceptional longevity through sustaining immune system function [154]. Single-cell differential composition, or scDC, is a specially developed tool to assist such composition analysis. It employs a bootstrap resampling process to determine the standard errors during estimating cell-type proportions, and generalized linear model (GLM) and mixed model (GLMM) analysis for subsequent significance testing to achieve statistical comparison [155].

One characteristic of compositional data is that they are subject to the constant-sum (i.e., unity) constraint [156]. This can lead to spuriously significant correlations among different cellular identities and results in negative bias. For instance, if a new cell type emerges due to activation, because of the constraint the relative proportions of other existing cell types will correspondingly decrease although their absolute numbers may not change. Standard univariate analysis methods that do not take this characteristic into consideration would lead to deflated estimation of existing cells. It is imperative, therefore, to develop novel methods that account for such inherent bias of single-cell compositional data. As an example, scCODA uses a Bayesian approach that takes into consideration the negative correlative bias as well as the uncertainty associated with determining cellular composition. It achieves this through modeling cell type proportions using a hierarchical Dirichlet-Multinomial distribution, and joint modeling of all detected cell type proportions instead of individual ones [157].

8.6 Differential Expression Analysis

The FindMarkers or other similar functions in Seurat, Cell Ranger, and other tools for marker gene identification, as introduced in Section 8.5.2, are in fact differential expression (DE) analysis methods that focus on finding genes that are highly expressed in a particular cluster of cells vs. all others. In this section, we focus on more general DE analysis to compare cells in two (or more) populations, samples, time points, or perturbation conditions. Methods developed for DE analysis of bulk RNA-seq data, such as DESeq2, edgeR, and limma, are also applicable to scRNA-seq data. As detailed in Chapter 7 (Section 7.3.5), DESeq2 and edgeR use the NB model under a GLM framework, and Limma employs the normal linear model, to capture the distribution of read counts. A core component of these methods is to assess gene expression variance from the rather limited number of samples usually used in bulk RNA-seq, but in applying to scRNA-seq data this is no longer a challenge because of the much larger number of single cells involved in the comparison. The substantial zero inflation, signal overdispersion, transcriptional bursting, and multimodality associated with scRNA-seq data, however, pose different challenges.

Methods developed for scRNA-seq DE analysis (see Table 8.3) use different approaches to address the specific challenges posed by single-cell data. Examples of these methods are SCDE [158], MAST [159], D^{3}E [160], scDD [161], BPSC [162], NBID [163], DEsingle [164], DECENT [165], and SwarnSeq

TABLE 8.3

Single-Cell Differential Expression Analysis Tools

Name	Description	Reference
Tools specifically developed for scRNA-seq data		
SCDE	Uses a mixture of negative-binomial distribution to model RNA signal amplification, and a Poisson distribution for signal dropout	[158]
MAST	Models scRNA-seq data using a mixture of Gaussian distribution to model gene expression level, and logistic regression to model drop-out events	[159]
BPSC	Performs DE analysis based on a beta-Poisson mixture model integrated into the GLM framework	[162]
DEsingle	Adopts ZINB model to describe read counts and excessive zeros, and define and detect three types of DE genes	[164]
D^{3}E	Uses non-parametric comparison of distributions for DE gene identification, and fits the transcriptional bursting model to explore gene expression change mechanisms	[160]
scDD	Uses a Bayesian framework to identify DE genes that are classified into different multimodal distributions	[161]
SwarnSeq	Integrates ZINB and a binomial model to model UMI counts to account for zero inflation and RNA capture rates, to identify and classify DE and differential zero-inflated genes	[166]
Monocle 3	Conducts regression analysis to find genes that change expression under different experimental conditions, and/or graph-autocorrelation analysis to identify genes that change with a trajectory or differ between clusters	[103]
tradeSeq	Performs trajectory based DE analysis using an NB generalized additive model	[167]
Tools originally developed for bulk RNA-seq data		
DESeq2	Models gene expression mean and variance using an NB distribution, and calculates each gene's p value using the observed sum of read counts of the two conditions	[173]
edgeR	Uses an NB model to model gene expression, and performs DE analysis using the GLM likelihood ratio test	[174]
limma	Fits a linear model for each gene, and use empirical Bayes procedures for borrowing information across genes for DE analysis	[175]
ROTS	A general test originally developed for microarray data that uses an adaptive reproducibility-optimized t-like test statistic for DE analysis	[176]

[166]. To deal with zero inflation, SCDE fits a mixture of two error models, with one using Poisson distribution to model the signal dropout process, with the other using the NB distribution to model the signal amplification process for detection of transcripts in correlation to their abundance in cells [158]. MAST uses a two-part generalized linear hurdle model, with one modeling the discrete expression rate of each gene across cells (i.e., how many cells express the gene) using logistic regression, and the other modeling the continuous positive expression level of each gene by Gaussian distribution [159]. To fit zero-inflated and overdispersed scRNA-seq data, SwarnSeq uses the zero-inflated negative binomial (ZINB) model to model the observed UMI counts of transcripts. In addition, through using a binomial model to adjust for cellular RNA capture rates, this method allows detection of DE genes, as well as differential zero-inflated genes, i.e., those that show significant difference in the number of cells that have zero expression between two groups [166]. To address the multimodality nature of scRNA-seq data, scDD employs Bayesian modeling to identify genes that display differential distributions across conditions, and the genes are then further classified into different multimodal expression patterns. SwarnSeq also classifies influential genes into various gene types based on their differential expression and zero inflation patterns. To address the issue of transcriptional bursting, D^{3}E has two modules, with one for DE gene identification, and the other for fitting a model for transcriptional bursting to help discover the mechanisms underlying the observed expression changes.

The same marker gene finding functions contained in most comprehensive pipeline toolkits as introduced earlier can also be generalized for DE analysis. For example, Seurat provides DE analysis from the same FindMarkers() function through specifying two groups of cells for comparison. Currently available differential test methods include Wilcoxon rank sum test, likelihood-ratio test, Student's *t*-test, negative binomial GLM, Poisson GLM, logistic regression, as well as the aforementioned MAST and DESeq2. SINCERA, as another example, offers DE analysis using one-tailed Welch's *t*-test if gene expression can be assumed to come from two independent normal distributions, or one-tailed Wilcoxon rank sum test in case of small sample sizes. To identify genes that are differentially expressed along a developmental lineage or trajectory (Trajectory Inference to be introduced next), methods such as Monocle 3 [103] and tradeSeq [167] can be used. Monocle 3, for example, employs two approaches: graph auto-correlation and regression analysis. The former is suitable to identify genes that change along a trajectory, or differ between clusters, while the latter is to find genes that change expression under different experimental conditions.

To evaluate the plethora of methods that are currently available for scRNA-seq DE analysis, several benchmarking studies were performed [168–171]. Based on these studies, methods that were originally developed for bulk RNA-seq data have been shown to perform as well as methods developed specifically for single-cell data, especially after applying some strategies such as prefiltering to remove lowly expressed genes [169] or weighting to deal with zero inflation [172]. As these bulk or scRNA-seq tools have different ways of dealing with signal sparsity, multimodality, and heterogeneity, there is a general lack of agreement in the DE genes they identify. In addition, these benchmarking studies also show a general tradeoff between precision and sensitivity, i.e., methods of high precision have low sensitivity, which leads to identification of less true positive genes but also introduces fewer false positives.

It should be noted that DE analysis is an integral step of an scRNA-seq analytical pipeline, and upstream data processing can have an effect on the overall performance of this step. Of the various upstream steps, normalization has been shown to have a significant impact on DE results by a systematic evaluative study conducted by Vieth et al. (2019). Based on this study, a good normalization before DE analysis, such as that provided by scran, can alleviate the need for complex DE methods [177]. Another note is that, just like in bulk RNA-seq analysis as detailed in Chapter 7 (Section 7.3.8), the identified DE genes can be subjected to further functional analysis, such as gene set enrichment analysis, to reveal what biological processes or pathways are enriched in them.

8.7 Trajectory Inference

Many biological processes, such as development, immune response, or tumorigenesis, are underlined by continuous dynamic cell changes across time. The path of changes that a cell undergoes in such a process is often called a trajectory. While it is not yet possible to monitor the continuous transcriptomic change of an individual cell over time, trajectory can be inferred from a population of cells that represent a continuum of transitional cellular states while cells undergo changes in an unsynchronized manner. Because trajectory inference (TI) is based off of a snapshot of gene expression of a population of cells at a certain point of time, it is also called pseudotemporal analysis. Methodologically, it is built on the premise that cells in the continuum share many common genes and their gene expression displays gradual change. In essence, to infer cellular trajectory is to find a path in the cellular gene expression space that connects cells of various transitional states by maximizing similarity between neighboring cells. The inferred cellular trajectory can then be validated with additional experimental evidence.

Trajectory inference is carried out on dimensionality-reduced data, often after the clustering step. General methods, such as minimum spanning tree (MST) that aims to connect all points (clustered cells) in a graph to a path that minimizes distance between points, can be directly used for TI [178].

Some of the methods specifically developed for TI are in fact based on MST. For example, the first version of Monocle, a pioneer method for inferring trajectory from single-cell sequencing data, first creates MST on cells projected in a dimensionality-reduced space, and then places cells along the longest path through the MST [92]. Slingshot [179], TSCAN [180], and Waterfall [181] build MST on cell cluster centroids, instead of cells, and then order cells onto the path through orthogonal projection. Besides these MST-based methods, some other commonly used methods are based on graph theory. For example, Diffusion Pseudotime (DPT) builds weighted kNN graph on cells, and then orders cells using random-walk-based distance [182]. Also based on the use of weighted kNN graph, PAGA performs graph partitioning and abstraction using the Louvain method to identify different cellular states or identities, and uses an extension of DPT for pseudotime calculation [110] (Figure 8.9 shows an example). Monocle 3 is built on PAGA and adds one step further to construct more fine-grained trajectory through learning a principal graph from the PAGA graph [103].

The methods mentioned above and listed on Table 8.4 are among an increasingly long list of TI methods available. Besides the different approaches these methods use to infer trajectories, they also differ in what trajectory topology they can infer, whether they require prior information, how scalable they are with increasing cell numbers, etc. Cellular trajectory topologies can be

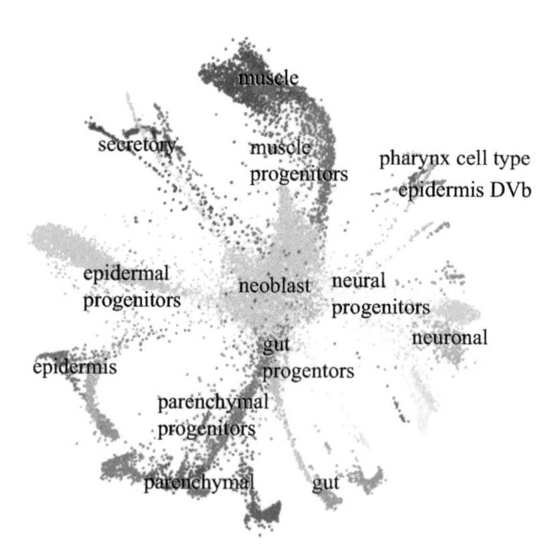

FIGURE 8.9
Cell trajectories inferred by PAGA to reconstruct a developmental lineage tree encompassing all cell types in the planarian body based on single-cell transcriptomic data. (Adapted from FA Wolf, FK Hamey, M Plass, J Solana, JS Dahlin, B Göttgens, N Rajewsky et al., PAGA: graph abstraction reconciles clustering with trajectory inference through a topology preserving map of single cells, *Genome Biology* 2019, 20(1):59. With permission.)

TABLE 8.4

Trajectory Inference Methods

Name	Description	Reference
PAGA	Constructs a weighted kNN graph, followed by graph partitioning and abstraction using the Louvain method. Applies an extended version of DPT for pseudotime calculation	[110]
Slingshot	Uses cell cluster-based MST to learn global lineage structure, and then orders cells along lineage curves through orthogonal projection	[179]
Monocle	Initial version (Monocle 1) builds an MST on cells in an ICA dimension reduced space and then orders cells along the longest path. Monocle 3 learns principal graph from PAGA graph to build more fine-grained trajectory	[92, 103]
TSCAN	Constructs MST to connect cell cluster centroids and projects cells to the MST backbone to build the pseudotime course	[180]
Diffusion Pseudotime (DPT)	Reduces dimensionality with diffusion maps, uses weighted kNN graph and orders cells using a random-walk-based distance	[182]
FateID	Uses random forests to quantify fate bias, which is then used to pseudo-temporally order cells	[191]
Waterfall	Builds MST on cell clusters, and use distance to cluster centers for orthogonal projection	[181]
Wanderlust	Selects the shortest path from an ensemble of kNN graphs, and order cells along the path based on their distance to user-defined start cell and random waypoint cells	[192]
Wishbone	An extension of Wanderlust for bifurcation branching topology	[193]

classified as linear, bifurcating, multifurcating, cyclic, tree-like, etc. While some methods (e.g., TSCAN, Wanderlust, and Waterfall) are designed to infer linear trajectories, many methods can infer multiple topologies, from linear, bifurcating, multifurcating, to even tree-like trajectories. Methods like PAGA can infer even more complicated topologies such as cyclical, connected, or disconnected graphs. Some of the methods, such as FateID, require prior information from user input in the form of specifying a starting or ending cell, or the number of branches, while others such as PAGA, Wanderlust, and Wishbone can take advantage of such information, but this is not mandatory. Besides using static snapshots of cellular states at a certain time point, some methods are developed to infer trajectories from time series data, the examples of which are TASIC [183], Waddington-OT [184], CSHMM [185], and Tempora [186]. Given the wide diversity of currently available TI tools, benchmarking studies may help guide their selection. One comprehensive evaluation of 45 currently available TI methods shows that (1) no one method applies to all scenarios, (2) there is considerable complementarity between the tested methods, and (3) the user should choose and use a variety of methods based on the expected trajectory topology [178]. Based on this comparison, the top performing methods include PAGA, Slingshot, different versions of Monocle, as well as generic methods such as MST.

After inference of cellular trajectory, the next questions to ask are what genes are associated with cell lineage development and what key genes underlie transitions between cellular states. As indicated in the last section, methods for DE gene analysis for trajectories are still limited. Besides Monocle 3 and tradeSeq as mentioned in the last section, other available trajectory DE methods include those employed by TSCAN, GPfates [187], and earlier versions of Monocle. Monocle 1 uses generalized additive models to test whether genes significantly change their expression as a function of pseudotime. TSCAN employs a similar approach. Monocle 2 uses a different approach called BEAM (branch expression analysis modeling) to test whether gene expression changes are associated with cell lineage branching along a trajectory. GPfates models gene expression-dependent cell fates as temporal mixtures of Gaussian processes. Similar to BEAM, it can identify gene expression changes associated with bifurcation points. Besides these tools that perform both trajectory inference and DE analysis, there are also tools that take as input pseudotemporal ordering of cells inferred by the TI tools detailed above, to conduct time-course DE analysis. LineagePulse, representing such an example, fits ZINB noise model to gene expression data collected from pseudotemporally ordered single cells [188].

Using static snapshots of cellular states, TI does not make predictions on the speed or direction of cell progression along the trajectory. To make such predictions, additional information is required. Change in cellular mRNA abundance inferred from the same static snapshots by a strategy called RNA velocity analysis [189] provides such information. The RNA velocity strategy is based on the detection and comparison of unspliced pre-mature transcripts that still contain introns, and spliced mature transcripts. In principle, this strategy is built on the premise that if there is a high ratio of unspliced to spliced mRNA molecules (called positive velocity) from a gene, it indicates that expression of the gene is upregulated from its steady state. Conversely, if the ratio of unspliced to spliced mRNA abundance is lower than its steady state ratio (i.e., negative velocity), it is indicative of downregulation for the gene. Based on aggregation of RNA velocities inferred across genes, this analysis then makes predictions on the future state of each cell in terms of the speed and direction of their movement along the trajectory. Currently available RNA velocity analysis tools include VeloCyto [189] and scVelo [190]. These RNA velocity tools are compatible and can be deployed with pipeline toolkits such as Seurat and Scanpy. Because it adds predictive information onto a trajectory about the direction and speed of cellular movement, RNA velocity analysis is often carried out in combination with TI.

8.8 Advanced Analyses

8.8.1 SNV/CNV Detection and Allele-Specific Expression Analysis

Besides transcriptomic profiles, scRNA-seq data also contains genotypic information specific for each cell, including single nucleotide and structural variants. Such additional information is especially helpful for studies that involve genome instability, such as cancer or other diseases related to aging. The genotypic information embedded in scRNA-seq data can help uncover functional variants in individual cells, and may also inform their specific gene expression pattern. Understandably, detection of these variants from scRNA-seq data is limited to expressed regions that have enough sequencing depth. While some single-cell sequencing platforms such as Smart-seq3 generate reads that cover the full length of transcripts, others such as 10× Chromium focus on the 3′ or 5′ end of transcripts. RNA editing may also add another layer of complication by revealing variants that may not be present at the DNA level, but occurrence of RNA editing is typically very rare. To detect SNVs, methods developed for calling variants from bulk RNA-seq data, such as MuTect2, Strelka2 [194], VarScan2, SAMtools [195], Pysam [196], FreeBayes, and BamBam, can be used on scRNA-seq data. The GATK RNA-seq short variant discovery best practices workflow, which uses HaplotyperCaller for variant calling followed by variant filtering using RNA-seq specific settings, is among the most used [197]. Monovar, a method developed for single-cell *DNA* sequencing data [198], can also be used for calling SNVs from scRNA-seq data [199]. There are currently a number of tools that have been developed for SNV detection from scRNA-seq data, including SSrGE [200], Trinity CTAT [201], and cellsnp-lite [202]. Among these methods, cellsnp-lite is a lightweight allelic reads pileup method with minimum filtering that can be applied to both 10× Chromium and Smart-seq3 data. Because of use of parallel processing it has improved running speed. Benchmarking comparison has shown that the performance of many current tools depends on sequencing depth, genomic context (such as high GC content), functional region, variant allele frequency, and platform (10x has more dropout events) [203]. It has also been shown that the main detection limitation is low sensitivity caused by low capture efficiency, sequencing depth, and signal dropout. Among the best performing tools so far are SAMtools, FreeBayes, Strelka2, and CTAT.

Deletion or duplication of a genomic region may lead to reduced or increased expression of genes located in the affected region. It is possible, therefore, to infer CNV information from scRNA-seq data. While it can be challenging due to uneven coverage of scRNA-seq signal across the genome, inferred CNVs do provide information on cellular heterogeneity at another dimension (genome instability), for instance, during cancer development [204]. To meet the challenges of calling CNVs from scRNA-seq data, a relatively small number of methods, including HoneyBADGER [205], CaSpER [206], inferCNV [207], and CONICS [208], are currently available. HoneyBADGER, for example, uses a probabilistic hidden Markov model to infer CNV from smoothed averaged gene expression profile of a set of cells in comparison to control cells. CaSpER identifies and visualizes CNV events from single-cell gene expression signals integrated with allelic shift signals, which quantifies loss of heterozygosity events across the genome facilitating CNV detection. InferCNV, a component of Trinity CTAT, searches for evidence of chromosomal segmental gains or losses by averaging expression over neighboring genes across the genome, in comparison to reference normal cells. CONICS (or CONICSmat when reference normal scRNA-seq data is not available) applies a two-component Gaussian mixture model to fit gene expression in a chromosomal region to make CNV calls.

To further connect genetic variants with gene expression regulation, allele-specific expression (ASE) analysis can be performed to determine potential imbalanced expression of transcripts from each allele. Imbalanced or preferential allelic expression leads to phenotypic variation, and is caused by epigenetic or genetic regulatory mechanisms. ASE analysis has been widely used to reveal allelic imbalance or allele-specific transcriptional bursting from bulk RNA-seq. Single-cell ASE analysis is still relatively new, and as a result available methods are still quite limited. Prashant et al. (2020) used a custom pipeline to estimate ASE from 10× Chromium scRNA-seq. This pipeline first uses STAR to align raw reads pooled from all cells, and then GATK to call all SNVs present in the data. Filtered high-quality, heterozygous SNVs are then used as input for a second round of STAR-based alignment, in which an SNV-aware option such as WASP [209] is employed on reads within each cell. To detect monoallelic and biallelic expression, VAF_{RNA}, i.e., the percentage of reads carrying the variant sequence, is then calculated. Besides directly calling from scRNA-seq data, the SNVs needed for the second round of alignment and VAF_{RNA} calculation can also be provided from bulk RNA-seq or DNA sequencing data [210]. Because of the inherent signal dropout issue and potential allele-mapping bias, interpretation of single-cell ASE results should use caution.

8.8.2 Alternative Splicing Analysis

As mentioned in Chapter 7 on bulk RNA-seq, alternative splicing plays an important role in regulating major biological processes, and abnormal splicing may lead to diseased states. At the single-cell level, alternative splicing can be analyzed despite the challenges of signal sparsity and background noise. Tools for this analysis include Outrigger [211], BRIE/BRIE2 [212, 213], ODEGR-NMF [214], ASCOT [215], Millefy [216], VALERIE [217], SCATS [218], scQuint [219], and DESJ-detection [220]. To illustrate how these tools work, Outrigger, a component of the Expedition suite, uses junction-spanning reads to create an exon-junction graph first, and then detects alternative splicing events through traversing the graph. Quantification of such events and identification of differential splicing between groups of cells are based on the use of "percent-spliced-in" or Psi (ψ). As another example, DESJ-detection first constructs a cell-splicing junction count matrix for each gene. Iterative K-means is then used to cluster cells; after removing clusters with low expression, a list of solid junctions is generated. The identification of DESJs, or differentially expressed splicing junctions, is achieved using limma. To help visualize differential splicing patterns across cells, Millefy and VALERIE can be used to uncover cellular heterogeneity and splicing differences between various cell groups. VALERIE, for example, is an R-based tool for using Psi values to display alternative splicing events. It can also be used to identify significant splicing difference between different cell populations, through performing statistical test, such as Kruskal–Wallis test, on the Psi values followed by multiple testing correction. Most of alternative splicing analysis tools use reads obtained from full-length scRNA-seq platforms such as Smart-seq3. Reads derived from the 3′ or 5′ end of transcripts, such as those generated from the 10× Chromium platform, cover limited number of splicing junctions at either end of genes. Despite this limitation, some tools such as SCATS can use 10× scRNA-seq data for alternative splicing analysis.

8.8.3 Gene Regulatory Network Inference

A gene regulatory network (GRN) is a graph representation of how genes interact with each other in a cell system. In a GRN, genes are represented as nodes, and their interactions constitute edges. Compared to the analytic steps above, GRN inference is relatively new and yet to be performed more widely. Currently there are a growing number of scRNA-seq based GRN inference methods available, including SCENIC [221], SCODE [222], SCOUP [223], PIDC [224], LEAP [225], NLNET [226], SCIMITAR [227], and GRISLI [228]. In addition, some of the methods originally developed for bulk RNA-seq data have also been applied to scRNA-seq data after some adjustment with good performance, including GENIE3 [229] and GRNBoost2 [230]. GENIE3 is a random forest regression-based algorithm that had the best performance in the DREAM4 *In Silico Multifactorial* challenge for assessment of GRN inference algorithms. GRNBoost2, based on the use of a stochastic gradient boosting machine regression model, has been developed as a faster alternative to GENIE3. These, and the methods specifically developed for scRNA-seq, use the general principle of gene co-expression to infer GRN. If two genes show co-expression in the context of other genes, they are considered to interact.

According to how they infer GRNs, the algorithms that are specifically developed for scRNA-seq data can be grouped into several categories. Methods such as SCENIC, PIDC, and NLNET are based on simple correlation analysis. SCENIC, for example, identifies sets of genes that are co-expressed with transcription factors with the use of GENIE3. To trim the large number of edges that represent potential gene-gene interactions to a shorter list of high-confidence edges, SCENIC performs a transcription factor-binding motif enrichment analysis to identify putative target genes. The output from SCENIC can then be imported into visualization tools such SCope for network visualization. Methods such as LEAP and SCIMITAR take into consideration effects of developmental stages on gene networking through incorporating pseudo-temporal information from trajectory inference and velocity analysis (Section 8.7). The pseudo-temporal ordering of cells helps establish directionality between an upstream gene and a downstream effector. For these methods, gene correlation is first calculated for each time window, and the multiple correlation matrices are then aggregated into one adjacency matrix to represent the overall gene-gene interactions. Other methods such as SCODE, SCOUP, and GRISLI use a similar approach with the application of pseudo-temporal information, but they use differential equations to estimate gene correlation and infer gene relationships. For example, SCODE relies on Monocle to provide pseudotime information, and uses ordinary differential equations to calculate gene correlation. There are also methods based on other approaches, such as SCNS [231] and BTR [232] that use Boolean models. With such models 0 or 1 represents deactivated or activated gene expression, and Boolean operations AND, OR, and NOT are used to capture relationships between two genes. Boolean models provide a simplistic presentation of the cell system through converting gene expression data into binary data, but this also leads to loss of gene-gene interaction information.

To pick an appropriate GRN inference method, besides having knowledge of how they are designed, it also helps to understand whether prior information is required and what is the basic characteristic of the cells under study. Some of the methods require prior information, in the form of pseudo-temporal ordering of cells or cell types, which can be revealed with trajectory inference. Such methods are more suitable for cells that are in different developmental stages. If the objective is to compare cellular composition or heterogeneity under different conditions (e.g., healthy vs. diseased), methods that employ static data are more appropriate. To provide systematic guidance to the selection of appropriate GRN inference methods, results from several benchmark studies [233–236] are available on currently available GRN inference methods. Overall these studies showed underperformance of current methods and call for development of better designed tools. For example, currently inferred networks still show poor agreement with ground truth. In the meantime, these studies also revealed the challenges of inferring GRN from scRNA-seq data, which can be technical (due mostly to signal heterogeneity and sparsity), biological (e.g., complex nature of molecular interactions), or computational (e.g., complexity of analysis). Validation of an inferred GRN is also a very challenging task. Most of the currently available methods still output one gene network for all cells, or for specific cell types, not individual cells. Some newer methods such as CSN [237] and c-CSN [238] allow building of cell-specific networks, i.e., one network per cell.

Despite the current challenges, the value of GRN analysis cannot be overemphasized. In-depth analysis of a GRN enables detection of network modules and key nodes (hub genes). A module refers to a group of genes that are highly connected to fulfill a cellular function. The overall topology of a module, or key nodes linking different modules, might change with development or cell differentiation, or differ under different conditions. Differential network analysis may reveal altered gene-gene interactions between conditions. These network analyses can be carried out using tools such as WGCNA [239].

References

1. Cui Y, Irudayaraj J. Inside single cells: quantitative analysis with advanced optics and nanomaterials. *Wiley Interdiscip Rev Nanomed Nanobiotechnol* 2015, 7(3):387–407.
2. Huang XT, Li X, Qin PZ, Zhu Y, Xu SN, Chen JP. Technical advances in single-cell RNA sequencing and applications in normal and malignant hematopoiesis. *Front Oncol* 2018, 8:582.
3. Shalek AK, Satija R, Shuga J, Trombetta JJ, Gennert D, Lu D, Chen P, Gertner RS, Gaublomme JT, Yosef N *et al.* Single-cell RNA-seq reveals dynamic paracrine control of cellular variation. *Nature* 2014, 510(7505):363–369.
4. Marinov GK, Williams BA, McCue K, Schroth GP, Gertz J, Myers RM, Wold BJ. From single-cell to cell-pool transcriptomes: stochasticity in gene expression and RNA splicing. *Genome Res* 2014, 24(3):496–510.
5. Zhang M, Zou Y, Xu X, Zhang X, Gao M, Song J, Huang P, Chen Q, Zhu Z, Lin W *et al.* Highly parallel and efficient single cell mRNA sequencing with paired picoliter chambers. *Nat Commun* 2020, 11(1):2118.
6. Hagemann-Jensen M, Ziegenhain C, Chen P, Ramskold D, Hendriks GJ, Larsson AJM, Faridani OR, Sandberg R. Single-cell RNA counting at allele and isoform resolution using Smart-seq3. *Nat Biotechnol* 2020, 38(6):708–714.
7. Macosko EZ, Basu A, Satija R, Nemesh J, Shekhar K, Goldman M, Tirosh I, Bialas AR, Kamitaki N, Martersteck EM *et al.* Highly Parallel Genome-wide Expression Profiling of Individual Cells Using Nanoliter Droplets. *Cell* 2015, 161(5):1202–1214.
8. Klein AM, Mazutis L, Akartuna I, Tallapragada N, Veres A, Li V, Peshkin L, Weitz DA, Kirschner MW. Droplet barcoding for single-cell transcriptomics applied to embryonic stem cells. *Cell* 2015, 161(5):1187–1201.
9. Cao J, Packer JS, Ramani V, Cusanovich DA, Huynh C, Daza R, Qiu X, Lee C, Furlan SN, Steemers FJ *et al.* Comprehensive single-cell transcriptional profiling of a multicellular organism. *Science* 2017, 357(6352):661–667.

10. Zheng GX, Terry JM, Belgrader P, Ryvkin P, Bent ZW, Wilson R, Ziraldo SB, Wheeler TD, McDermott GP, Zhu J *et al*. Massively parallel digital transcriptional profiling of single cells. *Nat Commun* 2017, 8:14049.
11. Zhang X, Li T, Liu F, Chen Y, Yao J, Li Z, Huang Y, Wang J. Comparative analysis of droplet-based ultra-high-throughput single-cell RNA-seq systems. *Mol Cell* 2019, 73(1):130–142 e135.
12. Ding J, Adiconis X, Simmons SK, Kowalczyk MS, Hession CC, Marjanovic ND, Hughes TK, Wadsworth MH, Burks T, Nguyen LT *et al*. Systematic comparison of single-cell and single-nucleus RNA-sequencing methods. *Nat Biotechnol* 2020, 38(6):737–746.
13. How many Cells (https://satijalab.org/howmanycells)
14. Svensson V, da Veiga Beltrame E, Pachter L. Quantifying the tradeoff between sequencing depth and cell number in single-cell RNA-seq. *bioRxiv* 2019, doi: https://doi.org/10.1101/762773
15. Zhang MJ, Ntranos V, Tse D. Determining sequencing depth in a single-cell RNA-seq experiment. *Nat Commun* 2020, 11(1):774.
16. Parekh S, Ziegenhain C, Vieth B, Enard W, Hellmann I. zUMIs – A fast and flexible pipeline to process RNA sequencing data with UMIs. *GigaScience* 2018, 7(6):giy059.
17. Heimberg G, Bhatnagar R, El-Samad H, Thomson M. Low dimensionality in gene expression data enables the accurate extraction of transcriptional programs from shallow sequencing. *Cell Syst* 2016, 2(4):239–250.
18. Wu AR, Neff NF, Kalisky T, Dalerba P, Treutlein B, Rothenberg ME, Mburu FM, Mantalas GL, Sim S, Clarke MF *et al*. Quantitative assessment of single-cell RNA-sequencing methods. *Nat Methods* 2014, 11(1):41–46.
19. Svensson V, Natarajan KN, Ly LH, Miragaia RJ, Labalette C, Macaulay IC, Cvejic A, Teichmann SA. Power analysis of single-cell RNA-sequencing experiments. *Nat Methods* 2017, 14(4):381–387.
20. Genomics X. Technical Note–Removal of Dead Cells from Single Cell Suspensions Improves Performance for 10× Genomics® Single Cell Applications. 2017.
21. van den Brink SC, Sage F, Vertesy A, Spanjaard B, Peterson-Maduro J, Baron CS, Robin C, van Oudenaarden A. Single-cell sequencing reveals dissociation-induced gene expression in tissue subpopulations. *Nat Methods* 2017, 14(10):935–936.
22. Wohnhaas CT, Leparc GG, Fernandez-Albert F, Kind D, Gantner F, Viollet C, Hildebrandt T, Baum P. DMSO cryopreservation is the method of choice to preserve cells for droplet-based single-cell RNA sequencing. *Sci Rep* 2019, 9(1):10699.
23. Cha J, Lee I. Single-cell network biology for resolving cellular heterogeneity in human diseases. *Exp Mol Med* 2020, 52(11):1798–1808.
24. Korrapati S, Taukulis I, Olszewski R, Pyle M, Gu S, Singh R, Griffiths C, Martin D, Boger E, Morell RJ *et al*. Single cell and single nucleus RNA-seq reveal cellular heterogeneity and homeostatic regulatory networks in adult mouse stria vascularis. *Front Mol Neurosci* 2019, 12:316.
25. Gao R, Kim C, Sei E, Foukakis T, Crosetto N, Chan LK, Srinivasan M, Zhang H, Meric-Bernstam F, Navin N. Nanogrid single-nucleus RNA sequencing reveals phenotypic diversity in breast cancer. *Nat Commun* 2017, 8(1):228.

26. Liang Q, Dharmat R, Owen L, Shakoor A, Li Y, Kim S, Vitale A, Kim I, Morgan D, Liang S *et al.* Single-nuclei RNA-seq on human retinal tissue provides improved transcriptome profiling. *Nat Commun* 2019, 10(1):5743.
27. Zhu YY, Machleder EM, Chenchik A, Li R, Siebert PD. Reverse transcriptase template switching: a SMART approach for full-length cDNA library construction. *BioTechniques* 2001, 30(4):892–897.
28. Bray NL, Pimentel H, Melsted P, Pachter L. Near-optimal probabilistic RNA-seq quantification. *Nat Biotechnol* 2016, 34(5):525–527.
29. Du Y, Huang Q, Arisdakessian C, Garmire LX. Evaluation of STAR and Kallisto on single cell RNA-seq data alignment. *G3* 2020, 10(5):1775–1783.
30. Melsted P, Ntranos V, Pachter L. The barcode, UMI, set format and BUStools. *Bioinformatics* 2019, 35(21):4472–4473.
31. Srivastava A, Malik L, Smith T, Sudbery I, Patro R. Alevin efficiently estimates accurate gene abundances from dscRNA-seq data. *Genome Biol* 2019, 20(1):65.
32. McGinnis CS, Murrow LM, Gartner ZJ. DoubletFinder: doublet detection in single-cell RNA sequencing data using artificial nearest neighbors. *Cell Syst* 2019, 8(4):329–337 e324.
33. Wolock SL, Lopez R, Klein AM. Scrublet: Computational identification of cell doublets in single-cell transcriptomic data. *Cell Syst* 2019, 8(4):281–291 e289.
34. Gayoso A, Shor J, Carr AJ, Sharma R, Pe'er D. DoubletDetection. In.: Zenodo; 2020: https://zenodo.org/record/2678042.
35. Bais AS, Kostka D. scds: computational annotation of doublets in single-cell RNA sequencing data. *Bioinformatics* 2020, 36(4):1150–1158.
36. Bernstein NJ, Fong NL, Lam I, Roy MA, Hendrickson DG, Kelley DR. Solo: doublet identification in single-cell RNA-seq via semi-supervised deep learning. *Cell Syst* 2020, 11(1):95–101 e105.
37. Xi NM, Li JJ. Benchmarking computational doublet-detection methods for single-cell RNA sequencing data. *Cell Syst* 2020, 12(2):176–194.
38. Stoeckius M, Zheng S, Houck-Loomis B, Hao S, Yeung BZ, Mauck WM, 3rd, Smibert P, Satija R. Cell Hashing with barcoded antibodies enables multiplexing and doublet detection for single cell genomics. *Genome Biol* 2018, 19(1):224.
39. Kang HM, Subramaniam M, Targ S, Nguyen M, Maliskova L, McCarthy E, Wan E, Wong S, Byrnes L, Lanata CM *et al.* Multiplexed droplet single-cell RNA-sequencing using natural genetic variation. *Nat Biotechnol* 2018, 36(1):89.
40. Xu J, Falconer C, Nguyen Q, Crawford J, McKinnon BD, Mortlock S, Senabouth A, Andersen S, Chiu HS, Jiang LD *et al.* Genotype-free demultiplexing of pooled single-cell RNA-seq. *Genome Biol* 2019, 20(1):290.
41. Heaton H, Talman AM, Knights A, Imaz M, Gaffney DJ, Durbin R, Hemberg M, Lawniczak MKN. Souporcell: robust clustering of single-cell RNA-seq data by genotype without reference genotypes. *Nat Methods* 2020, 17(6):615.
42. Huang YH, McCarthy DJ, Stegle O. Vireo: Bayesian demultiplexing of pooled single-cell RNA-seq data without genotype reference. *Genome Biol* 2019, 20(1):273.
43. Lun ATL, Riesenfeld S, Andrews T, Dao TP, Gomes T, participants in the 1st Human Cell Atlas J, Marioni JC. EmptyDrops: distinguishing cells from empty droplets in droplet-based single-cell RNA sequencing data. *Genome Biol* 2019, 20(1):63.

44. Heiser CN, Wang VM, Chen B, Hughey JJ, Lau KS. Automated quality control and cell identification of droplet-based single-cell data using dropkick. *Genome Res* 2021, 31(10):1742–1752 .
45. Ni Z, Chen S, Brown J, Kendziorski C. CB2 improves power of cell detection in droplet-based single-cell RNA sequencing data. *Genome Biol* 2020, 21(1):137.
46. Yang S, Corbett SE, Koga Y, Wang Z, Johnson WE, Yajima M, Campbell JD. Decontamination of ambient RNA in single-cell RNA-seq with DecontX. *Genome Biol* 2020, 21(1):57.
47. Young MD, Behjati S. SoupX removes ambient RNA contamination from droplet-based single-cell RNA sequencing data. *Gigascience* 2020, 9(12):giaa151.
48. Osorio D, Cai JJ. Systematic determination of the mitochondrial proportion in human and mice tissues for single-cell RNA sequencing data quality control. *Bioinformatics* 2020, 37(7):963–967.
49. Bacher R, Chu LF, Leng N, Gasch AP, Thomson JA, Stewart RM, Newton M, Kendziorski C. SCnorm: robust normalization of single-cell RNA-seq data. *Nat Methods* 2017, 14(6):584–586.
50. Yip SH, Wang P, Kocher JA, Sham PC, Wang J. Linnorm: improved statistical analysis for single cell RNA-seq expression data. *Nucleic Acids Res* 2017, 45(22):e179.
51. Vallejos CA, Marioni JC, Richardson S. BASiCS: Bayesian Analysis of Single-Cell Sequencing Data. *PLoS Comput Biol* 2015, 11(6):e1004333.
52. Qiu X, Hill A, Packer J, Lin D, Ma YA, Trapnell C. Single-cell mRNA quantification and differential analysis with Census. *Nat Methods* 2017, 14(3):309–315.
53. Risso D, Perraudeau F, Gribkova S, Dudoit S, Vert JP. A general and flexible method for signal extraction from single-cell RNA-seq data. *Nat Commun* 2018, 9(1):284.
54. Hafemeister C, Satija R. Normalization and variance stabilization of single-cell RNA-seq data using regularized negative binomial regression. *Genome Biol* 2019, 20(1):296.
55. Butler A, Hoffman P, Smibert P, Papalexi E, Satija R. Integrating single-cell transcriptomic data across different conditions, technologies, and species. *Nat Biotechnol* 2018, 36(5):411–420.
56. Wolf FA, Angerer P, Theis FJ. SCANPY: large-scale single-cell gene expression data analysis. *Genome Biol* 2018, 19(1):15.
57. Lun AT, Bach K, Marioni JC. Pooling across cells to normalize single-cell RNA sequencing data with many zero counts. *Genome Biol* 2016, 17:75.
58. Lopez R, Regier J, Cole MB, Jordan MI, Yosef N. Deep generative modeling for single-cell transcriptomics. *Nat Methods* 2018, 15(12):1053–1058.
59. Lytal N, Ran D, An L. Normalization methods on single-cell RNA-seq data: an empirical survey. *Front Genet* 2020, 11:41.
60. Cole MB, Risso D, Wagner A, DeTomaso D, Ngai J, Purdom E, Dudoit S, Yosef N. Performance Assessment and Selection of Normalization Procedures for Single-Cell RNA-Seq. *Cell Syst* 2019, 8(4):315–328 e318.
61. Buttner M, Miao Z, Wolf FA, Teichmann SA, Theis FJ. A test metric for assessing single-cell RNA-seq batch correction. *Nat Methods* 2019, 16(1):43–49.
62. Johnson WE, Li C, Rabinovic A. Adjusting batch effects in microarray expression data using empirical Bayes methods. *Biostatistics* 2007, 8(1):118–127.

63. Haghverdi L, Lun ATL, Morgan MD, Marioni JC. Batch effects in single-cell RNA-sequencing data are corrected by matching mutual nearest neighbors. *Nat Biotechnol* 2018, 36(5):421–427.
64. Stuart T, Satija R. Integrative single-cell analysis. *Nat Rev Genet* 2019, 20(5):257–272.
65. Welch JD, Kozareva V, Ferreira A, Vanderburg C, Martin C, Macosko EZ. Single-cell multi-omic integration compares and contrasts features of brain cell identity. *Cell* 2019, 177(7):1873–1887 e1817.
66. Hie B, Bryson B, Berger B. Efficient integration of heterogeneous single-cell transcriptomes using Scanorama. *Nat Biotechnol* 2019, 37(6):685–691.
67. Polanski K, Young MD, Miao Z, Meyer KB, Teichmann SA, Park JE. BBKNN: fast batch alignment of single cell transcriptomes. *Bioinformatics* 2020, 36(3):964–965.
68. Korsunsky I, Millard N, Fan J, Slowikowski K, Zhang F, Wei K, Baglaenko Y, Brenner M, Loh PR, Raychaudhuri S. Fast, sensitive and accurate integration of single-cell data with Harmony. *Nat Methods* 2019, 16(12):1289–1296.
69. Barkas N, Petukhov V, Nikolaeva D, Lozinsky Y, Demharter S, Khodosevich K, Kharchenko PV. Joint analysis of heterogeneous single-cell RNA-seq dataset collections. *Nat Methods* 2019, 16(8):695–698.
70. Tran HTN, Ang KS, Chevrier M, Zhang X, Lee NYS, Goh M, Chen J. A benchmark of batch-effect correction methods for single-cell RNA sequencing data. *Genome Biol* 2020, 21(1):12.
71. Rousseeuw PJ. Silhouettes: a graphical aid to the interpretation and validation of cluster analysis. *J Comput Appl Math* 1987, 20:53–65.
72. van Dijk D, Sharma R, Nainys J, Yim K, Kathail P, Carr AJ, Burdziak C, Moon KR, Chaffer CL, Pattabiraman D *et al.* Recovering gene interactions from single-cell data using data diffusion. *Cell* 2018, 174(3):716–729 e727.
73. Wagner F, Yan Y, Yanai I. K-nearest neighbor smoothing for high-throughput single-cell RNA-Seq data. *bioRxiv* 2018, doi: https://doi.org/10.1101/217737
74. Huang M, Wang J, Torre E, Dueck H, Shaffer S, Bonasio R, Murray JI, Raj A, Li M, Zhang NR. SAVER: gene expression recovery for single-cell RNA sequencing. *Nat Methods* 2018, 15(7):539–542.
75. Wang J, Agarwal D, Huang M, Hu G, Zhou Z, Ye C, Zhang NR. Data denoising with transfer learning in single-cell transcriptomics. *Nat Methods* 2019, 16(9):875–878.
76. Linderman GC, Zhao J, Roulis M, Bielecki P, Flavell RA, Nadler B, Kluger Y. Zero-preserving imputation of single-cell RNA-seq data. *Nat Commun* 2022, 13(1):192.
77. Lin PJ, Troup M, Ho JWK. CIDR: Ultrafast and accurate clustering through imputation for single-cell RNA-seq data. *Genome Biol* 2017, 18(1):59.
78. Eraslan G, Simon LM, Mircea M, Mueller NS, Theis FJ. Single-cell RNA-seq denoising using a deep count autoencoder. *Nat Commun* 2019, 10(1):390.
79. Li WV, Li JJ. An accurate and robust imputation method scImpute for single-cell RNA-seq data. *Nat Commun* 2018, 9(1):997.
80. Mongia A, Sengupta D, Majumdar A. McImpute: matrix completion based imputation for single cell RNA-seq data. *Front Genet* 2019, 10:9.
81. Gong W, Kwak IY, Pota P, Koyano-Nakagawa N, Garry DJ. DrImpute: imputing dropout events in single cell RNA sequencing data. *BMC Bioinformatics* 2018, 19(1):220.

82. Lahnemann D, Koster J, Szczurek E, McCarthy DJ, Hicks SC, Robinson MD, Vallejos CA, Campbell KR, Beerenwinkel N, Mahfouz A *et al.* Eleven grand challenges in single-cell data science. *Genome Biol* 2020, 21(1):31.
83. Andrews TS, Hemberg M. False signals induced by single-cell imputation. *F1000Research* 2018, 7:1740.
84. Hou W, Ji Z, Ji H, Hicks SC. A systematic evaluation of single-cell RNA-sequencing imputation methods. *Genome Biol* 2020, 21(1):218.
85. Li Y, Willer C, Sanna S, Abecasis G. Genotype imputation. *Annu Rev Genomics Hum Genet* 2009, 10:387–406.
86. Brennecke P, Anders S, Kim JK, Kolodziejczyk AA, Zhang X, Proserpio V, Baying B, Benes V, Teichmann SA, Marioni JC *et al.* Accounting for technical noise in single-cell RNA-seq experiments. *Nat Methods* 2013, 10(11):1093–1095.
87. Buettner F, Natarajan KN, Casale FP, Proserpio V, Scialdone A, Theis FJ, Teichmann SA, Marioni JC, Stegle O. Computational analysis of cell-to-cell heterogeneity in single-cell RNA-sequencing data reveals hidden subpopulations of cells. *Nat Biotechnol* 2015, 33(2):155–160.
88. Yip SH, Sham PC, Wang J. Evaluation of tools for highly variable gene discovery from single-cell RNA-seq data. *Brief Bioinform* 2019, 20(4):1583–1589.
89. Duo A, Robinson MD, Soneson C. A systematic performance evaluation of clustering methods for single-cell RNA-seq data. *F1000Research* 2018, 7:1141.
90. Townes FW, Hicks SC, Aryee MJ, Irizarry RA. Feature selection and dimension reduction for single-cell RNA-Seq based on a multinomial model. *Genome Biol* 2019, 20(1):295.
91. Andrews TS, Hemberg M. M3Drop: dropout-based feature selection for scRNASeq. *Bioinformatics* 2019, 35(16):2865–2867.
92. Trapnell C, Cacchiarelli D, Grimsby J, Pokharel P, Li S, Morse M, Lennon NJ, Livak KJ, Mikkelsen TS, Rinn JL. The dynamics and regulators of cell fate decisions are revealed by pseudotemporal ordering of single cells. *Nat Biotechnol* 2014, 32(4):381–386.
93. Shao C, Hofer T. Robust classification of single-cell transcriptome data by nonnegative matrix factorization. *Bioinformatics* 2017, 33(2):235–242.
94. Pierson E, Yau C. ZIFA: Dimensionality reduction for zero-inflated single-cell gene expression analysis. *Genome Biol* 2015, 16:241.
95. Buettner F, Pratanwanich N, McCarthy DJ, Marioni JC, Stegle O. f-scLVM: scalable and versatile factor analysis for single-cell RNA-seq. *Genome Biol* 2017, 18(1):212.
96. Mahfouz A, van de Giessen M, van der Maaten L, Huisman S, Reinders M, Hawrylycz MJ, Lelieveldt BP. Visualizing the spatial gene expression organization in the brain through non-linear similarity embeddings. *Methods* 2015, 73:79–89.
97. Senabouth A, Lukowski SW, Hernandez JA, Andersen SB, Mei X, Nguyen QH, Powell JE. ascend: R package for analysis of single-cell RNA-seq data. *GigaScience* 2019, 8(8):giz087.
98. Sun S, Zhu J, Ma Y, Zhou X. Accuracy, robustness and scalability of dimensionality reduction methods for single-cell RNA-seq analysis. *Genome Biol* 2019, 20(1):269.
99. Welch JD, Hartemink AJ, Prins JF. SLICER: inferring branched, nonlinear cellular trajectories from single cell RNA-seq data. *Genome Biol* 2016, 17(1):106.

100. Haghverdi L, Buettner F, Theis FJ. Diffusion maps for high-dimensional single-cell analysis of differentiation data. *Bioinformatics* 2015, 31(18):2989–2998.
101. Sun X, Liu Y, An L. Ensemble dimensionality reduction and feature gene extraction for single-cell RNA-seq data. *Nat Commun* 2020, 11(1):5853.
102. Becht E, McInnes L, Healy J, Dutertre CA, Kwok IWH, Ng LG, Ginhoux F, Newell EW. Dimensionality reduction for visualizing single-cell data using UMAP. *Nat Biotechnol* 2018, 37(1):38–44.
103. Cao J, Spielmann M, Qiu X, Huang X, Ibrahim DM, Hill AJ, Zhang F, Mundlos S, Christiansen L, Steemers FJ *et al.* The single-cell transcriptional landscape of mammalian organogenesis. *Nature* 2019, 566(7745):496–502.
104. Kobak D, Berens P. The art of using t-SNE for single-cell transcriptomics. *Nat Commun* 2019, 10(1):5416.
105. Hu Q, Greene CS. Parameter tuning is a key part of dimensionality reduction via deep variational autoencoders for single cell RNA transcriptomics. *Pac Symp Biocomput* 2019, 24:362–373.
106. Ding J, Condon A, Shah SP. Interpretable dimensionality reduction of single cell transcriptome data with deep generative models. *Nat Commun* 2018, 9(1):2002.
107. Deng Y, Bao F, Dai QH, Wu LF, Altschuler SJ. Scalable analysis of cell-type composition from single-cell transcriptomics using deep recurrent learning. *Nature Methods* 2019, 16(4):311.
108. Wang B, Zhu J, Pierson E, Ramazzotti D, Batzoglou S. Visualization and analysis of single-cell RNA-seq data by kernel-based similarity learning. *Nat Methods* 2017, 14(4):414–416.
109. Xiang R, Wang W, Yang L, Wang S, Xu C, Chen X. A comparison for dimensionality reduction methods of single-cell RNA-seq data. *Front Genet* 2021, 12:646936.
110. Wolf FA, Hamey FK, Plass M, Solana J, Dahlin JS, Gottgens B, Rajewsky N, Simon L, Theis FJ. PAGA: graph abstraction reconciles clustering with trajectory inference through a topology preserving map of single cells. *Genome Biol* 2019, 20(1):59.
111. Moon KR, van Dijk D, Wang Z, Gigante S, Burkhardt DB, Chen WS, Yim K, Elzen AVD, Hirn MJ, Coifman RR *et al.* Visualizing structure and transitions in high-dimensional biological data. *Nat Biotechnol* 2019, 37(12):1482–1492.
112. Anchang B, Hart TD, Bendall SC, Qiu P, Bjornson Z, Linderman M, Nolan GP, Plevritis SK. Visualization and cellular hierarchy inference of single-cell data using SPADE. *Nat Protoc* 2016, 11(7):1264–1279.
113. Weinreb C, Wolock S, Klein AM. SPRING: a kinetic interface for visualizing high dimensional single-cell expression data. *Bioinformatics* 2018, 34(7):1246–1248.
114. Kim T, Chen IR, Lin Y, Wang AY, Yang JYH, Yang P. Impact of similarity metrics on single-cell RNA-seq data clustering. *Brief Bioinform* 2019, 20(6):2316–2326.
115. Moussa M, Mandoiu, II. Single cell RNA-seq data clustering using TF-IDF based methods. *BMC Genomics* 2018, 19(Suppl 6):569.
116. Blondel VD, Guillaume JL, Lambiotte R, Lefebvre E. Fast unfolding of communities in large networks. *J Stat Mech-Theory E* 2008, doi:10.1088/1742-5468/2008/10/P10008

117. Traag VA, Waltman L, van Eck NJ. From Louvain to Leiden: guaranteeing well-connected communities. *Sci Rep* 2019, 9(1):5233.
118. Kiselev VY, Kirschner K, Schaub MT, Andrews T, Yiu A, Chandra T, Natarajan KN, Reik W, Barahona M, Green AR *et al.* SC3: consensus clustering of single-cell RNA-seq data. *Nat Methods* 2017, 14(5):483–486.
119. Zurauskiene J, Yau C. pcaReduce: hierarchical clustering of single cell transcriptional profiles. *BMC Bioinformatics* 2016, 17:140.
120. Grun D, Lyubimova A, Kester L, Wiebrands K, Basak O, Sasaki N, Clevers H, van Oudenaarden A. Single-cell messenger RNA sequencing reveals rare intestinal cell types. *Nature* 2015, 525(7568):251–255.
121. Xu C, Su Z. Identification of cell types from single-cell transcriptomes using a novel clustering method. *Bioinformatics* 2015, 31(12):1974–1980.
122. Lake BB, Chen S, Sos BC, Fan J, Kaeser GE, Yung YC, Duong TE, Gao D, Chun J, Kharchenko PV *et al.* Integrative single-cell analysis of transcriptional and epigenetic states in the human adult brain. *Nat Biotechnol* 2018, 36(1):70–80.
123. Guo M, Wang H, Potter SS, Whitsett JA, Xu Y. SINCERA: a pipeline for single-cell RNA-seq profiling analysis. *PLoS Comput Biol* 2015, 11(11):e1004575.
124. Freytag S, Tian L, Lonnstedt I, Ng M, Bahlo M. Comparison of clustering tools in R for medium-sized 10× Genomics single-cell RNA-sequencing data. *F1000Research* 2018, 7:1297.
125. Yu D, Huber W, Vitek O. Shrinkage estimation of dispersion in negative binomial models for RNA-seq experiments with small sample size. *Bioinformatics* 2013, 29(10):1275–1282.
126. Pliner HA, Shendure J, Trapnell C. Supervised classification enables rapid annotation of cell atlases. *Nat Methods* 2019, 16(10):983–986.
127. Domanskyi S, Szedlak A, Hawkins NT, Wang J, Paternostro G, Piermarocchi C. Polled Digital Cell Sorter (p-DCS): Automatic identification of hematological cell types from single cell RNA-sequencing clusters. *BMC Bioinformatics* 2019, 20(1):369.
128. Zhang Z, Luo D, Zhong X, Choi JH, Ma Y, Wang S, Mahrt E, Guo W, Stawiski EW, Modrusan Z *et al.* SCINA: A semi-supervised subtyping algorithm of single cells and bulk samples. *Genes* 2019, 10(7):531.
129. Zhang AW, O'Flanagan C, Chavez EA, Lim JLP, Ceglia N, McPherson A, Wiens M, Walters P, Chan T, Hewitson B *et al.* Probabilistic cell-type assignment of single-cell RNA-seq for tumor microenvironment profiling. *Nat Methods* 2019, 16(10):1007–1015.
130. Xu C, Lopez R, Mehlman E, Regier J, Jordan MI, Yosef N. Probabilistic harmonization and annotation of single-cell transcriptomics data with deep generative models. *Mol Syst Biol* 2021, 17(1):e9620.
131. Franzen O, Gan LM, Bjorkegren JLM. PanglaoDB: a web server for exploration of mouse and human single-cell RNA sequencing data. *Database (Oxford)* 2019, 2019.
132. Zhang X, Lan Y, Xu J, Quan F, Zhao E, Deng C, Luo T, Xu L, Liao G, Yan M *et al.* CellMarker: a manually curated resource of cell markers in human and mouse. *Nucleic Acids Res* 2019, 47(D1):D721–D728.
133. Zeisel A, Hochgerner H, Lonnerberg P, Johnsson A, Memic F, van der Zwan J, Haring M, Braun E, Borm LE, La Manno G *et al.* Molecular architecture of the mouse nervous system. *Cell* 2018, 174(4):999–1014 e1022.

134. Ecker JR, Geschwind DH, Kriegstein AR, Ngai J, Osten P, Polioudakis D, Regev A, Sestan N, Wickersham IR, Zeng H. The BRAIN Initiative Cell Census Consortium: Lessons Learned toward Generating a Comprehensive Brain Cell Atlas. *Neuron* 2017, 96(3):542–557.
135. Saunders A, Macosko EZ, Wysoker A, Goldman M, Krienen FM, de Rivera H, Bien E, Baum M, Bortolin L, Wang S *et al.* Molecular diversity and specializations among the cells of the adult mouse brain. *Cell* 2018, 174(4):1015–1030 e1016.
136. Regev A, Teichmann SA, Lander ES, Amit I, Benoist C, Birney E, Bodenmiller B, Campbell P, Carninci P, Clatworthy M *et al.* The Human Cell Atlas. *Elife* 2017, 6:e27041.
137. Tabula Muris C, Overall c, Logistical c, Organ c, processing, Library p, sequencing, Computational data a, Cell type a, Writing g *et al.* Single-cell transcriptomics of 20 mouse organs creates a Tabula Muris. *Nature* 2018, 562(7727):367–372.
138. Kiselev VY, Yiu A, Hemberg M. scmap: projection of single-cell RNA-seq data across data sets. *Nat Methods* 2018, 15(5):359–362.
139. Tan Y, Cahan P. SingleCellNet: a computational tool to classify single cell RNA-seq data across platforms and across species. *Cell Syst* 2019, 9(2):207–213 e202.
140. Ma F, Pellegrini M. ACTINN: automated identification of cell types in single cell RNA sequencing. *Bioinformatics* 2020, 36(2):533–538.
141. Alquicira-Hernandez J, Sathe A, Ji HP, Nguyen Q, Powell JE. scPred: accurate supervised method for cell-type classification from single-cell RNA-seq data. *Genome Biol* 2019, 20(1):264.
142. Pedregosa F, Varoquaux G, Gramfort A, Michel V, Thirion B, Grisel O, Blondel M, Prettenhofer P, Weiss R, Dubourg V. Scikit-learn: Machine learning in Python. *J Mach Learn Res* 2011, 12:2825–2830.
143. Bakken T, Cowell L, Aevermann BD, Novotny M, Hodge R, Miller JA, Lee A, Chang I, McCorrison J, Pulendran B *et al.* Cell type discovery and representation in the era of high-content single cell phenotyping. *BMC Bioinformatics* 2017, 18(Suppl 17):559.
144. Wang S, Pisco AO, McGeever A, Brbic M, Zitnik M, Darmanis S, Leskovec J, Karkanias J, Altman RB. Leveraging the cell ontology to classify unseen cell types. *Nat Commun* 2021, 12(1):5556.
145. Bernstein MN, Ma Z, Gleicher M, Dewey CN. CellO: comprehensive and hierarchical cell type classification of human cells with the Cell Ontology. *iScience* 2021, 24(1):101913.
146. Abdelaal T, Michielsen L, Cats D, Hoogduin D, Mei H, Reinders MJT, Mahfouz A. A comparison of automatic cell identification methods for single-cell RNA sequencing data. *Genome Biol* 2019, 20(1):194.
147. Huang Q, Liu Y, Du Y, Garmire LX. Evaluation of cell type Annotation R Packages on Single-cell RNA-seq Data. *Genomics Proteomics Bioinformatics* 2021, 19(2):267–281.
148. Aran D, Looney AP, Liu L, Wu E, Fong V, Hsu A, Chak S, Naikawadi RP, Wolters PJ, Abate AR *et al.* Reference-based analysis of lung single-cell sequencing reveals a transitional profibrotic macrophage. *Nat Immunol* 2019, 20(2):163–172.

149. de Kanter JK, Lijnzaad P, Candelli T, Margaritis T, Holstege FCP. CHETAH: a selective, hierarchical cell type identification method for single-cell RNA sequencing. *Nucleic Acids Res* 2019, 47(16):e95.
150. Cao ZJ, Wei L, Lu S, Yang DC, Gao G. Searching large-scale scRNA-seq databases via unbiased cell embedding with Cell BLAST. *Nat Commun* 2020, 11(1):3458.
151. Hu J, Li X, Hu G, Lyu Y, Susztak K, Li M. Iterative transfer learning with neural network for clustering and cell type classification in single-cell RNA-seq analysis. *Nat Mach Intell* 2020, 2(10):607–618.
152. Lieberman Y, Rokach L, Shay T. CaSTLe – Classification of single cells by transfer learning: Harnessing the power of publicly available single cell RNA sequencing experiments to annotate new experiments. *PLoS One* 2018, 13(10):e0205499.
153. Haber AL, Biton M, Rogel N, Herbst RH, Shekhar K, Smillie C, Burgin G, Delorey TM, Howitt MR, Katz Y *et al.* A single-cell survey of the small intestinal epithelium. *Nature* 2017, 551(7680):333–339.
154. Hashimoto K, Kouno T, Ikawa T, Hayatsu N, Miyajima Y, Yabukami H, Terooatea T, Sasaki T, Suzuki T, Valentine M *et al.* Single-cell transcriptomics reveals expansion of cytotoxic CD4 T cells in supercentenarians. *Proc Natl Acad Sci U S A* 2019, 116(48):24242–24251.
155. Cao Y, Lin Y, Ormerod JT, Yang P, Yang JYH, Lo KK. scDC: single cell differential composition analysis. *BMC Bioinformatics* 2019, 20(Suppl 19):721.
156. Jackson DA. Compositional data in community ecology: the paradigm or peril of proportions? *Ecology* 1997, 78(3):929–940.
157. Büttner M, Ostner J, Müller C, Theis F, Schubert B. scCODA is a Bayesian model for compositional single-cell data analysis. *Nat Commun* 2021, 12(1):6876.
158. Kharchenko PV, Silberstein L, Scadden DT. Bayesian approach to single-cell differential expression analysis. *Nat Methods* 2014, 11(7):740–742.
159. Finak G, McDavid A, Yajima M, Deng J, Gersuk V, Shalek AK, Slichter CK, Miller HW, McElrath MJ, Prlic M *et al.* MAST: a flexible statistical framework for assessing transcriptional changes and characterizing heterogeneity in single-cell RNA sequencing data. *Genome Biol* 2015, 16:278.
160. Delmans M, Hemberg M. Discrete distributional differential expression (D3E) – a tool for gene expression analysis of single-cell RNA-seq data. *BMC Bioinformatics* 2016, 17:110.
161. Korthauer KD, Chu LF, Newton MA, Li Y, Thomson J, Stewart R, Kendziorski C. A statistical approach for identifying differential distributions in single-cell RNA-seq experiments. *Genome Biol* 2016, 17(1):222.
162. Vu TN, Wills QF, Kalari KR, Niu N, Wang L, Rantalainen M, Pawitan Y. Beta-Poisson model for single-cell RNA-seq data analyses. *Bioinformatics* 2016, 32(14):2128–2135.
163. Chen W, Li Y, Easton J, Finkelstein D, Wu G, Chen X. UMI-count modeling and differential expression analysis for single-cell RNA sequencing. *Genome Biol* 2018, 19(1):70.
164. Miao Z, Deng K, Wang X, Zhang X. DEsingle for detecting three types of differential expression in single-cell RNA-seq data. *Bioinformatics* 2018, 34(18):3223–3224.

165. Ye C, Speed TP, Salim A. DECENT: differential expression with capture efficiency adjustmeNT for single-cell RNA-seq data. *Bioinformatics* 2019, 35(24):5155–5162.
166. Das S, Rai SN. SwarnSeq: An improved statistical approach for differential expression analysis of single-cell RNA-seq data. *Genomics* 2021, 113(3):1308–1324.
167. Van den Berge K, Roux de Bezieux H, Street K, Saelens W, Cannoodt R, Saeys Y, Dudoit S, Clement L. Trajectory-based differential expression analysis for single-cell sequencing data. *Nat Commun* 2020, 11(1):1201.
168. Wang T, Li B, Nelson CE, Nabavi S. Comparative analysis of differential gene expression analysis tools for single-cell RNA sequencing data. *BMC Bioinformatics* 2019, 20(1):40.
169. Soneson C, Robinson MD. Bias, robustness and scalability in single-cell differential expression analysis. *Nat Methods* 2018, 15(4):255–261.
170. Dal Molin A, Baruzzo G, Di Camillo B. Single-cell RNA-sequencing: assessment of differential expression analysis methods. *Front Genet* 2017, 8:62.
171. Jaakkola MK, Seyednasrollah F, Mehmood A, Elo LL. Comparison of methods to detect differentially expressed genes between single-cell populations. *Brief Bioinform* 2017, 18(5):735–743.
172. Van den Berge K, Perraudeau F, Soneson C, Love MI, Risso D, Vert JP, Robinson MD, Dudoit S, Clement L. Observation weights unlock bulk RNA-seq tools for zero inflation and single-cell applications. *Genome Biol* 2018, 19(1):24.
173. Love MI, Huber W, Anders S. Moderated estimation of fold change and dispersion for RNA-seq data with DESeq2. *Genome Biol* 2014, 15(12):550.
174. Robinson MD, McCarthy DJ, Smyth GK. edgeR: a Bioconductor package for differential expression analysis of digital gene expression data. *Bioinformatics* 2010, 26(1):139–140.
175. Ritchie ME, Phipson B, Wu D, Hu Y, Law CW, Shi W, Smyth GK. limma powers differential expression analyses for RNA-sequencing and microarray studies. *Nucleic Acids Res* 2015, 43(7):e47.
176. Seyednasrollah F, Rantanen K, Jaakkola P, Elo LL. ROTS: reproducible RNA-seq biomarker detector-prognostic markers for clear cell renal cell cancer. *Nucleic Acids Res* 2016, 44(1):e1.
177. Vieth B, Parekh S, Ziegenhain C, Enard W, Hellmann I. A systematic evaluation of single cell RNA-seq analysis pipelines. *Nat Commun* 2019, 10(1):4667.
178. Saelens W, Cannoodt R, Todorov H, Saeys Y. A comparison of single-cell trajectory inference methods. *Nat Biotechnol* 2019, 37(5):547–554.
179. Street K, Risso D, Fletcher RB, Das D, Ngai J, Yosef N, Purdom E, Dudoit S. Slingshot: cell lineage and pseudotime inference for single-cell transcriptomics. *BMC Genomics* 2018, 19(1):477.
180. Ji Z, Ji H. TSCAN: Pseudo-time reconstruction and evaluation in single-cell RNA-seq analysis. *Nucleic Acids Res* 2016, 44(13):e117.
181. Shin J, Berg DA, Zhu Y, Shin JY, Song J, Bonaguidi MA, Enikolopov G, Nauen DW, Christian KM, Ming GL *et al.* Single-cell RNA-seq with waterfall reveals molecular cascades underlying adult neurogenesis. *Cell Stem Cell* 2015, 17(3):360–372.

182. Haghverdi L, Buttner M, Wolf FA, Buettner F, Theis FJ. Diffusion pseudotime robustly reconstructs lineage branching. *Nat Methods* 2016, 13(10):845–848.
183. Rashid S, Kotton DN, Bar-Joseph Z. TASIC: determining branching models from time series single cell data. *Bioinformatics* 2017, 33(16):2504–2512.
184. Schiebinger G, Shu J, Tabaka M, Cleary B, Subramanian V, Solomon A, Gould J, Liu S, Lin S, Berube P *et al.* Optimal-Transport Analysis of Single-Cell Gene Expression identifies developmental trajectories in reprogramming. *Cell* 2019, 176(4):928–943 e922.
185. Lin C, Bar-Joseph Z. Continuous-state HMMs for modeling time-series single-cell RNA-Seq data. *Bioinformatics* 2019, 35(22):4707–4715.
186. Tran TN, Bader GD. Tempora: cell trajectory inference using time-series single-cell RNA sequencing data. *PLoS Comput Biol* 2020, 16(9):e1008205.
187. Lonnberg T, Svensson V, James KR, Fernandez-Ruiz D, Sebina I, Montandon R, Soon MS, Fogg LG, Nair AS, Liligeto U *et al.* Single-cell RNA-seq and computational analysis using temporal mixture modelling resolves Th1/Tfh fate bifurcation in malaria. *Sci Immunol* 2017, 2(9):eaal2192.
188. LineagePulse (https://github.com/YosefLab/LineagePulse)
189. La Manno G, Soldatov R, Zeisel A, Braun E, Hochgerner H, Petukhov V, Lidschreiber K, Kastriti ME, Lonnerberg P, Furlan A *et al.* RNA velocity of single cells. *Nature* 2018, 560(7719):494–498.
190. Bergen V, Lange M, Peidli S, Wolf FA, Theis FJ. Generalizing RNA velocity to transient cell states through dynamical modeling. *Nat Biotechnol* 2020, 38(12):1408–1414.
191. Herman JS, Sagar, Grun D. FateID infers cell fate bias in multipotent progenitors from single-cell RNA-seq data. *Nat Methods* 2018, 15(5):379–386.
192. Bendall SC, Davis KL, Amir el AD, Tadmor MD, Simonds EF, Chen TJ, Shenfeld DK, Nolan GP, Pe'er D. Single-cell trajectory detection uncovers progression and regulatory coordination in human B cell development. *Cell* 2014, 157(3):714–725.
193. Setty M, Tadmor MD, Reich-Zeliger S, Angel O, Salame TM, Kathail P, Choi K, Bendall S, Friedman N, Pe'er D. Wishbone identifies bifurcating developmental trajectories from single-cell data. *Nat Biotechnol* 2016, 34(6):637–645.
194. Kim S, Scheffler K, Halpern AL, Bekritsky MA, Noh E, Kallberg M, Chen X, Kim Y, Beyter D, Krusche P *et al.* Strelka2: fast and accurate calling of germline and somatic variants. *Nat Methods* 2018, 15(8):591–594.
195. Rodriguez-Meira A, Buck G, Clark SA, Povinelli BJ, Alcolea V, Louka E, McGowan S, Hamblin A, Sousos N, Barkas N *et al.* Unravelling Intratumoral Heterogeneity through High-Sensitivity Single-Cell Mutational Analysis and Parallel RNA Sequencing. *Mol Cell* 2019, 73(6):1292–1305 e1298.
196. Pysam (https://github.com/pysam-developers/pysam)
197. Fasterius E, Uhlen M, Al-Khalili Szigyarto C. Single-cell RNA-seq variant analysis for exploration of genetic heterogeneity in cancer. *Sci Rep* 2019, 9(1):9524.
198. Zafar H, Wang Y, Nakhleh L, Navin N, Chen K. Monovar: single-nucleotide variant detection in single cells. *Nat Methods* 2016, 13(6):505–507.
199. Schnepp PM, Chen M, Keller ET, Zhou X. SNV identification from single-cell RNA sequencing data. *Hum Mol Genet* 2019, 28(21):3569–3583.

200. Poirion O, Zhu X, Ching T, Garmire LX. Using single nucleotide variations in single-cell RNA-seq to identify subpopulations and genotype-phenotype linkage. *Nat Commun* 2018, 9(1):4892.
201. Fangal VD. CTAT Mutations: A Machine Learning Based RNA-Seq Variant Calling Pipeline Incorporating Variant Annotation, Prioritization, and Visualization. 2020.
202. Huang X, Huang Y. Cellsnp-lite: an efficient tool for genotyping single cells. *Bioinformatics* 2021, 37(23):4569–4571.
203. Liu F, Zhang Y, Zhang L, Li Z, Fang Q, Gao R, Zhang Z. Systematic comparative analysis of single-nucleotide variant detection methods from single-cell RNA sequencing data. *Genome Biol* 2019, 20(1):242.
204. Chung W, Eum HH, Lee HO, Lee KM, Lee HB, Kim KT, Ryu HS, Kim S, Lee JE, Park YH *et al.* Single-cell RNA-seq enables comprehensive tumour and immune cell profiling in primary breast cancer. *Nat Commun* 2017, 8:15081.
205. Fan J, Lee HO, Lee S, Ryu DE, Lee S, Xue C, Kim SJ, Kim K, Barkas N, Park PJ *et al.* Linking transcriptional and genetic tumor heterogeneity through allele analysis of single-cell RNA-seq data. *Genome Res* 2018, 28(8):1217–1227.
206. Serin Harmanci A, Harmanci AO, Zhou X. CaSpER identifies and visualizes CNV events by integrative analysis of single-cell or bulk RNA-sequencing data. *Nat Commun* 2020, 11(1):89.
207. inferCNV of the Trinity CTAT Project (https://github.com/broadinstitute/inferCNV)
208. Muller S, Cho A, Liu SJ, Lim DA, Diaz A. CONICS integrates scRNA-seq with DNA sequencing to map gene expression to tumor sub-clones. *Bioinformatics* 2018, 34(18):3217–3219.
209. van de Geijn B, McVicker G, Gilad Y, Pritchard JK. WASP: allele-specific software for robust molecular quantitative trait locus discovery. *Nat Methods* 2015, 12(11):1061–1063.
210. Borel C, Ferreira PG, Santoni F, Delaneau O, Fort A, Popadin KY, Garieri M, Falconnet E, Ribaux P, Guipponi M *et al.* Biased allelic expression in human primary fibroblast single cells. *Am J Hum Genet* 2015, 96(1):70–80.
211. Song Y, Botvinnik OB, Lovci MT, Kakaradov B, Liu P, Xu JL, Yeo GW. Single-cell alternative splicing analysis with expedition reveals splicing dynamics during neuron differentiation. *Mol Cell* 2017, 67(1):148–161 e145.
212. Huang Y, Sanguinetti G. BRIE: transcriptome-wide splicing quantification in single cells. *Genome Biol* 2017, 18(1):123.
213. Huang Y, Sanguinetti G. BRIE2: computational identification of splicing phenotypes from single-cell transcriptomic experiments. *Genome Biol* 2021, 22(1):251.
214. Matsumoto H, Hayashi T, Ozaki H, Tsuyuzaki K, Umeda M, Iida T, Nakamura M, Okano H, Nikaido I. An NMF-based approach to discover overlooked differentially expressed gene regions from single-cell RNA-seq data. *NAR Genom Bioinform* 2019, 2(1):lqz020.
215. Ling JP, Wilks C, Charles R, Leavey PJ, Ghosh D, Jiang L, Santiago CP, Pang B, Venkataraman A, Clark BS *et al.* ASCOT identifies key regulators of neuronal subtype-specific splicing. *Nat Commun* 2020, 11(1):137.
216. Ozaki H, Hayashi T, Umeda M, Nikaido I. Millefy: visualizing cell-to-cell heterogeneity in read coverage of single-cell RNA sequencing datasets. *BMC Genomics* 2020, 21(1):177.

217. Wen WX, Mead AJ, Thongjuea S. VALERIE: Visual-based inspection of alternative splicing events at single-cell resolution. *PLoS Comput Biol* 2020, 16(9):e1008195.
218. Hu Y, Wang K, Li M. Detecting differential alternative splicing events in scRNA-seq with or without unique molecular identifiers. *PLoS Comput Biol* 2020, 16(6):e1007925.
219. Benegas G, Fischer J, Song YS. Robust and annotation-free analysis of alternative splicing across diverse cell types in mice. *Elife* 2022, 11:e73520.
220. Liu S, Zhou B, Wu L, Sun Y, Chen J, Liu S. Single-cell differential splicing analysis reveals high heterogeneity of liver tumor-infiltrating T cells. *Sci Rep* 2021, 11(1):5325.
221. Aibar S, Gonzalez-Blas CB, Moerman T, Huynh-Thu VA, Imrichova H, Hulselmans G, Rambow F, Marine JC, Geurts P, Aerts J *et al.* SCENIC: single-cell regulatory network inference and clustering. *Nat Methods* 2017, 14(11):1083–1086.
222. Matsumoto H, Kiryu H, Furusawa C, Ko MSH, Ko SBH, Gouda N, Hayashi T, Nikaido I. SCODE: an efficient regulatory network inference algorithm from single-cell RNA-Seq during differentiation. *Bioinformatics* 2017, 33(15):2314–2321.
223. Matsumoto H, Kiryu H. SCOUP: a probabilistic model based on the Ornstein-Uhlenbeck process to analyze single-cell expression data during differentiation. *BMC Bioinformatics* 2016, 17(1):232.
224. Chan TE, Stumpf MPH, Babtie AC. Gene Regulatory Network Inference from Single-Cell Data Using Multivariate Information Measures. *Cell Syst* 2017, 5(3):251–267 e253.
225. Specht AT, Li J. LEAP: constructing gene co-expression networks for single-cell RNA-sequencing data using pseudotime ordering. *Bioinformatics* 2017, 33(5):764–766.
226. Liu H, Li P, Zhu M, Wang X, Lu J, Yu T. Nonlinear Network Reconstruction from Gene Expression Data Using Marginal Dependencies Measured by DCOL. *PLoS One* 2016, 11(7):e0158247.
227. Cordero P, Stuart JM. Tracing Co-Regulatory Network Dynamics in Noisy, Single-Cell Transcriptome Trajectories. *Pac Symp Biocomput* 2017, 22:576–587.
228. Aubin-Frankowski PC, Vert JP. Gene regulation inference from single-cell RNA-seq data with linear differential equations and velocity inference. *Bioinformatics* 2020, 36(18):4774–4780.
229. Huynh-Thu VA, Irrthum A, Wehenkel L, Geurts P. Inferring regulatory networks from expression data using tree-based methods. *PLoS One* 2010, 5(9):e12776.
230. Moerman T, Aibar Santos S, Bravo Gonzalez-Blas C, Simm J, Moreau Y, Aerts J, Aerts S. GRNBoost2 and Arboreto: efficient and scalable inference of gene regulatory networks. *Bioinformatics* 2019, 35(12):2159–2161.
231. Woodhouse S, Piterman N, Wintersteiger CM, Gottgens B, Fisher J. SCNS: a graphical tool for reconstructing executable regulatory networks from single-cell genomic data. *BMC Syst Biol* 2018, 12(1):59.
232. Lim CY, Wang H, Woodhouse S, Piterman N, Wernisch L, Fisher J, Gottgens B. BTR: training asynchronous Boolean models using single-cell expression data. *BMC Bioinformatics* 2016, 17(1):355.

233. Pratapa A, Jalihal AP, Law JN, Bharadwaj A, Murali TM. Benchmarking algorithms for gene regulatory network inference from single-cell transcriptomic data. *Nat Methods* 2020, 17(2):147–154.
234. Chen S, Mar JC. Evaluating methods of inferring gene regulatory networks highlights their lack of performance for single cell gene expression data. *BMC Bioinformatics* 2018, 19(1):232.
235. Nguyen H, Tran D, Tran B, Pehlivan B, Nguyen T. A comprehensive survey of regulatory network inference methods using single-cell RNA sequencing data. *Brief Bioinform* 2020, 22(3):bbaa190.
236. Kang Y, Thieffry D, Cantini L. Evaluating the Reproducibility of Single-Cell Gene Regulatory Network Inference Algorithms. *Front Genet* 2021, 12:617282.
237. Dai H, Li L, Zeng T, Chen L. Cell-specific network constructed by single-cell RNA sequencing data. *Nucleic Acids Res* 2019, 47(11):e62.
238. Li L, Dai H, Fang Z, Chen L. c-CSN: Single-cell RNA Sequencing Data Analysis by Conditional Cell-specific Network. *Genomics Proteomics Bioinformatics* 2021, 19(2):319–329.
239. Langfelder P, Horvath S. WGCNA: an R package for weighted correlation network analysis. *BMC Bioinformatics* 2008, 9:559.

9

Small RNA Sequencing

Small RNAs play an important role in regulating gene expression in both the cytoplasm and the nucleus through inducing both post-transcriptional and transcriptional gene silencing mechanisms. In addition to RNAi, some studies also show that some small RNAs can increase gene expression via a mechanism called RNA activation (or RNAa) [1]. Through these regulatory activities, small RNAs are involved in many cellular processes, affect growth and development, and if their own expression goes awry, lead to diseases such as cancer and Alzheimer's disease.

As introduced in Chapter 3, the major categories of small RNAs in cells include miRNAs, siRNAs, and piRNAs. Among these three types of small RNAs, miRNAs are so far the most studied. A total of 38,589 miRNA loci have been catalogued in 271 species at the time of this writing in miRBase (release 22.1), the gold-standard database for miRNAs. It has been estimated that a typical mammalian cell contains hundreds of miRNA species, each of which regulates transcripts from multiple genes. The expression of these miRNAs is cell- and tissue-specific, and dynamically regulated based on cellular state. Mutations or methylations in miRNA genes often lead to dysregulation in their expression. Studying the expression of miRNAs and other small RNAs is an important aspect of studying their roles in biological processes and diseases. Compared to other small RNA expression analysis methods, such as microarray and qPCR, NGS has a broader dynamic range for measuring small RNAs even at extremely high or low levels, single-base resolution to differentiate closely related small RNA molecules, the ability to study organisms without a currently available genome assembly, and the capability to discover novel small RNA species.

On new small RNA discovery, although from human and other model organisms the community has catalogued thousands of miRNAs and other small RNA species, more remain to be found. For less studied species, the number of known small RNAs is still low. Many *in silico* miRNA prediction algorithms have been developed, but their predictions have to be validated with experimental evidence. Small-RNA sequencing, through interrogating

DOI: 10.1201/9780429329180-12

the entire pool of small RNAs, provides an excellent tool for novel miRNA discovery and experimental validation of computational predictions. Furthermore, small RNA sequencing offers an assumption-free, comprehensive analysis of the small RNA transcriptome in biological targets, including differential expression between conditions. In general, small RNA sequencing data analysis shares much commonality with the analysis of RNA-seq data (Chapter 7). In the meantime, some aspects of small RNA sequencing data analysis are unique and mostly focused on in this chapter.

9.1 Small RNA NGS Data Generation and Upstream Processing

9.1.1 Data Generation

Since sequencing analysis of small RNAs in the transcriptome is similar to mRNA analysis, the experimental aspects detailed in Chapter 7 on factorial design, replication and randomization, and sample collection equally apply here and are therefore not repeated. Mature miRNA species, generated as a result of Dicer and Argonaute processing (Figure 9.1), have an average size of 22 nucleotides. Small RNA molecules can be purified from cells or tissues, while total RNA extracts that retain small RNA species work equally well and are often used. A size selection step in the sequencing library construction process removes larger RNA molecules in total RNA extracts. Furthermore, the small RNA sequencing library construction process takes advantage of the particular end structure on small RNAs, which are absent on mRNAs.

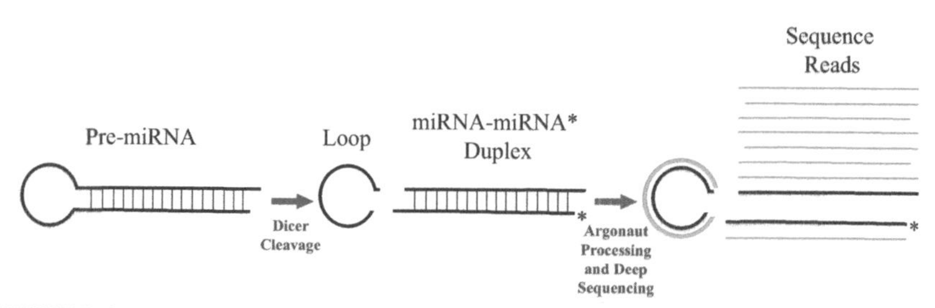

FIGURE 9.1
Deep sequencing of mature miRNAs after Dicer and Argonaute processing. Dicer cleaves a short stem-loop structure out of pre-miRNA to form the miRNA:miRNA* duplex. Upon loading into RISC, Argonaute unwinds the duplex and uses one strand as guide for gene silencing while discards the other strand (the star strand). While the short stem-loop and star strand sequences are usually degraded, they may still generate sequencing signals, because of undegraded residues or the fact that they may exist to perform other functions (e.g., the star strand is sometimes functional).

Canonical mature small RNAs have a monophosphate group at the 5′ end and a hydroxyl group at the 3′ end, which is derived from the action of small RNA processing enzymes such as Dicer.

Many small RNA sequencing library construction processes start with ligation of adapter sequences to their 3′ and 5′ ends. The universal adapter sequences provide anchoring for subsequent reverse transcription, and then PCR amplification. Among these steps, the initial ligation step has been found to introduce most bias to the process [2]. To counter the biases introduced during ligation as well as subsequent PCR amplification, alternative strategies have been devised and employed. The strategies for mitigation of the ligation bias include use of modified adapters that carry randomized nucleotides at the ligation boundary [3], and ligation-free procedures such as the SMART polyadenylation and template switch technology followed by PCR amplification, or probe capture-based target miRNA sequencing. To mitigate PCR bias, unique molecular identifiers (UMIs) can be used. To select an appropriate library construction procedure, the specific needs of the project need to be considered, as all currently available protocols have strengths and weaknesses based on currently available benchmark studies [2, 4].

Because of their short length, constructed small RNA sequencing libraries do not need to be sequenced very long. The actual read length depends on the configuration of library constructs and whether the index sequences are read in the same pass or as a separate reading step. In the current version of the Illumina small RNA sequencing protocol that reads index sequences in a second pass, 50 cycles of sequencing can be enough. Sequencing depth is another key factor in the data generation process that determines the power of differential expression analysis and novel small RNA discovery. While this depends on sample source as small RNA amount and composition vary greatly with cell type and species, in general 2–3 million aligned reads (or 4–5 million raw unmapped reads) should offer enough confidence for most studies. A study has shown that coverage higher than 5 million reads contributes little to the detection of new small RNA species [5].

9.1.2 Preprocessing

After obtaining sequencing reads and demultiplexing, the reads generated from each sample need to be first checked for quality using the QC tools introduced in Chapter 5 such as FastQC, NGS QC Toolkit, and fastp, or specifically developed miRNA-seq data QC tools including miRTrace [6] and mirnaQC [7]. Besides the typical NGS data QC metrics, specific miRNA-seq QC tools often provide additional features. For example, mirnaQC provides quality measures on miRNA yield and the fraction of putative degradation products (e.g., rRNA fragments) in both absolute values and relative ranks in comparison to a reference collection of 36,000 published datasets. Because small RNA libraries are usually sequenced longer than the actual lengths of the small RNA inserts, the 3′ adapter sequence is often part of the generated sequence reads and therefore should also be trimmed off. The trimming can be carried out with stand-alone tools such as Cutadapt and Trimmomatic, or utilities in the NGS QC Toolkit or fastp. Adapter trimming can also be conducted coincidentally with mapping, as some mappers provide such an option, or using data preprocessing modules within some small RNA data analysis tools (to be covered next).

9.1.3 Mapping

For mapping small RNA sequencing reads to a reference genome, short read aligners introduced in Chapter 5, such as Bowtie/Bowtie2, BWA, Novoalign, or SOAP/SOAP2, or RNA-seq aligner in Chapter 7, such as STAR, can be used. Among these aligners, Novoalign offers the option of stripping off adapter sequences in the mapping command. As for the reference genome, the most recent assembly should always be used. Because of the short target read length, the number of allowed mismatches should be set as 1. To speed up the mapping process, a multi-threading parameter, which enables the use of multiple CPU cores, can be used if the aligner supports it. After mapping, reads that are aligned to unique regions are then searched against small RNA databases to establish their identities (see next section), while those that are mapped to a large number (e.g., >5,000) of genomic locations should be removed from further analysis.

Besides the aforementioned general tools for small RNA reads preprocessing and mapping, tools have also been developed specially for small RNA-seq analysis, such as miRDeep/miRDeep2 [8], sRNAtoolbox [9], ShortStack [10], sRNAnalyzer [11], miRge [12], and miRMaster [13]. Among these tools, sRNAtoolbox is a collection of small RNA-seq data tools for differential expression analysis and other downstream analyses. Its center piece is sRNAbench, which replaces the previously widely used miRanalyzer. It provides functions such as data preprocessing, genome mapping using Bowtie, visualization of genome mapped reads, expression profiling, etc.

While the mapping of small RNA reads to a reference genome is similar to the mapping in RNA-seq as covered in Chapter 7, some characteristics of small RNAs, mostly their short length and post-transcriptional editing (see next), present different challenges to the small RNA read mapping process. Because of their short length, sizeable numbers of small RNA reads are usually mapped to more than one genomic region. In comparison, this issue is minimal for RNA-seq data, as longer and sometimes paired-end reads greatly increase specificity. The easiest way to deal with multi-mapped small RNA reads is to simply ignore them, but this leads to the loss of great amounts of data. A more commonly used approach is to assign them to one of the mapped positions randomly, while an alternative approach is to report them to all possible positions. More sophisticated algorithms have also been developed in an effort to avoid the precision or sensitivity pitfalls of these approaches. For example, ShortStack employs local weighting, which uses local genomic context to guide placement of multi-mapped miRNA-seq reads.

Post-transcriptional editing, on the other hand, leads to the generation of isomiRs [14], i.e., isoforms of canonical miRNAs that resemble but nevertheless vary from the reference miRNA annotated in miRBase. The isomiRs have various forms of variations from the canonical sequence, including alternative 3′ (more often) and 5′ termini, and nucleotide substitutions in the body sequence. Since their discovery, which itself is attributed to small RNA sequencing, isomiRs have been shown to have functional importance [15]. While the discovery of isomiRs is more recent, many newer tools cover isomiRs, including miRge, sRNAnalyzer, and sRNAbench.

9.1.4 Identification of Known and Putative Small RNA Species

To identify currently known small RNA species, the mapped reads need to be searched against the most recent version of miRBase or other small RNA databases (such as piRNABank, piRBase, and piRNAclusterDB for piRNAs). Reads with no matches in these databases can then be searched against other databases (Rfam, repeat, and mRNA) to determine if they are degradation products of ncRNAs, genomic repeats, and mRNAs. The previously mentioned small RNA-seq data analysis tools all provide these database search capabilities.

To discover potentially novel miRNA species, mapped reads that do not match known miRNAs and sequences in the other databases are submitted to algorithms such as sRNAbench and miRDeep2, which are designed to search for putative miRNAs. The approach used by miRDeep2 takes into consideration the biogenic process of miRNAs. It first identifies potential miRNA precursor coding regions out of the genomic regions that are clustered with the mapped reads. RNA secondary structures are then predicted on these identified regions using RNA folding software, and examined to see if they resemble a typical miRNA hairpin structure seen in pri-miRNA molecules and if they are thermodynamically stable. Putative miRNA species are called if the reads fall into stable hairpins in an expected manner, along with other evidences such as reads from the star strand. Similarly, sRNAbench also uses both structural and biogenic features to predict novel miRNAs [16].

9.1.5 Normalization

Before identifying differentially expressed small RNAs, read counts for each small RNA species in the samples need to be normalized. The goal of normalization is to make the samples directly comparable by removing unwanted sample-specific variations, which are usually due to differences in library size and therefore sequencing depth. The normalization approaches used in bulk

RNA-seq as detailed in Chapter 7 can be similarly applied here. The general assumption for most of the normalization approaches, that the majority of small RNAs stay constant between conditions, seems to hold. For the total read-count-based normalization, since all small RNAs are similarly short in size, the RPKM normalization can be simplified as RPM (reads per million). This method, however, has been found to be inadequate in some benchmark studies [17–19]. Other more sophisticated normalization approaches introduced in Chapter 7, including RLE used by DESeq/DESeq2 and TMM used by edgeR, have been found to work well for identification of differentially expressed small RNAs.

9.2 Identification of Differentially Expressed Small RNAs

The packages and tests introduced for RNA-seq differential expression analysis in Chapter 7 can also be directly used for small RNA analysis. For example, DESeq2, edgeR, limma, DEGseq, and NOISeq have been widely employed on the identification of differentially expressed small RNAs. In terms of implementation, these methods can be either paired with or integrated into the tools specifically designed for small RNA-seq data analysis as detailed in Section 9.1.3. For example, miRDeep2 can be used first to identify and quantify known and novel miRNAs, and the raw counts of identified miRNAs can then be input into edgeR or limma for DE analysis. To compute differential expression, miRge integrates with DESeq2. The sRNAtoolbox has a module called sRNAde specially developed for DE analysis. Using sRNAbench-generated or user-provided expression matrix as input, this module detects differentially expressed small RNAs with the use of five methods: edgeR, DESeq, DESeq2, NOISeq, and Student's *t*-test. Besides DE results generated from the individual methods, sRNAde also provides consensus DE results from the five methods. FDR is provided by all of these methods for multiple testing correction.

9.3 Functional Analysis of Identified Known Small RNAs

To perform functional analysis of differentially expressed small RNAs, their gene targets need to be predicted first. A number of tools are available for this task, including TargetScan [20], miRanda [21], mirSVR [22], miRDB [23], miRWalk [24], PicTar [25], PITA [26], RNA22 [27], RNAhybrid [28], DeepMirTar [29], and the DNA intelligent analysis (DIANA) applications microT-CDS and microT [30]. These tools predict target genes based on two general approaches, with one based on characteristics of miRNA-mRNA interaction, and the other on inference from machine learning models. The first approach makes predictions on the basis of seed pairing, thermodynamic stability, sequence conservation, and 3′-UTR structural accessibility. Many of the methods above use this approach. For example, miRanda, one of the earliest developed and widely used methods, makes predictions based on miRNA-mRNA complementarity pattern, location of the binding site in the mRNA, binding energy, and miRNA evolutionary conservation. The second machine-learning-based approach applies artificial intelligence to learn from known miRNA-mRNA duplexes to make new predictions. This approach uses features extracted from validated miRNA-mRNA interactions, plus negative dataset, to train a classifier model that can make distinctions and predict new miRNA targets. This approach has the ability to automatically improve based on new results. Examples of methods using this approach include DIANA-microT-CDS, mirSVR, and DeepMirTar.

Despite the progress made on miRNA target gene prediction, miRNA target gene prediction is still no easy task because of the small size of the miRNA-mRNA binding area, often imperfect complementarity of the binding, and sometimes lack of conservation [31]. Once a list of potential target genes is generated, functional analysis, such as Gene Ontology (GO) and pathway analysis, can be conducted using the approaches detailed in Chapter 7. In addition, for pathway analysis, a list of miRNAs can also be directly uploaded to the DIANA miRPath web server to a generate a list of biological pathways that are significantly enriched with the miRNAs′ target genes, which are predicted with DIANA-microT-CDS or documented with existing experimental evidence [32].

References

1. Huang V, Qin Y, Wang J, Wang X, Place RF, Lin G, Lue TF, Li LC. RNAa is conserved in mammalian cells. *PLoS One* 2010, 5(1):e8848.
2. Androvic P, Benesova S, Rohlova E, Kubista M, Valihrach L. Small RNA-sequencing for analysis of circulating mirnas: benchmark study. *J Mol Diagn* 2022, 24(4):386–394.
3. Baran-Gale J, Kurtz CL, Erdos MR, Sison C, Young A, Fannin EE, Chines PS, Sethupathy P. Addressing bias in small RNA library preparation for sequencing: a new protocol recovers microRNAs that evade capture by current methods. *Front Genet* 2015, 6:352.
4. Benesova S, Kubista M, Valihrach L. Small RNA-sequencing: approaches and considerations for miRNA analysis. *Diagnostics (Basel)* 2021, 11(6):964.
5. Metpally RP, Nasser S, Malenica I, Courtright A, Carlson E, Ghaffari L, Villa S, Tembe W, Van Keuren-Jensen K. Comparison of analysis tools for miRNA high throughput sequencing using nerve crush as a model. *Front Genet* 2013, 4:20.

6. Kang W, Eldfjell Y, Fromm B, Estivill X, Biryukova I, Friedlander MR. miRTrace reveals the organismal origins of microRNA sequencing data. *Genome Biol* 2018, 19(1):213.
7. Aparicio-Puerta E, Gomez-Martin C, Giannoukakos S, Medina JM, Marchal JA, Hackenberg M. mirnaQC: a webserver for comparative quality control of miRNA-seq data. *Nucleic Acids Res* 2020, 48(W1):W262–W267.
8. Friedlander MR, Mackowiak SD, Li N, Chen W, Rajewsky N. miRDeep2 accurately identifies known and hundreds of novel microRNA genes in seven animal clades. *Nucleic Acids Res* 2012, 40(1):37–52.
9. Aparicio-Puerta E, Gomez-Martin C, Giannoukakos S, Medina JM, Scheepbouwer C, Garcia-Moreno A, Carmona-Saez P, Fromm B, Pegtel M, Keller A *et al.* sRNAbench and sRNAtoolbox 2022 update: accurate miRNA and sncRNA profiling for model and non-model organisms. *Nucleic Acids Res* 2022, 50(W1):W710–W717.
10. Johnson NR, Yeoh JM, Coruh C, Axtell MJ. Improved placement of multi-mapping small RNAs. *G3* 2016, 6(7):2103–2111.
11. Wu X, Kim TK, Baxter D, Scherler K, Gordon A, Fong O, Etheridge A, Galas DJ, Wang K. sRNAnalyzer-a flexible and customizable small RNA sequencing data analysis pipeline. *Nucleic Acids Res* 2017, 45(21):12140–12151.
12. Patil AH, Halushka MK. miRge3.0: a comprehensive microRNA and tRF sequencing analysis pipeline. *NAR Genom Bioinform* 2021, 3(3):lqab068.
13. Fehlmann T, Kern F, Laham O, Backes C, Solomon J, Hirsch P, Volz C, Muller R, Keller A. miRMaster 2.0: multi-species non-coding RNA sequencing analyses at scale. *Nucleic Acids Res* 2021, 49(W1):W397–W408.
14. Morin RD, O'Connor MD, Griffith M, Kuchenbauer F, Delaney A, Prabhu AL, Zhao Y, McDonald H, Zeng T, Hirst M *et al.* Application of massively parallel sequencing to microRNA profiling and discovery in human embryonic stem cells. *Genome Res* 2008, 18(4):610–621.
15. Tomasello L, Distefano R, Nigita G, Croce CM. The MicroRNA Family Gets Wider: The IsomiRs Classification and Role. *Front Cell Dev Biol* 2021, 9:668648.
16. Barturen G, Rueda A, Hamberg M, Alganza A, Lebron R, Kotsyfakis M, Shi B-J, Koppers-Lalic D, Hackenberg M. sRNAbench: profiling of small RNAs and its sequence variants in single or multi-species high-throughput experiments. Methods Next-Generation Seq. 2014, 1:21–31.
17. Garmire LX, Subramaniam S. Evaluation of normalization methods in mammalian microRNA-Seq data. *RNA* 2012, 18(6):1279–1288.
18. Dillies MA, Rau A, Aubert J, Hennequet-Antier C, Jeanmougin M, Servant N, Keime C, Marot G, Castel D, Estelle J *et al.* A comprehensive evaluation of normalization methods for Illumina high-throughput RNA sequencing data analysis. *Brief Bioinform* 2013, 14(6):671–683.
19. Tam S, Tsao MS, McPherson JD. Optimization of miRNA-seq data preprocessing. *Brief Bioinform* 2015, 16(6):950–963.
20. Agarwal V, Bell GW, Nam JW, Bartel DP. Predicting effective microRNA target sites in mammalian mRNAs. *Elife* 2015, 4:e05005.
21. Enright AJ, John B, Gaul U, Tuschl T, Sander C, Marks DS. MicroRNA targets in Drosophila. *Genome Biol* 2003, 5(1):R1.
22. Betel D, Koppal A, Agius P, Sander C, Leslie C. Comprehensive modeling of microRNA targets predicts functional non-conserved and non-canonical sites. *Genome Biol* 2010, 11(8):R90.

23. Chen Y, Wang X. miRDB: an online database for prediction of functional microRNA targets. *Nucleic Acids Res* 2020, 48(D1):D127–D131.
24. Sticht C, De La Torre C, Parveen A, Gretz N. miRWalk: An online resource for prediction of microRNA binding sites. *PLoS One* 2018, 13(10):e0206239.
25. Krek A, Grun D, Poy MN, Wolf R, Rosenberg L, Epstein EJ, MacMenamin P, da Piedade I, Gunsalus KC, Stoffel M *et al.* Combinatorial microRNA target predictions. *Nat Genet* 2005, 37(5):495–500.
26. Kertesz M, Iovino N, Unnerstall U, Gaul U, Segal E. The role of site accessibility in microRNA target recognition. *Nat Genet* 2007, 39(10):1278–1284.
27. Miranda KC, Huynh T, Tay Y, Ang YS, Tam WL, Thomson AM, Lim B, Rigoutsos I. A pattern-based method for the identification of microRNA binding sites and their corresponding heteroduplexes. *Cell* 2006, 126(6):1203–1217.
28. Rehmsmeier M, Steffen P, Hochsmann M, Giegerich R. Fast and effective prediction of microRNA/target duplexes. *RNA* 2004, 10(10):1507–1517.
29. Wen M, Cong P, Zhang Z, Lu H, Li T. DeepMirTar: a deep-learning approach for predicting human miRNA targets. *Bioinformatics* 2018, 34(22):3781–3787.
30. Paraskevopoulou MD, Georgakilas G, Kostoulas N, Vlachos IS, Vergoulis T, Reczko M, Filippidis C, Dalamagas T, Hatzigeorgiou AG. DIANA-microT web server v5.0: service integration into miRNA functional analysis workflows. *Nucleic Acids Res* 2013, 41(Web Server issue):W169–173.
31. Ritchie W, Flamant S, Rasko JE. Predicting microRNA targets and functions: traps for the unwary. *Nat Methods* 2009, 6(6):397–398.
32. Vlachos IS, Zagganas K, Paraskevopoulou MD, Georgakilas G, Karagkouni D, Vergoulis T, Dalamagas T, Hatzigeorgiou AG. DIANA-miRPath v3.0: deciphering microRNA function with experimental support. *Nucleic Acids Res* 2015, 43(W1):W460–466.

10

Genotyping and Variation Discovery by Whole Genome/Exome Sequencing

Detection of genomic variation among individuals of a population is among the most frequent applications of NGS. Genome sequence heterogeneity is prevalent in a naturally occurring population, which cannot be captured by the current use of a single reference genome for a species. Genomic variant cataloging projects in many countries, such as the All of Us Program in the United States that aims to sequence one million genomes, underscore the importance of genomic variation discovery. Locating genomic sequence variations that correlate with disease predisposition or drug response, and establishing genotypic basis of various phenotypes, have become common focuses of many NGS studies in biomedical and life science research. Besides variations carried through the germline for generations, NGS has also been applied to identify *de novo* germline and somatic mutations, which occur more frequently than previously expected and underlie numerous human diseases including cancer and neurodegenerative diseases [1–3].

Detecting from NGS data the various forms of genomic variations/mutations detailed in Chapter 2, including SNVs, indels, and SVs, is not an easy task. The primary challenge is to differentiate true sequence variations/mutations from false positives caused by sequencing errors and artifacts generated in basecalling and sequence alignment. It is, therefore, important to generate high-quality sequencing data before performing data analysis. Equally importantly, sensitive and yet specific variant/mutant calling algorithms are required to achieve high accuracy in genomic variation and mutation discovery. This chapter first provides details on data preprocessing, alignment, realignment, and recalibration, then focuses on methods for the detection of germline and somatic SNVs/indels and SVs, followed by annotation of identified variants. Figure 10.1 shows an overview of the data analysis pipeline.

DOI: 10.1201/9780429329180-13

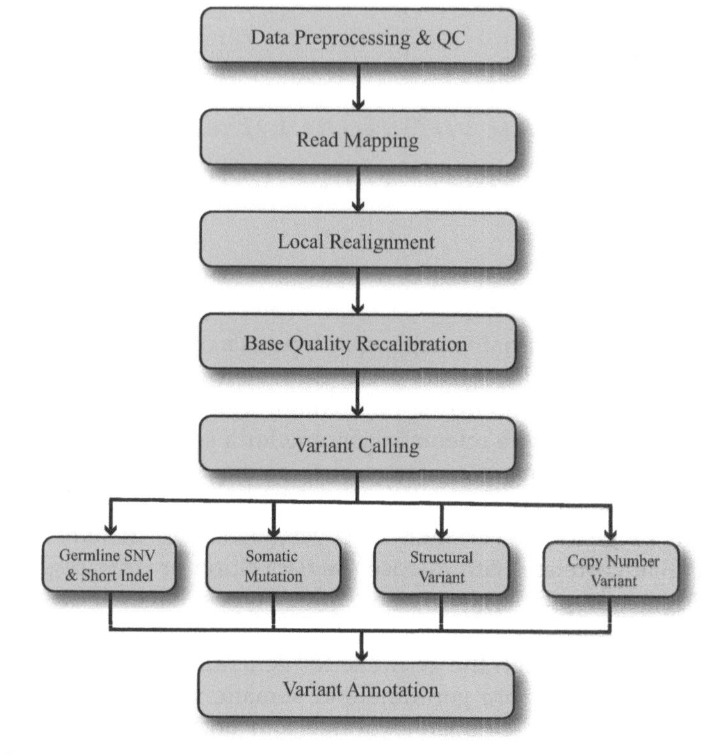

FIGURE 10.1
General workflow for genotyping and variation discovery from resequencing data.

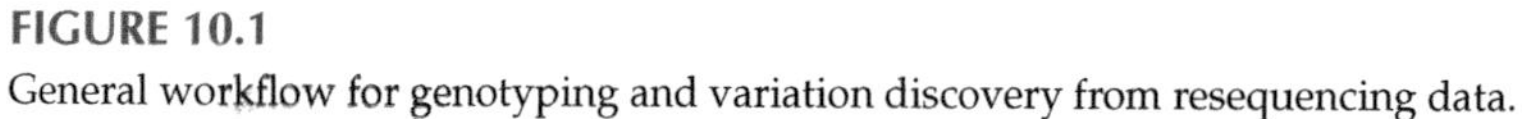

10.1 Data Preprocessing, Mapping, Realignment, and Recalibration

Besides the general data preprocessing and QC steps introduced in Chapter 5, such as examining sequencing data quality, removing low-quality and duplicate reads, additional steps are needed for variant calling. The read mapping step requires the use of highly sensitive alignment algorithm such as BWA-MEM, Bowtie2, and minimap2. After examining mapping quality, reads with low-quality mapping scores need to be filtered out. For paired-end reads, they should map to the reference genome as pairs at the expected interval and those that do not show the expected pattern should be filtered out as well.

After the initial alignment, realignment around indels usually leads to improvement in mapping results. This is usually due to the fact that short indels, especially those at the ends of reads, often cause problems in the initial alignment process. To realign around the indel regions, the original BAM file is first processed to identify where realignment is needed using tools such as the GATK RealignerTargetCreator. In this process, using a known set of indels (such as those in dbSNP) can speed up the process and improve accuracy. After the target regions for realignment are identified, programs such as the GATK IndelRealigner can be employed to conduct the realignment. At the end of this process, a new BAM file is generated containing realigned reads.

Prior to variant calling, the original basecall quality scores should also be recalibrated to further improve data quality. This base quality score recalibration can be conducted with tools such as the GATK BaseRecalibrator and ApplyBQSR. These tools recalibrate raw quality values using covariate-aware base quality score recalibration algorithms, which adjust for covariates, such as machine sequencing cycle and local sequence context, that are known to affect sequencing signal and basecall quality. To carry out the recalibration, a model of covariation is first analyzed and built by BaseRecalibrator when using GATK, which is then used by ApplyBQSR to recalibrate the data. Variant calling based on the recalibrated data has higher accuracy and cuts down on the number of false positives.

10.2 Single Nucleotide Variant (SNV) and Short Indel Calling

10.2.1 Germline SNV and Indel Calling

In general, variant calls are affected by a number of factors (Figure 10.2). These factors include: 1) basecall quality, 2) mapping quality, 3) single vs. paired-end sequencing, 4) read length, 5) depth of coverage, and 6) sequence context. Because of errors or uncertainties that occur in the steps of sequencing/basecalling/mapping, there are almost always certain levels of uncertainty associated with each variant call. To minimize this uncertainty, basecalling algorithms use statistical models, heuristics, or more recently, deep learning. By modeling the errors and biases, and sometimes incorporating other related prior information, variant callers that use statistical models significantly reduces the probability of miscalling variants. Methods that are based on the heuristic approach, on the other hand, call variants based on a number of heuristic factors, such as minimum read depth, base quality, and allele frequency. Deep learning is a subfield of machine learning that uses artificial neural networks (ANNs) to learn from training data with known truth, to make predictions or classifications from new unknown data. To call variants from NGS data using deep learning, pre-labeled variants are first used as training data to build a model for subsequent variant calling. Among the algorithms that use these three different approaches, those based on statistical models are currently more widely used than those based on

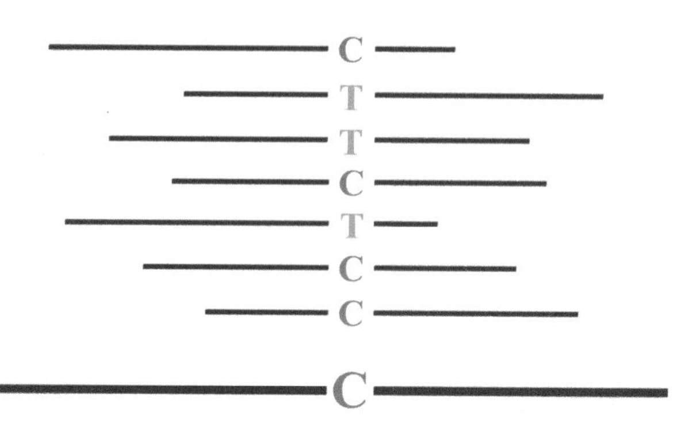

FIGURE 10.2
The variant calling process is usually affected by various factors. In this illustration, a number of reads are aligned against a reference sequence (bottom). At the illustrated site, the reference sequence has a C while the reads have C and T. Depending on the factors mentioned in the text and prior information, this site can be called as heterozygous (C/T), or no variation (C/C) if the T's are treated as errors. It is also possible to be called as a homozygous T/T, if the C's are regarded as errors.

heuristics and deep learning. It should be noted, however, that statistical models are usually based on certain assumptions. Under circumstances when the assumptions are violated, the heuristic and deep learning-based methods can be more robust.

Among the variant callers that are based on statistical models, GATK [4] is currently among the most widely used. GATK mostly uses HaplotypeCaller to call germline SNVs and short indels. HaplotypeCaller, as the name suggests, considers the linkage between nearby variants and calls SNPs and indels simultaneously through local *de novo* haplotype assembly. More specifically, the variant calling process used by HaplotypeCaller is composed of four main steps: (1) identification of active genomic regions that most likely harbor variants based on evidence from reads; (2) determination of haplotypes for each active region via reassembly of the region using a De Bruijn-like graph, followed by realignment of the haplotypes against the reference to identify potential variant sites; (3) calculation of the likelihoods of the haplotypes through aligning each read against each haplotype using a Pair-Hidden Markov Model (PairHMM); and (4) application of Bayes' rule to find the most likely genotypes in the sample. This variant calling process is highly accurate, but it is also computationally intensive and relatively slow. BCFtools [5] uses a similar genotype likelihood model for variant calling, which is achieved in two steps, namely "bcftools mpileup" and "bcftools call." In the mpileup step, it collects summary information from input BAM files and computes the likelihoods of possible genotypes, which are stored in BCF files (to be detailed next). The subsequent call step uses the likelihood information in the BCF files to conduct variant calling.

Besides GATK and BCFtools, other model-based variant callers include Strelka2 [6], freebayes [7], and 16GT [8]. A commonly used heuristics-based variant caller is VarScan2 [9], which works more robustly on data confounded by factors such as extreme read depth, pooled samples, and contaminated or impure samples. Many of these tools, such as GATK, BCFtools, Strelka2, and VarScan2, work for both single- and multiple-sample data. Multiple-sample analysis usually has increased detection power than single-sample analysis [10], because with multiple samples it is more likely to call a variant when more than one sample shows the same variation.

Google's open-source DeepVariant tool was the first attempt to use deep learning for variant calling. It is based on the use of convolutional neural network (CNN), a class of ANN that is often applied to image classification problems [11]. With the use of CNN, the variant calling process is converted into an image classification process [12]. With the increased use of long reads from the ONT and PacBio platforms, many of the variant callers specifically developed for long reads are based on the use of deep learning. These tools include Clair3 [13], Longshot [14], NanoCaller [15], and Medaka [16]. NanoCaller, for example, combines long-range haplotype information and deep CNN in an effort to improve variant calling accuracy. DeepVariant can also be used on long reads besides short reads, when coupled with algorithms such as PEPPER that uses a recurrent neural network to find variant candidates for subsequent variant calling by DeepVariant [17]. Based on currently available benchmarking studies [18–20], DeepVariant has been shown to have similar or even better performance than the often used GATK pipeline.

10.2.2 Somatic Mutation Detection

Most of the currently available variant calling methods are designed to identify germline variations that can be passed from generation to generation. While these variants are major detection targets, somatic mutations (Figure 10.3) also play important roles in many diseases such as cancer. To identify such acquired somatic mutations, some of the germline variant callers mentioned above also provide utilities or options designed for calling somatic mutations, such as GATK Mutect2 [21], Strelka2, and VarScan2. Tools specifically designed for somatic mutation detection include SomaticSniper [22], VarDict [23], NeuSomatic [24], JointSNVmix [25], and Lancet [26]. Mechanistically, while some of these algorithms (such as Mutect2 and VarScan2) carry out mutation calling on each of the contrasting samples (e.g., normal vs. cancer tissues carrying somatic mutations from the same patient) separately against a reference genome, others (such as Strelka2, JointSNVmix, and SomaticSniper) directly compare the contrasting samples. In the former approach, sequence reads generated from contrasting samples are independently aligned to and variants called against a reference genome. The called

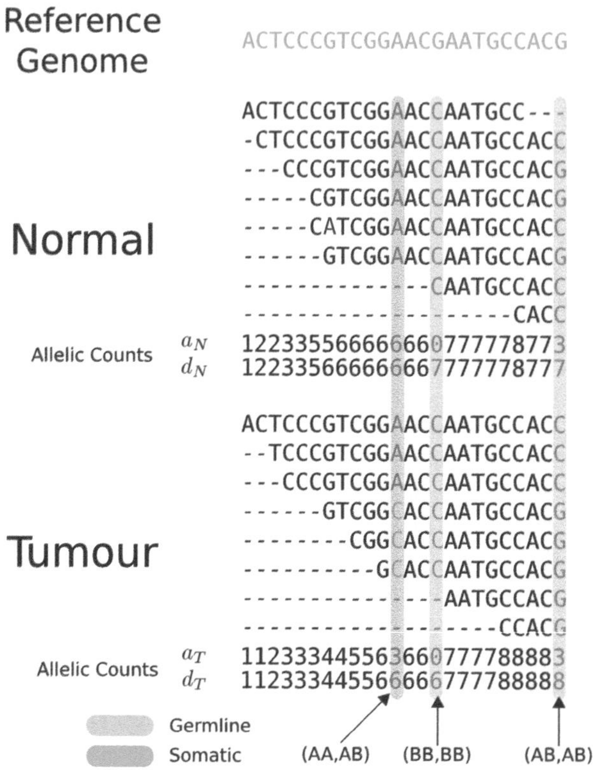

Reference Genome ACTCCCGTCGGAACGAATGCCACG

Normal
ACTCCCGTCGGAACCAATGCC---
-CTCCCGTCGGAACCAATGCCACC
---CCCGTCGGAACCAATGCCACG
-----CGTCGGAACCAATGCCACG
-----CATCGGAACCAATGCCACC
------GTCGGAACCAATGCCACG
---------------CAATGCCACC
---------------------CACC

Allelic Counts a_N 122335566666660777778773
d_N 122335666666667777778777

Tumour
ACTCCCGTCGGAACCAATGCCACC
--TCCCGTCGGAACCAATGCCACC
---CCCGTCGGAACCAATGCCACC
------GTCGGCACCAATGCCACG
--------CGGCACCAATGCCACG
----------GCACCAATGCCACG
----------------AATGCCACG
--------------------CCACG

Allelic Counts a_T 112333445563660777788883
d_T 112333445566666777788888

Germline
Somatic
(AA,AB) (BB,BB) (AB,AB)

FIGURE 10.3
Detection of somatic mutations vs. germline variations. In this example, sequence reads from normal and tumor tissues are aligned to the reference genome (shown at the top in green). The allelic counts, i.e., the number of matches (a_N and a_T) and depth of reads (d_N and d_T), at each base position are shown. The blue sites show germline positions, while the red shows a position where a somatic mutation occurred in some tumor cells. Also shown at the bottom are the predicted genotypes for the normal and tumor tissues. (Modified from Roth A. et al. JointSNVMix: a probabilistic model for accurate detection of somatic mutations in normal/tumour paired next-generation sequencing data. *Bioinformatics*, 2012, 28 (7): 907–13, by permission of Oxford University Press.)

variants in the contrasting samples are then compared to each other to locate somatic mutations in the cancer tissue. In the latter approach, the samples are directly compared to each other using statistical tests on the basis of joint probability. NeuSomatic represents the first attempt to use deep learning (CNN) for somatic variant detection. The currently developed Mutect3 also uses machine learning in an effort to improve somatic mutation detection accuracy. To help evaluate the performance of these various somatic variant callers, several benchmarking studies [27–30] are available showing that Mutect2 and Strelka2 are among the top performers so far.

10.2.3 Variant Calling from RNA Sequencing Data

While variant calling is mostly carried out from DNA sequencing data, RNA-seq can also be used to call variants from transcriptionally active regions of the genome. RNA-seq-based variant calling is more challenging due to the inherent heterogeneity in the abundance of reads transcribed from different regions, differential splicing of exons, allele-specific expression, RNA editing, etc. Variant calling from RNA-seq data offers certain advantages, however, as it does not incur additional cost beyond collecting the original transcriptomic data, and it directly interrogates transcriptionally active regions of the genome. In addition, RNA-seq-based variant discovery can be used to validate variants called from whole-genome or –exome sequencing. Methods for RNA-seq-based variant calling are still limited. Currently available tools/pipelines include the GATK best practices workflow for RNA-seq short variant calling [31], SNPiR [32], eSNV-Detect [33], SNVMix [34], and Opossum-Platypus [35]. Most of these tools are based on the use of variant callers developed for germline DNA sequencing, such as the GATK HaplotypeCaller or BCFtools. The often used GATK best practices workflow for RNA-seq data, for example, centers around HaplotypeCaller and comprises multiple steps including: 1) alignment of raw RNA-seq reads to the reference genome using STAR; 2) cleanup of data through identification and removal of duplicate reads; 3) reformatting of alignments that span intronic regions as a preparatory step for variant calling; 4) recalibration of base quality scores to improve variant calling accuracy; 5) variant calling using HaplotypeCaller; and 6) variant filtering to produce the final list of variants. SNVMix, on the other hand, employs a probabilistic binomial mixture model to call variants from pre-mapped RNA-seq reads.

10.2.4 Variant Call Format (VCF)

VCF is a text-based standard file format for storing sequence variations, including SNVs, short indels, and SVs (to be detailed next) [36]. This format is designed to be scalable to encompass millions of sites from thousands of samples. Originally developed for the 1000 Genomes Project, it is designed for fast data retrieval. Besides reporting variants and their genomic positions, it offers fields to store additional information such as variant call quality score, and allows users to add their own custom tags to describe new sequence variations. BCF is the binary version of VCF, providing speed and efficiency through compression of the variant data.

Figure 10.4 provides an example of the VCF format. It contains meta-information lines at the front, a header line, and data lines each of which describes a variant position. The metainfo lines start with "##" and describe related analysis information, such as species, file date, and assembly version. In addition, abbreviations used in the user definable data columns are also defined in the metainfo lines. The subsequent header line lists the names of

```
##fileformat=VCFv4.3
##fileDate=20090805
##source=myImputationProgramV3.1
##reference=file:///seq/references/1000GenomesPilot-NCBI36.fasta
##contig=<ID=20,length=62435964,assembly=B36,md5=f126cdf8a6e0c7f379d618ff66beb2da,species="Homo sapiens",taxonomy=x>
##phasing=partial
##INFO=<ID=NS,Number=1,Type=Integer,Description="Number of Samples With Data">
##INFO=<ID=DP,Number=1,Type=Integer,Description="Total Depth">
##INFO=<ID=AF,Number=A,Type=Float,Description="Allele Frequency">
##INFO=<ID=AA,Number=1,Type=String,Description="Ancestral Allele">
##INFO=<ID=DB,Number=0,Type=Flag,Description="dbSNP membership, build 129">
##INFO=<ID=H2,Number=0,Type=Flag,Description="HapMap2 membership">
##FILTER=<ID=q10,Description="Quality below 10">
##FILTER=<ID=s50,Description="Less than 50% of samples have data">
##FORMAT=<ID=GT,Number=1,Type=String,Description="Genotype">
##FORMAT=<ID=GQ,Number=1,Type=Integer,Description="Genotype Quality">
##FORMAT=<ID=DP,Number=1,Type=Integer,Description="Read Depth">
##FORMAT=<ID=HQ,Number=2,Type=Integer,Description="Haplotype Quality">
#CHROM POS     ID        REF    ALT     QUAL FILTER INFO                              FORMAT      NA00001        NA00002        NA00003
20     14370   rs6054257 G      A       29   PASS   NS=3;DP=14;AF=0.5;DB;H2           GT:GQ:DP:HQ 0|0:48:1:51,51 1|0:48:8:51,51 1/1:43:5:.,.
20     17330   .         T      A       3    q10    NS=3;DP=11;AF=0.017               GT:GQ:DP:HQ 0|0:49:3:58,50 0|1:3:5:65,3   0/0:41:3
20     1110696 rs6040355 A      G,T     67   PASS   NS=2;DP=10;AF=0.333,0.667;AA=T;DB GT:GQ:DP:HQ 1|2:21:6:23,27 2|1:2:0:18,2   2/2:35:4
20     1230237 .         T      .       47   PASS   NS=3;DP=13;AA=T                   GT:GQ:DP:HQ 0|0:54:7:56,60 0|0:48:4:51,51 0/0:61:2
20     1234567 microsat1 GTC    G,GTCT  50   PASS   NS=3;DP=9;AA=G                    GT:GQ:DP    0/1:35:4       0/2:17:2       1/1:40:3
```

FIGURE 10.4
The VCF format (version 4.3). The format is currently managed by the Large Scale Genomics Work Stream, part of the Global Alliance for Genomics & Health (GA4GH). BCF is the binary counterpart of VCF. (From http://samtools.github.io/hts-specs/.)

TABLE 10.1

Mandatory Fields in a VCF File

Col	Field	Type	Description
1	#CHROM	String	Chromosome number
2	POS	Integer	Start position of the variation
3	ID	String	Database identifier
4	REF	String	Reference allele
5	ALT	String	Alternate allele(s)
6	QUAL	Numeric	Quality score (Phred-style)
7	FILTER	String	Filter status
8	INFO	String	User extensible information

the 8 mandatory columns (Table 10.1). In the QUAL column, a Phred-like quality score for the alternative allele (ALT) call is given (e.g., a QUAL value of 30 means the probability of the ALT call being wrong is 0.001). In the FILTER column, "PASS" means this position has passed all filters, while a value of "q10" as shown in Figure 10.4 indicates that the variant call quality at this site is below 10. The data lines, containing variant calls for a list of genomic positions, make the body of a VCF file.

VCF/BCF files can be parsed, manipulated, and visualized using tools such BCFtools and vcfR [37]. BCFtools, beyond its capability to call variants, offers a set of utilities for VCF/BCF files on a long range of operations, such as file format conversion, summarizing variant statistics, variant filtering, sorting, concatenation or merging of multiple files, finding how variants in multiple files intersect, detection of sample swaps and contamination, etc. Besides built-in commands for such operations, it also supports the use of plugins for specific single-purpose tasks. Examples of such plugins and tasks are *frameshifts* to annotate frameshift indels, *guess-ploidy* to determine sample sex, *split-vep* to query and extract from variants' VEP annotations (see Section 10.4 for VEP variant annotation), *trio-dnm2* to determine *de novo* mutations in parents-offspring trios, and *gvcfz* to compress gVCF (or genomic VCF) files. The gVCF format is an extended version of VCF with the same format specification, but it also contains information for the rest of non-variant genomic regions with confidence estimates that they match the reference genome. The goal to include all sites of the genome in a gVCF is to facilitate joint genotyping of a cohort of samples when needed.

10.2.5 Evaluating VCF Results

SNVs and indels reported in VCF files need to be evaluated to identify false positives. Visualization of called variants and supporting reads in a genome browser, such as IGV or Savant, provides an initial examination of the variant call result. Further evaluation should be based on criteria such as deviation from Hardy–Weinberg equilibrium, systematic call quality difference

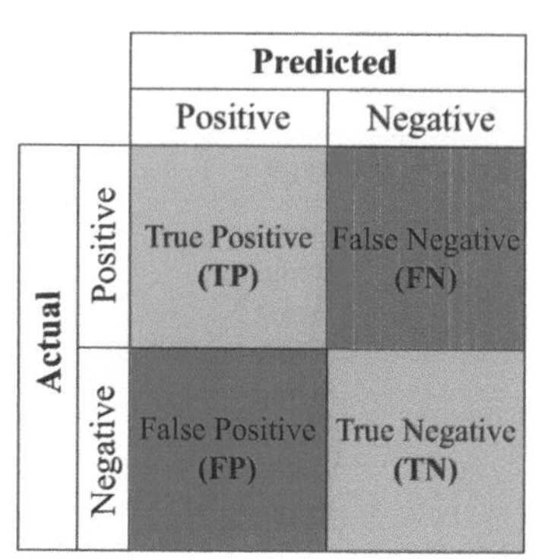

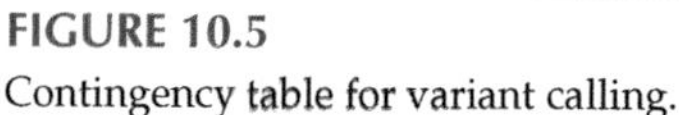
FIGURE 10.5
Contingency table for variant calling.

between major and minor alleles, extreme depth of coverage, or strand bias. The ratio of transitions and transversions (Ti/Tv) is an additional indicator of variant call specificity and quality. The theoretical ratio of Ti/Tv is 0.5, because purely from the point of statistical probability the chance of producing transitions is half that of transversions. However, due to biochemical mechanisms involved in these nucleotide substitution processes, the frequency of having transitions is higher than that of transversions. Based on existing NGS data from multiple species, the expected values of Ti/Tv for whole genome and exome datasets are usually in the ranges of 2.0–2.1, 3.0–3.5, respectively [38]. Variants that do not pass these QC criteria are then filtered out. Besides such filtering using preset criteria, low quality variants may also be identified for removal using machine learning approaches such as ForestQC [39] and VQSR (part of GATK).

As different variant callers employ different approaches, the variants they identify usually only partially overlap. It is advisable, therefore, to examine closely on the specifics of an experiment to decide on more appropriate variant caller(s). If more than one method is used, it is advisable to compare their outputs and analyze how they intersect. Use of convergent variants is an effective way to reduce rates of miscalled variants. Alternatively, ensemble methods such as VariantMetaCaller [40] and BAYSIC [41] can also be used.

To further compare results from multiple variant callers side-by-side, precision and recall are key metrics often used to measure the performance of variant callers. Precision refers to the ability to not detect false positives, while recall to not detect false negatives (Figure 10.5). Mathematically, precision and recall are calculated as

$$Precision = \frac{TP}{TP + FP\left(i.e., \text{ } all \text{ } predicted \text{ } positives\right)}$$

$$Recall = \frac{TP}{TP + FN\left(i.e.,\ all\ actual\ positives\right)}$$

To combine precision and recall into a single metric to simplify evaluation of variant callers, F1 score is often used. Mathematically, F1 is the harmonic mean of the two, calculated as

$$F1 = 2 \times \frac{Precision \times Recall}{Precision + Recall}$$

By using the harmonic mean instead of the arithmetic mean, F1 provides a more balanced way to measure variant calling performance. For example, if either precision or recall is low, the F1 score is low. To have a high F1 score, both precision and recall need to be high. Utilities such as RTG Tools [42] can help compare results from multiple callers.

10.3 Structural Variant (SV) Calling

10.3.1 Short-Read-Based SV Calling

As covered in Chapter 2, structural variants in the genome involve insertions, deletions, duplications, inversions, and translocations of sequences at least 50 bp in size. SVs are largely the basis of genome evolution and diversification, and produce more genomic differences than SNVs between individuals [43]. Earlier experimental methods on the detection of SVs were mostly based on comparative genome hybridization and SNP whole genome arrays. The advent of NGS, especially the use of paired-end short reads and more recently long reads, has greatly pushed SV detection forward. As illustrated in Figure 10.6, the basic approaches to detect SVs using short reads are based on: (1) the change in read depth (RD) as compared to the rest of the genome; (2) the use of paired-reads (PR) to detect unexpected change in orientation, distance between them, or their localization to different chromosomes; (3) the existence of split-reads (SR) that span disjoint regions of the genome; and (4) the deviation of *de novo* assembled (AS) sequence from the reference sequence.

Among these approaches, the RD approach is to be detailed in Section 10.3.3. Figure 10.7 illustrates the main steps in the PR approach. The first step is to separate read pairs into concordant or discordant groups, defined by the distance between a read pair matching or deviating from the expected distance based on the reference genome. The discordant read pairs are then assembled into different clusters based on the genomic region they cover to generate candidate SV calling regions. In the last step, the candidate SV

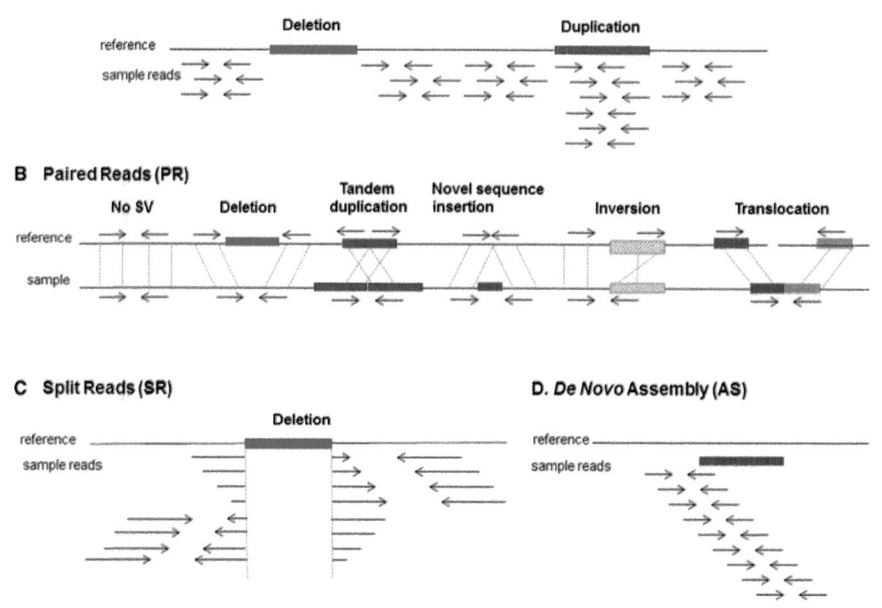

FIGURE 10.6
Four approaches for SV detection. (From Escaramís, G. & Docampo, E. A decade of structural variants: description, history and methods to detect structural variation. *Briefings in Functional Genomics*, 2015, 14 (5): 305–14, by permission of Oxford University Press.)

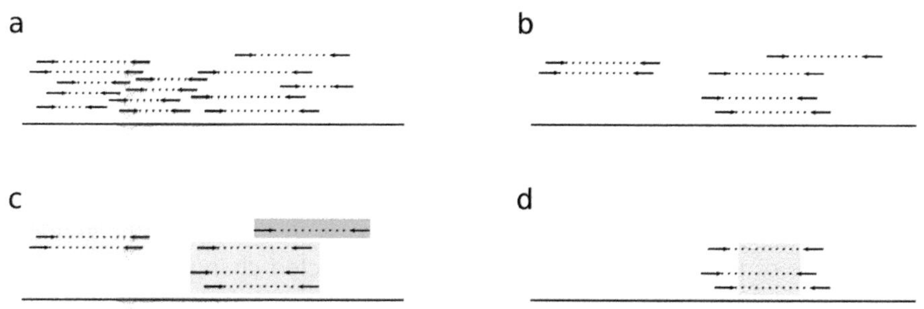

FIGURE 10.7
General steps of calling SVs using paired-end reads. (Used with permission from Whelan, Christopher, "Detecting and Analyzing Genomic Structural Variation Using Distributed Computing" (2014). Scholar Archive. Paper 3482.)

clusters are filtered based on statistical assessment so that only clusters that are covered by multiple read pairs are reported as SVs. The boundaries of possible break points in the region are also identified in this step (indicated by the shaded area in Figure 10.7, panel d). Among currently available SV detection algorithms, PEMer (or Paired-End Mapper) [44], BreakDancer [45],

SVDetect [46], and 1-2-3-SV [47] apply this paired-reads based approach. Pindel [48] provides an example for the SR approach. It first searches for read pairs in which one read aligns to the reference genome but the other does not. Based on the assumption that the second read contains a breakpoint, it uses the aligned read as anchor to scan the surrounding regions for split mapping of the second read. While it can locate breakpoints at single base resolution, this approach is computationally expensive because of the challenge associated with aligning read sub-sequences to different genomic regions with gaps in between. Cortex [49] and AsmVar [50] are examples of the DS approach. In this approach, the genome is first assembled from reads, and subsequently SVs are called through alignment and statistical analysis of the *de novo* genome assembly against the reference.

To improve detection accuracy, many currently available SV detection algorithms use a combination of these approaches. For example, DELLY [51], Meerkat [52], SoftSV [53], and Wham [54] combine PR and SR, while GASV/GASVPro [55, 56], Genome STRiP [57], and inGAP-sv [58] combine RD and PR. HYDRA [59] is an example of combining RD and AS. As examples of combining three approaches, MANTA [60], GRIDSS [61], SvABA [62], and CREST [63] combine PR, SR, and AS. LUMPY [64] and TIDDIT [65], on the other hand, combine RD, PR, and SR.

10.3.2 Long-Read-Based SV Calling

SV detection from the use of short reads has high miscalling rates because of the limitations of short-read sequencing. Long-read technologies such as PacBio and Oxford Nanopore sequencing overcome such inherent limitations caused by short read length. Mechanistically, long-read-based SV callers are mostly built on the use of the SR and/or AS approaches. These callers include pbsv [66], Sniffles [67], Phased Assembly Variant (PAV) [68], MELT [69], NanoVar [70], NanoSV [71], PALMER [72], SVIM [73], and Picky [74]. Some callers, such as Dysgu [75], are developed to use both long and short reads. Besides long reads generated from long-read sequencers, synthetic long reads obtained using technologies such as linked-read sequencing [76–79] can also be used for SV detection by deploying tools such as Long Ranger, an open-source pipeline developed by 10× Genomics [80].

10.3.3 CNV Detection

CNVs, caused by duplications, insertions, or deletions, are an important subtype of structural variation. Among the four basic approaches outlined in Section 10.3.1, CNV detection algorithms are often based on RD. These algorithms are based on the assumption that the number of reads obtained from a region is proportional to its copy number in the genome. If a genomic segment is repeated multiple times, for example, a significantly higher number of reads will be observed from the segment compared to other non-repeated regions. If a segment is deleted, on the other hand, there will be no read coverage for it (see Figure 10.6, panel A). These RD-based algorithms include CNVnator [81], CNV-seq [82], CNVkit [83], cn.MOPS (Copy Number estimation by a Mixture of PoissonS) [84], CNAseg [85], ERDS [86], RDXplorer (Read Depth eXplorer) [87], Control-FREEC [88], mrFAST [89], SegSeq [90], readDepth [91], Canvas [92], and iCopyDAV [93]. Among these algorithms, CNVnator, built on the use of a mean-shift approach originally designed for image processing, is one of the most commonly used. As factors such as GC content may also affect local read density, a normalization step is often conducted when deploying these algorithms to account for these compounding factors. In studies that involve comparison of samples from the same genetic background, e.g., a diseased tissue vs. a healthy tissue from the same patient, these compounding factors are often cancelled out.

10.3.4 Integrated SV Analysis

The different software tools introduced above are based on different algorithmic design, and as a result show varying performance for detecting particular types (or aspects) of SVs [94, 95]. In order to improve call performance for the full range of SVs, there have been efforts to take an integrated approach towards comprehensive SV calling using the different but often complementary tools. SVMerge, being one of these efforts, integrates SV calling results from different callers [96]. It first feeds BAM files into a number of SV callers including those introduced above to generate BED files, and then the SV calls in the BED files are merged. After computational validation and breakpoint refinement by local *de novo* alignment, a final list of SVs is generated. Other efforts that take a similarly integrated approach include Parliament2 [97], FusorSV [98], SURVIVOR [99], MetaSV [100], and CNVer [101].

10.4 Annotation of Called Variants

To gain biological insights from identified SNVs, indels, or SVs, annotation of the variants is needed. For example, if an SNV is annotated to be nonsynonymous in a gene, it may impair protein function if the affected amino acid is located within the active site of the protein. Through examination of their annotations, called variants can be filtered and prioritized for more in-depth analysis. Because of the large number of variants usually called from an experiment, an automatic pipeline is usually preferred. To meet this demand a number of variant annotation tools have been developed. ANNOVAR [102] is one such tool among the most widely used. It takes SNVs, indels, and CNVs as input, and as output, it reports their functional impacts and provides significance scores to help with filtering and prioritization. Its TABLE_ANNOVAR script can quickly turn a variant list into an Excel-compatible file containing many annotation fields that can help researcher evaluate the function importance of the variants. ANNOVAR offers flexibility and extensibility, e.g., it can identify variants located in conserved genomic regions, or find variants that overlap with those from the 1000 genomes project or dbSNP. Other widely used variant annotation tools include SnpEff [103], VEP (Variant Effect Predictor) [104], UCSC Genome Browser's Variant Annotation Integrator [105], and SeattleSeq [106].

On the back end of these annotation tools are various types of annotation databases. These databases provide reference information on what genes or regions of the genome they are at (e.g., transcriptional factor binding sites, DNase I hypersensitivity sites, or highly conserved regions), what functional consequence(s) they might cause based on known phenotype or computational inference, whether an identified variant has been observed before, how frequently they are observed in different populations if previously known, whether there is evidence for their connection with a human disease (especially relevant for human patient samples), etc. Annotational information from these databases is essential for filtering out non-target variants, and ranking/prioritizing the remaining variants for reporting purpose and further analyses. Many of these databases, as well as the detailed steps of variant filtering, ranking, and prioritizing, are covered in detail in the next chapter.

References

1. Acuna-Hidalgo R, Veltman JA, Hoischen A. New insights into the generation and role of de novo mutations in health and disease. *Genome Biol* 2016, 17(1):241.
2. Miller MB, Reed HC, Walsh CA. Brain Somatic Mutation in Aging and Alzheimer's Disease. *Annu Rev Genomics Hum Genet* 2021, 22:239–256.
3. Martincorena I, Campbell PJ. Somatic mutation in cancer and normal cells. *Science* 2015, 349(6255):1483–1489.
4. McKenna A, Hanna M, Banks E, Sivachenko A, Cibulskis K, Kernytsky A, Garimella K, Altshuler D, Gabriel S, Daly M *et al.* The Genome Analysis Toolkit: a MapReduce framework for analyzing next-generation DNA sequencing data. *Genome Res* 2010, 20(9):1297–1303.
5. Danecek P, Bonfield JK, Liddle J, Marshall J, Ohan V, Pollard MO, Whitwham A, Keane T, McCarthy SA, Davies RM *et al.* Twelve years of SAMtools and BCFtools. *GigaScience* 2021, 10(2).
6. Kim S, Scheffler K, Halpern AL, Bekritsky MA, Noh E, Kallberg M, Chen X, Kim Y, Beyter D, Krusche P *et al.* Strelka2: fast and accurate calling of germline and somatic variants. *Nat Methods* 2018, 15(8):591–594.

7. Garrison E, Marth G. Haplotype-based variant detection from short-read sequencing. *arXiv:12073907*, 2012.
8. Luo R, Schatz MC, Salzberg SL. 16GT: a fast and sensitive variant caller using a 16-genotype probabilistic model. *GigaScience* 2017, 6(7):1–4.
9. Koboldt DC, Zhang Q, Larson DE, Shen D, McLellan MD, Lin L, Miller CA, Mardis ER, Ding L, Wilson RK. VarScan 2: somatic mutation and copy number alteration discovery in cancer by exome sequencing. *Genome Res* 2012, 22(3):568–576.
10. Liu J, Shen Q, Bao H. Comparison of seven SNP calling pipelines for the next-generation sequencing data of chickens. *PLoS One* 2022, 17(1):e0262574.
11. Valueva MV, Nagornov N, Lyakhov PA, Valuev GV, Chervyakov NI. Application of the residue number system to reduce hardware costs of the convolutional neural network implementation. *Math Comput Simul* 2020, 177:232–243.
12. Poplin R, Chang PC, Alexander D, Schwartz S, Colthurst T, Ku A, Newburger D, Dijamco J, Nguyen N, Afshar PT *et al.* A universal SNP and small-indel variant caller using deep neural networks. *Nat Biotechnol* 2018, 36(10):983–987.
13. Zheng Z, Li S, Su J, Leung AW-S, Lam T-W, Luo R. Symphonizing pileup and full-alignment for deep learning-based long-read variant calling. *Nat Comput Sci* 2022, 2(12):797–803.
14. Edge P, Bansal V. Longshot enables accurate variant calling in diploid genomes from single-molecule long read sequencing. *Nat Commun* 2019, 10(1):4660.
15. Ahsan MU, Liu Q, Fang L, Wang K. NanoCaller for accurate detection of SNPs and indels in difficult-to-map regions from long-read sequencing by haplotype-aware deep neural networks. *Genome Biol* 2021, 22(1):261.
16. (https://github.com/nanoporetech/medaka)
17. Shafin K, Pesout T, Chang PC, Nattestad M, Kolesnikov A, Goel S, Baid G, Kolmogorov M, Eizenga JM, Miga KH *et al.* Haplotype-aware variant calling with PEPPER-Margin-DeepVariant enables high accuracy in nanopore long-reads. *Nat Methods* 2021, 18(11):1322–1332.
18. Supernat A, Vidarsson OV, Steen VM, Stokowy T. Comparison of three variant callers for human whole genome sequencing. *Sci Rep* 2018, 8(1):17851.
19. Pei S, Liu T, Ren X, Li W, Chen C, Xie Z. Benchmarking variant callers in next-generation and third-generation sequencing analysis. *Brief Bioinform* 2021, 22(3):bbaa148.
20. Lin YL, Chang PC, Hsu C, Hung MZ, Chien YH, Hwu WL, Lai F, Lee NC. Comparison of GATK and DeepVariant by trio sequencing. *Sci Rep* 2022, 12(1):1809.
21. Benjamin D, Sato T, Cibulskis K, Getz G, Stewart C, Lichtenstein L. Calling somatic SNVs and indels with Mutect2. *bioRxiv* 2019, doi: https://doi.org/10.1101/861054
22. Larson DE, Harris CC, Chen K, Koboldt DC, Abbott TE, Dooling DJ, Ley TJ, Mardis ER, Wilson RK, Ding L. SomaticSniper: identification of somatic point mutations in whole genome sequencing data. *Bioinformatics* 2012, 28(3):311–317.
23. Lai Z, Markovets A, Ahdesmaki M, Chapman B, Hofmann O, McEwen R, Johnson J, Dougherty B, Barrett JC, Dry JR. VarDict: a novel and versatile variant caller for next-generation sequencing in cancer research. *Nucleic Acids Res* 2016, 44(11):e108.

24. Sahraeian SME, Liu R, Lau B, Podesta K, Mohiyuddin M, Lam HYK. Deep convolutional neural networks for accurate somatic mutation detection. *Nat Commun* 2019, 10(1):1041.
25. Roth A, Ding J, Morin R, Crisan A, Ha G, Giuliany R, Bashashati A, Hirst M, Turashvili G, Oloumi A *et al.* JointSNVMix: a probabilistic model for accurate detection of somatic mutations in normal/tumour paired next-generation sequencing data. *Bioinformatics* 2012, 28(7):907–913.
26. Narzisi G, Corvelo A, Arora K, Bergmann EA, Shah M, Musunuri R, Emde AK, Robine N, Vacic V, Zody MC. Genome-wide somatic variant calling using localized colored de Bruijn graphs. *Commun Biol* 2018, 1:20.
27. Cai L, Yuan W, Zhang Z, He L, Chou KC. In-depth comparison of somatic point mutation callers based on different tumor next-generation sequencing depth data. *Sci Rep* 2016, 6:36540.
28. Kroigard AB, Thomassen M, Laenkholm AV, Kruse TA, Larsen MJ. Evaluation of nine somatic variant callers for detection of somatic mutations in exome and targeted deep sequencing data. *PLoS One* 2016, 11(3):e0151664.
29. Chen Z, Yuan Y, Chen X, Chen J, Lin S, Li X, Du H. Systematic comparison of somatic variant calling performance among different sequencing depth and mutation frequency. *Sci Rep* 2020, 10(1):3501.
30. Zhao S, Agafonov O, Azab A, Stokowy T, Hovig E. Accuracy and efficiency of germline variant calling pipelines for human genome data. *Sci Rep* 2020, 10(1):20222.
31. GATK Best Practices Workflow for RNAseq Short Variant Discovery (SNPs + Indels) (https://gatk.broadinstitute.org/hc/en-us/articles/360035531192-RNAseq-short-variant-discovery-SNPs-Indels-)
32. Piskol R, Ramaswami G, Li JB. Reliable identification of genomic variants from RNA-seq data. *Am J Hum Genet* 2013, 93(4):641–651.
33. Tang X, Baheti S, Shameer K, Thompson KJ, Wills Q, Niu N, Holcomb IN, Boutet SC, Ramakrishnan R, Kachergus JM *et al.* The eSNV-detect: a computational system to identify expressed single nucleotide variants from transcriptome sequencing data. *Nucleic Acids Res* 2014, 42(22):e172.
34. Goya R, Sun MG, Morin RD, Leung G, Ha G, Wiegand KC, Senz J, Crisan A, Marra MA, Hirst M *et al.* SNVMix: predicting single nucleotide variants from next-generation sequencing of tumors. *Bioinformatics* 2010, 26(6):730–736.
35. Oikkonen L, Lise S. Making the most of RNA-seq: Pre-processing sequencing data with Opossum for reliable SNP variant detection. *Wellcome Open Res* 2017, 2:6.
36. Danecek P, Auton A, Abecasis G, Albers CA, Banks E, DePristo MA, Handsaker RE, Lunter G, Marth GT, Sherry ST *et al.* The variant call format and VCFtools. *Bioinformatics* 2011, 27(15):2156–2158.
37. Knaus BJ, Grunwald NJ. vcfr: a package to manipulate and visualize variant call format data in R. *Mol Ecol Resour* 2017, 17(1):44–53.
38. Freudenberg-Hua Y, Freudenberg J, Kluck N, Cichon S, Propping P, Nothen MM. Single nucleotide variation analysis in 65 candidate genes for CNS disorders in a representative sample of the European population. *Genome Res* 2003, 13(10):2271–2276.
39. Li J, Jew B, Zhan L, Hwang S, Coppola G, Freimer NB, Sul JH. ForestQC: Quality control on genetic variants from next-generation sequencing data using random forest. *PLoS Comput Biol* 2019, 15(12):e1007556.

40. Gezsi A, Bolgar B, Marx P, Sarkozy P, Szalai C, Antal P. VariantMetaCaller: automated fusion of variant calling pipelines for quantitative, precision-based filtering. *BMC Genomics* 2015, 16:875.
41. Cantarel BL, Weaver D, McNeill N, Zhang J, Mackey AJ, Reese J. BAYSIC: a Bayesian method for combining sets of genome variants with improved specificity and sensitivity. *BMC Bioinformatics* 2014, 15:104.
42. RTG Tools: Utilities for accurate VCF comparison and manipulation (https://github.com/RealTimeGenomics/rtg-tools)
43. Sudmant PH, Rausch T, Gardner EJ, Handsaker RE, Abyzov A, Huddleston J, Zhang Y, Ye K, Jun G, Fritz MH *et al*. An integrated map of structural variation in 2,504 human genomes. *Nature* 2015, 526(7571):75–81.
44. Korbel JO, Abyzov A, Mu XJ, Carriero N, Cayting P, Zhang Z, Snyder M, Gerstein MB. PEMer: a computational framework with simulation-based error models for inferring genomic structural variants from massive paired-end sequencing data. *Genome Biol* 2009, 10(2):R23.
45. Chen K, Wallis JW, McLellan MD, Larson DE, Kalicki JM, Pohl CS, McGrath SD, Wendl MC, Zhang Q, Locke DP *et al*. BreakDancer: an algorithm for high-resolution mapping of genomic structural variation. *Nat Methods* 2009, 6(9):677–681.
46. Zeitouni B, Boeva V, Janoueix-Lerosey I, Loeillet S, Legoix-ne P, Nicolas A, Delattre O, Barillot E. SVDetect: a tool to identify genomic structural variations from paired-end and mate-pair sequencing data. *Bioinformatics* 2010, 26(15):1895–1896.
47. 1-2-3-SV (https://github.com/Vityay/1-2-3-SV)
48. Ye K, Guo L, Yang X, Lamijer EW, Raine K, Ning Z. Split-read indel and structural variant calling using PINDEL. *Methods Mol Biol* 2018, 1833:95–105.
49. Iqbal Z, Caccamo M, Turner I, Flicek P, McVean G. De novo assembly and genotyping of variants using colored de Bruijn graphs. *Nat Genet* 2012, 44(2):226–232.
50. Liu S, Huang S, Rao J, Ye W, Genome Denmark Consortium II, Krogh A, Wang J. Discovery, genotyping and characterization of structural variation and novel sequence at single nucleotide resolution from de novo genome assemblies on a population scale. *GigaScience* 2015, 4 :64.
51. Rausch T, Zichner T, Schlattl A, Stutz AM, Benes V, Korbel JO. DELLY: structural variant discovery by integrated paired-end and split-read analysis. *Bioinformatics* 2012, 28(18):i333–i339.
52. Yang L, Luquette LJ, Gehlenborg N, Xi R, Haseley PS, Hsieh CH, Zhang C, Ren X, Protopopov A, Chin L *et al*. Diverse mechanisms of somatic structural variations in human cancer genomes. *Cell* 2013, 153(4):919–929.
53. Bartenhagen C, Dugas M. Robust and exact structural variation detection with paired-end and soft-clipped alignments: SoftSV compared with eight algorithms. *Brief Bioinform* 2016, 17(1):51–62.
54. Kronenberg ZN, Osborne EJ, Cone KR, Kennedy BJ, Domyan ET, Shapiro MD, Elde NC, Yandell M. Wham: Identifying structural variants of biological consequence. *PLoS Comput Biol* 2015, 11(12):e1004572.
55. Sindi S, Helman E, Bashir A, Raphael BJ. A geometric approach for classification and comparison of structural variants. *Bioinformatics* 2009, 25(12):i222–230.

56. Sindi SS, Onal S, Peng LC, Wu HT, Raphael BJ. An integrative probabilistic model for identification of structural variation in sequencing data. *Genome Biol* 2012, 13(3):R22.
57. Handsaker RE, Van Doren V, Berman JR, Genovese G, Kashin S, Boettger LM, McCarroll SA. Large multiallelic copy number variations in humans. *Nat Genet* 2015, 47(3):296–303.
58. Qi J, Zhao F. inGAP-sv: a novel scheme to identify and visualize structural variation from paired end mapping data. *Nucleic Acids Res* 2011, 39(Web Server issue):W567–575.
59. Quinlan AR, Clark RA, Sokolova S, Leibowitz ML, Zhang Y, Hurles ME, Mell JC, Hall IM. Genome-wide mapping and assembly of structural variant breakpoints in the mouse genome. *Genome Res* 2010, 20(5):623–635.
60. Chen X, Schulz-Trieglaff O, Shaw R, Barnes B, Schlesinger F, Kallberg M, Cox AJ, Kruglyak S, Saunders CT. Manta: rapid detection of structural variants and indels for germline and cancer sequencing applications. *Bioinformatics* 2016, 32(8):1220–1222.
61. Cameron DL, Schroder J, Penington JS, Do H, Molania R, Dobrovic A, Speed TP, Papenfuss AT. GRIDSS: sensitive and specific genomic rearrangement detection using positional de Bruijn graph assembly. *Genome Res* 2017, 27(12):2050–2060.
62. Wala JA, Bandopadhayay P, Greenwald NF, O'Rourke R, Sharpe T, Stewart C, Schumacher S, Li Y, Weischenfeldt J, Yao X *et al.* SvABA: genome-wide detection of structural variants and indels by local assembly. *Genome Res* 2018, 28(4):581–591.
63. Wang J, Mullighan CG, Easton J, Roberts S, Heatley SL, Ma J, Rusch MC, Chen K, Harris CC, Ding L *et al.* CREST maps somatic structural variation in cancer genomes with base-pair resolution. *Nat Methods* 2011, 8(8):652–654.
64. Layer RM, Chiang C, Quinlan AR, Hall IM. LUMPY: a probabilistic framework for structural variant discovery. *Genome Biol* 2014, 15(6):R84.
65. Eisfeldt J, Vezzi F, Olason P, Nilsson D, Lindstrand A. TIDDIT, an efficient and comprehensive structural variant caller for massive parallel sequencing data. *F1000Research* 2017, 6:664.
66. pbsv (https://github.com/PacificBiosciences/pbsv)
67. Sedlazeck FJ, Rescheneder P, Smolka M, Fang H, Nattestad M, von Haeseler A, Schatz MC. Accurate detection of complex structural variations using single-molecule sequencing. *Nat Methods* 2018, 15(6):461–468.
68. Ebert P, Audano PA, Zhu Q, Rodriguez-Martin B, Porubsky D, Bonder MJ, Sulovari A, Ebler J, Zhou W, Serra Mari R *et al.* Haplotype-resolved diverse human genomes and integrated analysis of structural variation. *Science* 2021, 372(6537):eabf7117.
69. Gardner EJ, Lam VK, Harris DN, Chuang NT, Scott EC, Pittard WS, Mills RE, Genomes Project C, Devine SE. The Mobile Element Locator Tool (MELT): population-scale mobile element discovery and biology. *Genome Res* 2017, 27(11):1916–1929.
70. Tham CY, Tirado-Magallanes R, Goh Y, Fullwood MJ, Koh BTH, Wang W, Ng CH, Chng WJ, Thiery A, Tenen DG *et al.* NanoVar: accurate characterization of patients' genomic structural variants using low-depth nanopore sequencing. *Genome Biol* 2020, 21(1):56.

71. Cretu Stancu M, van Roosmalen MJ, Renkens I, Nieboer MM, Middelkamp S, de Ligt J, Pregno G, Giachino D, Mandrile G, Espejo Valle-Inclan J *et al*. Mapping and phasing of structural variation in patient genomes using nanopore sequencing. *Nat Commun* 2017, 8(1):1326.
72. Zhou W, Emery SB, Flasch DA, Wang Y, Kwan KY, Kidd JM, Moran JV, Mills RE. Identification and characterization of occult human-specific LINE-1 insertions using long-read sequencing technology. *Nucleic Acids Res* 2020, 48(3):1146–1163.
73. Heller D, Vingron M. SVIM: structural variant identification using mapped long reads. *Bioinformatics* 2019, 35(17):2907–2915.
74. Gong L, Wong CH, Cheng WC, Tjong H, Menghi F, Ngan CY, Liu ET, Wei CL. Picky comprehensively detects high-resolution structural variants in nanopore long reads. *Nat Methods* 2018, 15(6):455–460.
75. Cleal K, Baird DM. Dysgu: efficient structural variant calling using short or long reads. *Nucleic Acids Res* 2022, 50(9):e53.
76. Zheng GX, Lau BT, Schnall-Levin M, Jarosz M, Bell JM, Hindson CM, Kyriazopoulou-Panagiotopoulou S, Masquelier DA, Merrill L, Terry JM *et al*. Haplotyping germline and cancer genomes with high-throughput linked-read sequencing. *Nat Biotechnol* 2016, 34(3):303–311.
77. Zhang F, Christiansen L, Thomas J, Pokholok D, Jackson R, Morrell N, Zhao Y, Wiley M, Welch E, Jaeger E *et al*. Haplotype phasing of whole human genomes using bead-based barcode partitioning in a single tube. *Nat Biotechnol* 2017, 35(9):852–857.
78. Wang O, Chin R, Cheng X, Wu MKY, Mao Q, Tang J, Sun Y, Anderson E, Lam HK, Chen D *et al*. Efficient and unique cobarcoding of second-generation sequencing reads from long DNA molecules enabling cost-effective and accurate sequencing, haplotyping, and de novo assembly. *Genome Res* 2019, 29(5):798–808.
79. Chen Z, Pham L, Wu TC, Mo G, Xia Y, Chang PL, Porter D, Phan T, Che H, Tran H *et al*. Ultralow-input single-tube linked-read library method enables short-read second-generation sequencing systems to routinely generate highly accurate and economical long-range sequencing information. *Genome Res* 2020, 30(6):898–909.
80. Long Ranger (https://github.com/10XGenomics/longranger)
81. Abyzov A, Urban AE, Snyder M, Gerstein M. CNVnator: an approach to discover, genotype, and characterize typical and atypical CNVs from family and population genome sequencing. *Genome Res* 2011, 21(6):974–984.
82. Xie C, Tammi MT. CNV-seq, a new method to detect copy number variation using high-throughput sequencing. *BMC bioinformatics* 2009, 10:80.
83. Talevich E, Shain AH, Botton T, Bastian BC. CNVkit: Genome-wide copy number detection and visualization from targeted DNA sequencing. *PLoS Comput Biol* 2016, 12(4):e1004873.
84. Klambauer G, Schwarzbauer K, Mayr A, Clevert DA, Mitterecker A, Bodenhofer U, Hochreiter S. cn.MOPS: mixture of Poissons for discovering copy number variations in next-generation sequencing data with a low false discovery rate. *Nucleic Acids Res* 2012, 40(9):e69.
85. Ivakhno S, Royce T, Cox AJ, Evers DJ, Cheetham RK, Tavare S. CNAseg--a novel framework for identification of copy number changes in cancer from second-generation sequencing data. *Bioinformatics* 2010, 26(24):3051–3058.

86. Zhu M, Need AC, Han Y, Ge D, Maia JM, Zhu Q, Heinzen EL, Cirulli ET, Pelak K, He M *et al.* Using ERDS to infer copy-number variants in high-coverage genomes. *Am J Hum Genet* 2012, 91(3):408–421.
87. Yoon S, Xuan Z, Makarov V, Ye K, Sebat J. Sensitive and accurate detection of copy number variants using read depth of coverage. *Genome Res* 2009, 19(9):1586–1592.
88. Boeva V, Popova T, Bleakley K, Chiche P, Cappo J, Schleiermacher G, Janoueix-Lerosey I, Delattre O, Barillot E. Control-FREEC: a tool for assessing copy number and allelic content using next-generation sequencing data. *Bioinformatics* 2012, 28(3):423–425.
89. Alkan C, Kidd JM, Marques-Bonet T, Aksay G, Antonacci F, Hormozdiari F, Kitzman JO, Baker C, Malig M, Mutlu O *et al.* Personalized copy number and segmental duplication maps using next-generation sequencing. *Nat Genet* 2009, 41(10):1061–1067.
90. Chiang DY, Getz G, Jaffe DB, O'Kelly MJ, Zhao X, Carter SL, Russ C, Nusbaum C, Meyerson M, Lander ES. High-resolution mapping of copy-number alterations with massively parallel sequencing. *Nat Methods* 2009, 6(1):99–103.
91. Miller CA, Hampton O, Coarfa C, Milosavljevic A. ReadDepth: a parallel R package for detecting copy number alterations from short sequencing reads. *PLoS One* 2011, 6(1):e16327.
92. Roller E, Ivakhno S, Lee S, Royce T, Tanner S. Canvas: versatile and scalable detection of copy number variants. *Bioinformatics* 2016, 32(15):2375–2377.
93. Dharanipragada P, Vogeti S, Parekh N. iCopyDAV: Integrated platform for copy number variations-Detection, annotation and visualization. *PLoS One* 2018, 13(4):e0195334.
94. Cameron DL, Di Stefano L, Papenfuss AT. Comprehensive evaluation and characterisation of short read general-purpose structural variant calling software. *Nat Commun* 2019, 10(1):3240.
95. Kosugi S, Momozawa Y, Liu X, Terao C, Kubo M, Kamatani Y. Comprehensive evaluation of structural variation detection algorithms for whole genome sequencing. *Genome Biol* 2019, 20(1):117.
96. Wong K, Keane TM, Stalker J, Adams DJ. Enhanced structural variant and breakpoint detection using SVMerge by integration of multiple detection methods and local assembly. *Genome Biol* 2010, 11(12):R128.
97. Zarate S, Carroll A, Mahmoud M, Krasheninina O, Jun G, Salerno WJ, Schatz MC, Boerwinkle E, Gibbs RA, Sedlazeck FJ. Parliament2: Accurate structural variant calling at scale. *GigaScience* 2020, 9(12):giaa145.
98. Becker T, Lee WP, Leone J, Zhu Q, Zhang C, Liu S, Sargent J, Shanker K, Mil-Homens A, Cerveira E *et al.* FusorSV: an algorithm for optimally combining data from multiple structural variation detection methods. *Genome Biol* 2018, 19(1):38.
99. Jeffares DC, Jolly C, Hoti M, Speed D, Shaw L, Rallis C, Balloux F, Dessimoz C, Bahler J, Sedlazeck FJ. Transient structural variations have strong effects on quantitative traits and reproductive isolation in fission yeast. *Nat Commun* 2017, 8:14061.
100. Mohiyuddin M, Mu JC, Li J, Bani Asadi N, Gerstein MB, Abyzov A, Wong WH, Lam HY. MetaSV: an accurate and integrative structural-variant caller for next generation sequencing. *Bioinformatics* 2015, 31(16):2741–2744.

101. Medvedev P, Fiume M, Dzamba M, Smith T, Brudno M. Detecting copy number variation with mated short reads. *Genome Res* 2010, 20(11):1613–1622.
102. Wang K, Li M, Hakonarson H. ANNOVAR: functional annotation of genetic variants from high-throughput sequencing data. *Nucleic Acids Res* 2010, 38(16):e164.
103. Cingolani P, Platts A, Wang le L, Coon M, Nguyen T, Wang L, Land SJ, Lu X, Ruden DM. A program for annotating and predicting the effects of single nucleotide polymorphisms, SnpEff: SNPs in the genome of Drosophila melanogaster strain w1118; iso-2; iso-3. *Fly* 2012, 6(2):80–92.
104. McLaren W, Gil L, Hunt SE, Riat HS, Ritchie GR, Thormann A, Flicek P, Cunningham F. The Ensembl Variant Effect Predictor. *Genome Biol* 2016, 17(1):122.
105. Hinrichs AS, Raney BJ, Speir ML, Rhead B, Casper J, Karolchik D, Kuhn RM, Rosenbloom KR, Zweig AS, Haussler D *et al.* UCSC Data Integrator and Variant Annotation Integrator. *Bioinformatics* 2016, 32(9):1430–1432.
106. SeattleSeq (http://snp.gs.washington.edu/SeattleSeqAnnotation138/)

11

Clinical Sequencing and Detection of Actionable Variants

Next-generation sequencing has not only altered the landscape of life science research, its impact on clinical diagnosis, prognosis, and intervention selection has also become increasingly evident. The launch of precision or personalized health initiatives worldwide is a testament to the power of NGS in improving human health, and also a key driver for integrating NGS into medical practice. From the rapid development of clinical sequencing, it is apparent that personal genome information guided medicine is the future of medical practice. Compared to research-oriented NGS, clinical sequencing is subjected to more regulations as required for other clinical tests of patient samples to ensure accurate and reliable results. In the United States, clinical sequencing is mostly regulated by the Food and Drug Administration (FDA) and Centers for Medicare & Medicaid Services through the Clinical Laboratory Improvement Amendments (CLIA). Many countries around the world and international organizations such as ISO have similar regulations.

Diagnosis, prognosis, and treatment of oncologic and pediatric diseases are two exemplary areas that have seen great benefits from clinical sequencing. As cancer is a disease of the genome, NGS is well suited to unravel tumor heterogeneity and classify tumors into different types or subtypes based on what genomic variants they possess [1]. Sequencing of various oncological gene panels, whole exome, and increasingly whole genome has become more and more commonplace, and provided much needed guidance for clinical actions. Tumor mutation burden, an overall index of the total amount of nonsynonymous mutations in a genome measured by NGS, serves as a good indicator of immunotherapy efficacy [2]. For pediatric patients, especially those in the neonatal intensive care unit (NICU), speedy diagnosis and treatment are essential, which require rapid sequencing and data processing. The use of NGS in the NICU setting is a test to its speed, accuracy, and overall utility in meeting clinical needs. The development of rapid genome sequencing pipelines, including bioinformatics, has been proven to decrease infant morbidity and at the same time lead to cost savings [3].

Often different from a research setting, the immediate goal of clinical sequencing is to identify disease-causing variant(s) based on which a treatment plan can be decided. The major challenge to achieve this goal is how to identify the causal variant(s) for the primary indication from thousands or even millions

DOI: 10.1201/9780429329180-14

of called variants. As focused on in this chapter, this requires a multi-faceted approach and corresponding tools to filter, rank, and prioritize the variants. This chapter starts with patient sample collection and sequencing approaches. After the steps on how to find actionable variants are detailed, the current variant classification system based on their pathogenicity is presented. This is followed by conduct of clinical review and generation of clinical reports. Validation of a bioinformatics pipeline for clinical sequencing is presented at the end.

11.1 Clinical Sequencing Data Generation

11.1.1 Patient Sample Collection

For the diagnosis of Mendelian disorders, i.e., those caused by inherited mutation(s) in a single gene, various types of patient samples may be used. Among the most widely used are peripheral blood (including dried blood spots), saliva, buccal swab, etc. For diseases that require identification of somatic mutations, such as solid tumors, tissue biopsies collected for pathological examinations, such as hematoxylin & eosin staining, immunohistochemistry, or fluorescent *in situ* hybridization, can be used. To detect somatic mutations, besides DNA extracted from such pathogenic tissues (or cells), matched normal control DNA is also needed. Such control DNA is often prepared from peripheral blood for most diseases, including solid tumors. For hematologic malignancies, however, control DNA may come from buccal swab, saliva, nail, hair follicle, skin biopsy, etc. In cases where normal control DNA is not available, the use of DNA from pathogenic tissues/cells alone can lead to overestimation of somatic mutations [4], and/or missing of germline mutations that may underlie genetic predisposition to other conditions.

Among the most abundant and readily available tissue biopsies are those that are chemically fixed with traditional fixatives such as formaldehyde. Such fixatives, while preserving tissue morphology for pathological examination, cause DNA damage ranging from chemical modifications (such as formation of adducts, intra- and inter-strand cross-links, as well as crosslinks with proteins), fragmentation, and strand separation [5]. This damage may introduce up to 30% artifactual mutation profiles that were not found in the original tissue [6]. Therefore, caution must be used when starting with such previously fixed archival tissues in order to minimize sequence artifacts [7]. Newer, alternative fixatives that have been shown to preserve DNA quality, protein antigenicity, and at the same time morphology [5] should be used for new collections. If archival formalin-fixed paraffin-embedded (FFPE) tissues need to be used, strategies developed to obtain the best NGS sample quality possible, including deparaffinization and FFPE-compatible nucleic acid extraction [8], can help decrease the frequency of such false-positive discoveries.

Other surgical biopsy materials, including fine needle aspiration and core needle biopsy, are also regularly used to sample discrete site(s) of diseased tissue. A potential pitfall here is that they may not be representative of the remaining, unsampled parts of the tissue. Some diseases, with cancer being the best known, are characterized by clonality and cellular heterogeneity. For these diseases, the sampling with tissue biopsies may not fully reveal the entire set of mutations. Also because of its invasiveness, obtaining tissue biopsies is not suitable for sample collection at regular intervals for the purpose of tracking disease progression and monitoring treatment outcome. Liquid biopsy represents a more recently developed sample type that is less invasive and therefore better suited for real-time, longitudinal clinical tracking and monitoring. Instead of sampling diseased tissue directly, liquid biopsy collects cells or DNA that are shed from diseased tissue into the blood or other bodily fluid (such as urine). For example, circulating tumor cells (CTCs), or circulating cell-free DNA (cfDNA), are increasingly used as input. Because the amount of cfDNA in the plasma or other bodily fluid is rather low, of which circulating tumor DNA (ctDNA) only constitutes a minor fraction (<0.1–10%), it is crucial to use a cfDNA extraction method that provides high efficiency and recovery [9, 10]. While cancer patients usually have more cfDNA than healthy individuals [11], the extraction yield is generally within the range from below 10 ng to 100 ng cfDNA per mL plasma (usually below 20 ng). The fragment size of cfDNA is mostly within the range of 160–200 bp [12]. Side-by-side comparisons have demonstrated high levels of concordance on the detection of mutations between cfDNA and matched tissue biopsies [11, 13]. Because of its low invasiveness and high accuracy, liquid biopsy has been used for multiple clinical applications, including detection of minimal residual disease for patients that are in remission, or early screening of healthy individuals before a disease manifests itself [14].

Prior to sample collection, patients need to counseled and informed consent must be obtained. Besides the affected individual (proband), often the proband's parents and/or other family members may also need to be sequenced. This is especially required to determine whether a mutation is passed on from parents or formed *de novo*, for which samples are collected from the proband and their biological parents for trio sequencing. During counseling, the purpose of performing genetic testing and the types of results anticipated are conveyed to the patient and their family members. In addition, the patient and family members are also informed of the test's limitations and potential risks. For example, the test may not reveal a genetic link, the interpretation may involve uncertainty, or the findings may be distressing instead of reassuring. Further, interpretation of the results may change over time, and the test results may have implications to other untested family members and their lives.

Post sample collection, to avoid potential sample mix-up, patient sample transfer, storage, and sequencing library preparation (detailed below) should be systematically tracked. This tracking can be achieved through performing routine sample genotyping using microarray or STR (Short Tandem Repeats) marker profiling, and subsequently comparing the genotyping result to that directly generated from the sequencing data. When proband-parent trios as well as other family members are sequenced, parentage and relatedness need to be checked using standard kinship determination methods [15].

11.1.2 Library Preparation and Sequencing Approaches

What type of NGS library to make from the DNA extracted from collected patient samples, and how to make them to best inform clinical decision making, require consideration of multiple factors. These factors include patient sample type, quality and quantity of the extracted DNA, genomic coverage needed, type(s) of variant to be detected, and limit of detection (LOD) to be achieved. For example, if liquid biopsy is used as input, only limited quantity of target DNA is available, which requires a sequencing library preparation procedure with high sensitivity. If archival FFPE tissue is used, the quality of the extracted DNA might be low and needs to be evaluated first using fragment analysis and/or quantitative PCR. If low-quality FFPE-derived DNA sample has to be used, more input DNA and/or increased PCR cycle number are usually needed for library prep. To mitigate the effects of fixative-caused DNA damage, DNA repair may be conducted prior to library making with the use of uracil-DNA glycosylase, or a mixture of multiple DNA repair enzymes.

Depending on the quantity of starting DNA, the library preparatory procedure may involve PCR amplification or be PCR-free. A PCR-free procedure can avoid amplification-caused artifacts, including PCR stochasticity, polymerase errors, and amplification biases (such as different amplification efficiencies due to variation in fragment length and GC content), but does require larger amounts of DNA to start, e.g., 1–2 μg. For lower input quantity, e.g., at the ng or even pg level, PCR amplification is needed to boost library yield. With continuous advancements in NGS library construction technology, generating enough library molecules from ultra-low amounts of DNA, such as those extracted from very limited patient biopsy materials or liquid biopsy, is no longer a major challenge. With commercial library prep kits currently available (as of 2022), libraries can be prepared from as little as 10 pg of DNA. With whole genome amplification using approaches such as multiple displacement amplification, libraries can even be prepared from a single cell that has only two copies of genomic DNA (~6–7 pg) [16].

Depending on genomic coverage needed, sequencing libraries for whole genome, exome, or selected target genes (gene panel) may be prepared. Whole genome sequencing enables generation of the most comprehensive set of patient genetic variant information, encompassing coding and non-coding single nucleotide variants, indels, copy number variants, and structural variants. Exome sequencing, in comparison, focuses on detecting SNVs and small indels in coding regions, while it also allows detection of CNVs and SVs although at reduced power. With the continuous drop in sequencing cost, there is an ongoing debate on the use of whole genome vs. exome sequencing for patient sample sequencing. While WGS generates the most comprehensive variant information for a patient, it requires significantly more data storage and processing capabilities, and a significant number of identified variants (especially those located in non-coding regions) is currently uninterpretable. Exome sequencing needs significantly less sequencing power and informatics resources, generates more interpretable SNV and indel information, but this comes at the cost of much reduced power for CNV and SV detection. Targeted gene panel sequencing uses a different approach interrogating only genes that are known to be associated with a disease of interest. By focusing on selected genes, the amount of sequencing required is further reduced and, as a result, high sequencing depth can be readily accomplished thereby leading to increased detection sensitivity and specificity. This is particularly helpful for making heterozygous calls, detecting somatic mutations at low frequency, and analyzing heteroplasmic mitochondrial DNA variants for diseases linked to dysfunctional mitochondrial metabolism. Because only disease-associated genes are analyzed, the demand on bioinformatics power is further reduced, and the result is more likely to be interpretable. Among the widely available disease-specific gene panels are those built for precision oncology, including the Memorial Sloan Kettering–Integrated Mutation Profiling of Actionable Cancer Targets (MSK-IMPACT) and the FoundationOne CDx (F1CDx) panels, both of which are approved by the FDA. These panels use hybridization probes to capture their gene targets and are designed to measure SNVs, indels, CNVs, and SVs in the genes they target, as well as microsatellite instability (MSI) and tumor mutation burden (TMB).

Targeted gene panel sequencing is particularly suitable for detecting low-frequency variant alleles for demanding applications such as cancer screening, minimal residual disease assessment, prenatal diagnostics, and infectious agent detection. To reach the low levels of LOD needed in these applications (e.g., variant allele frequency or VAF <5%), sequencing needs to be performed at greater depth than that needed to detect germline mutations, or somatic mutations from tissues enriched with pathogenic cells. For example, based on calculation using the binomial distribution, a sequencing depth of 1,650× is needed to detect mutations at ≥3% frequency with at least 30 mutation-supporting reads [17]. In comparison, for detection of germline mutations using WGS, the typical sequencing depth is 30×. To detect somatic mutations from tissues that contain at least 20% tumor cells using WGS, a depth of 100× is needed along with at least 10 variant-supporting reads, with the matched control DNA sequenced to ≥30× [18]. Sequencing depth is largely determined by the accuracy of sequencing (e.g., 0.1–1% of error rate on Illumina sequencers, refer to Chapter 4 for more details), and other molecular steps such as PCR amplification during library preparation. To reach even lower LOD (e.g., VAF <1%), strategies to improve NGS accuracy and reduce PCR errors have been deployed, which include employment of signal-to-noise correction methodologies and single-molecule consensus sequencing schemes [19]. One of such strategies is called Duplex Sequencing built on the use of molecular barcoding. Molecular barcoding utilizes the so-called unique molecular identifiers (or UMIs) to label single molecules prior to PCR amplification. From sequences generated from PCR duplicates that carry the same UMI, i.e., those derived from the same molecule, consensus sequence is reached that corrects errors introduced during amplification (except the first cycle) as well as random sequencing errors. In Duplex Sequencing, the two strands of the original DNA duplex generate two separate consensus sequences, and comparing them leads to generation of duplex consensus sequence, further removing errors introduced during the first PCR cycle. With Duplex Sequencing, somatic mutations that occur at a frequency of 10^{-5} or lower can be detected with high confidence [20]. Because it relies on single molecule consensus sequence generation through sequencing multiple amplified copies of the same original molecule in a strand-specific fashion and then later collapsing them, this technique requires a lot more reads than conventional NGS to achieve lower LOD levels.

For QC/QA during library preparation and sequencing, sample and data quality needs to be checked at multiple steps. Prior to proceeding to library preparation, input DNA quality and quantity need to be assessed against pre-defined criteria, which may vary depending on the intended detection targets of the sequencing assay. For example, for detection of structural variants, which needs long-range genomic information, FFPE samples are not recommended even with the use of remedial measures mentioned above. Once an appropriate library preparatory workflow is chosen, it needs to be standardized to ensure consistent performance, and multiple QC/QA steps should be specified to monitor results of the workflow at key junctures, eventually library yield and fragment size range. To help assess workflow performance, reference DNA samples with known variants can be used as positive control. Such reference samples include well-characterized reference DNA from the Genome in a Bottle (GIAB) consortium supported by the U.S. National Institute of Standards and Technology [21], engineered DNA that contains clinically relevant synthetic variants at pre-defined allele frequencies (available from commercial sources), and clinical samples that have been analyzed by another CLIA-accredited NGS lab. Sequencing run quality metrics, such as the percentage of reads over Q30 and overall error rate, need to be monitored and must pass a pre-defined quality threshold.

11.2 Read Mapping and Variant Calling

Identification of disease-causing variants from sequencing data is a multi-step process. Chapters 5 and 10 cover many of the upstream steps in this process, including read mapping and variant calling. As these steps also apply to clinical samples, they are not to be repeated here. In the clinical setting, however, speed and turnaround time are often of essence for clinicians to make diagnosis, prognosis, and treatment decisions. Various hardware or algorithmic implementation strategies have been devised to speed up the read mapping and variant calling process. These strategies include: (1) deployment of specialized hardware, such as field programmable gate arrays (FPGAs), as used by the DRAGEN (Dynamic Read Analysis for GENomics) platform for significantly accelerated mapping and variant calling [22]; (2) optimization for speed of highly efficient workflows such as Genalice [23] and Sentieon [24]; (3) real-time [25] or pulsed [26] read mapping and variant calling while sequencing is still underway; or (4) employment of a sequencing technology that has concurrent analysis capabilities while sequencing, like Oxford Nanopore Technology. With these strategies, it has become a reality to receive provisional diagnosis and prognosis from clinical sample collection within 20 hours [27]. After mapping of reads to the reference genome, median (or mean) depth of coverage, uniformity of coverage, and on-target capture rate (for exome and gene panel sequencing) need to be examined and compared to pre-defined QC threshold. Next the called variants are subjected to an extensive process of filtering and annotation to produce a short list of actionable variants, which as detailed next is the major focus of this chapter.

11.3 Variant Filtering

The procedure from a long list of called variants to a clinical testing report (Figure 11.1) is the major focus of this chapter. Among the large number of called variants, most are benign and do not have an impact on human health. For example, a typical WGS of an individual's germline DNA usually identifies 5 million or more variants, of which only 30,000 or more are in protein-coding regions. Of these variants, ~10,000 represent missense amino acid substitutions, aberrant splicing sites, or small indels [28]. These variants are further narrowed down to produce a short list of clinically relevant and actionable variants, to assist clinicians in disease diagnosis, prognosis, and treatment. This multi-step process requires a multitude of tools and databases to screen the called variants based on their frequency, functional consequence, known linkage to human disease(s), and match of clinical phenotype, as well

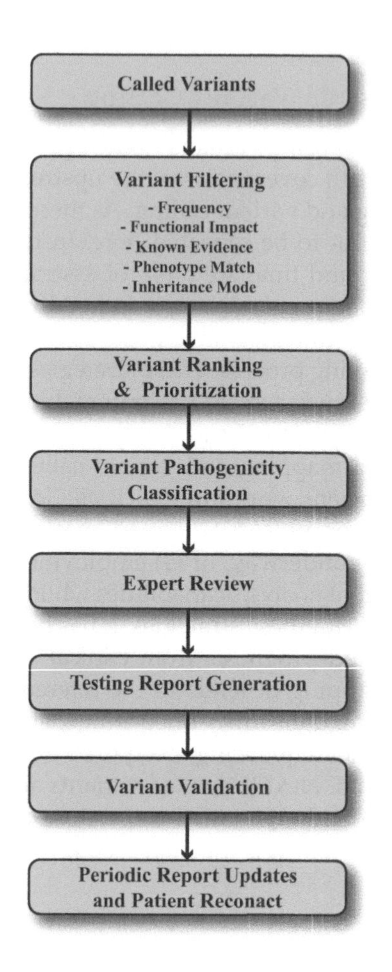

FIGURE 11.1
Clinical sequencing general data analytic workflow starting from variant calls.

as mode of inheritance. Besides these in silico variant filtering steps, wet lab strategies such as pedigree sequencing can also help to reliably and significantly reduce the number of potential candidate variants.

Presented below are major steps of the variant filtering workflow. Prior to performing these steps, preliminary variant screening is needed to filter out variants that do not pass a predefined variant call quality threshold. While the following filtering steps are usually performed on all variants that pass the threshold, these steps can also be applied to a pre-selected list of genes that are known to be associated with the phenotype/disease of the patient, for the purpose of minimizing incidental findings and reducing analytic burden. Each of the filtering steps detailed below focuses on one relevant aspect of a variant to its potential role in underlying the disease or phenotype observed in the patient. In combination, the multi-faceted filtering provided by these steps will become the basis for generating a list of top candidates for additional vetting toward clinical action.

11.3.1 Frequency of Occurrence

Many called variants are too common to be consistent with the low incidence of a genetic disease. To identify disease-causing variant(s), such common variants need to be filtered out. While the threshold for occurrence frequency can be set at different levels, an MAF (minor allele frequency) of less than 1% is often used. To determine the occurrence frequency of a called variant in the general population, large databases of human genetic variations, such as gnomAD [29], the 1000 Genomes Project (1KGP) database [30], TOPMed [31], UK10K [32], and NHLBI Exome Sequencing Project (ESP) [33], are often used (Table 11.1, Page 249). These databases contain mostly SNVs and short indels. Some of these databases, such as gnomAD and 1KGP, also contain common structural variants, but most SVs are catalogued by specific databases including the Database of Genomic Variants (DGV) [34], Database of Chromosomal Imbalance and Phenotype in Humans using Ensembl Resources (DECIPHER) [35], and dbVar [36]. It should be noted that variant allele frequency is often population specific. For example, the minor allele of an SNV in the *EML6* gene, rs17046386 (A>G), is not rare in African populations, but rare in non-African populations. Therefore, the ancestral background of the affected individual should be taken into consideration in this step.

11.3.2 Functional Consequence

Variants that change amino acid residues in the active site of a protein may significantly affect its function. Variants located at other conservative base positions, such as those that affect gene transcript splicing or gene transcription initiation, may also exert significant effects on gene product. On the other hand, functional significance of variants that fall into intergenic regions are often hard to assess. To sort variants based on their genomic locations, e.g., those in protein-coding, regulatory (e.g., intron, splicing site, 5′ or 3′ UTR, promoter, etc.), or intergenic non-coding regions, variant annotation tools such as ANNOVAR, VEP, or VariantAnnotation [37] can be used. Intergenic or non-coding variants are usually filtered out, unless they are predicted to have regulatory functions such as affecting gene splicing. To predict potential variant pathogenicity caused by altered splicing, SpliceAI [38], MaxEntScan [39], and NNSplice [40] are among the best performing methods, based on currently available benchmarking studies [41–44]. For amino acid-altering variants, a variety of tools are available to predict their potential impacts to help determine whether they should be filtered out. Such tools, based on their underlying algorithm, can be divided into three groups: function prediction, evolutionary conservation, and ensemble. Function prediction methods, typically based on biochemical properties of amino acids, include PolyPhen [45], PROVEAN [46], SIFT [47], MutationTaster [48], MutationAssessor [49], VEST [50], FATHMM [51], MutPred [52], and LRT [53]. Methods that are based on evolutionary sequence conservation include phyloP [54], phastCons [54], GERP [55], and SiPhy [56]. Those that employ the ensemble approach aggregate and integrate information about a variant from different sources for variant prioritization to help identify disease-causing genes. Examples of these methods are CADD [57], M-CAP [58], MetaRNN [59], REVEL [60], Eigen [61], and VAAST [62].

11.3.3 Existing Evidence of Relationship to Human Disease

There are multiple community efforts to catalog the relationship of human diseases to individual genes and the variants they harbor. For example, ClinVar [63], Human Gene Mutation Database (HGMD) [64], and Online Mendelian Inheritance in Man (OMIM) [65] are databases that aggregate genes/variants and their relationship to human health. Besides these comprehensive databases, there are also efforts to collect variants as they relate to particular classes of diseases. For example, COSMIC (Catalogue Of Somatic Mutations In Cancer) [66], National Cancer Institute's GDC (Genomic Data Commons) [67], and American Association for Cancer Research (AACR)'s GENIE (Genomics Evidence Neoplasia Information Exchange) [68], catalog variants that appear in various cancers. Among variants identified in a patient, if certain variant(s) have already been cataloged by a relevant database, the chance of the variant(s) bearing a relationship to the patient's disease becomes higher. Most of these databases can be queried through web interface providing user-friendliness and most up-to-date information. For efficiency and consistency, the variant-disease relationship contained in these databases can also be accessed through the use of Application Programming Interface (API), or downloaded for local deployment.

11.3.4 Clinical Phenotype Match

Identification of actionable variants starts even before sequencing, as a patient's clinical phenotype can greatly aid variant filtering. To standardize clinical phenotyping, the Human Phenotype Ontology (HPO) project provides structured vocabulary to describe phenotypic abnormalities manifested in human diseases, thereby enabling integrated analysis of semantic phenotypic information with NGS genotypic data. Once appropriate HPO terms are identified to capture the clinical phenotype of a patient, a number of tools can then be used to filter variants based on such HPO terms. These tools include Exomiser [69], eXtasy [70], Genomiser [71], Phenolyzer [53], Phenomizer [72], PhenIX [28], Phevor [73], Phen-Gen [74], VarElect [75], and DeepPVP [76].

These tools draw on current knowledge of the relationships between disease phenotypes and genetic variants to rank variants from a patient. Besides using highly specific phenotypic terms, sometimes it also helps to expand the search to include synonyms or more general terms at higher levels up the Disease Ontology (DO) hierarchy. While the phenotype – genetic variant relationship is typically extracted from human disease databases such as OMIM, Orphanet, DECIPHER, and ClinVar, they can also be extracted from databases of model organisms, such as MGI (Mouse Genome Informatics) and ZFIN (Zebrafish Information Network). Phenotypic information extracted from these databases can be translated to human diseases using tools such as PhenomeNET [77] and PHIVE (for mouse) or hiPHIVE (human/interaction PHIVE, for both mouse and zebrafish), both as parts of Exomiser [69].

11.3.5 Mode of Inheritance

Family medical history, if available, can also greatly aid the variant filtering process. For a Mendelian disorder, the five basic modes of inheritance are autosomal dominant, autosomal recessive, sex-linked dominant, sex-linked recessive, and mitochondrial. Traditional pedigree analysis can lead to revealing of the mode of inheritance of such a disorder. From variants called from a proband and their pedigree (most commonly trio sequencing), those that do not conform to the inheritance pattern are filtered out.

11.4 Variant Ranking and Prioritization

Using the multiple filters above, to eventually get down to a short list of clinically actionable variants for reporting, filtered variants need to be ranked and prioritized. To speed up this process, systems are needed to integrate the large number of tools and databases above. Publicly available open-source systems, as well as commercially developed proprietary systems, are available to meet this need. Examples of publicly available tools include VarFish [83], KGGSeq [84], MutationTaster2021 [48], VAAST Variant Prioritizer (VVP) [85], VaRank [86], Variant Ranker [87], and Variant Prioritization Ordering Tool (VPOT) [88]. Using as input VCF files and often patient's clinical phenotype, these tools offer the capabilities to annotate, filter, rank, and prioritize the imported variants. Many of the tools offer highly interactive filtering affording users the flexibility to use different permutations of the various filters on allele frequency, functional consequence, phenotypic association, existing disease connection, and inheritance pattern. Some of these tools, such as KGGSeq, MutationTaster2021, and Variant Ranker, also allow users to identify variants in genes associated with a particular biological pathway,

Gene Ontology term, gene network, or displaying a particular expression profile such as tissue specificity. To facilitate variants ranking, some of the tools, exemplified by VaRank, Variant Ranker, and VPOT, generate an aggregated score from the various scores computed by many of the tools summarized on Table 11.1. Each of these scores provides information on one aspect of a variant's potential contribution to disease development, such as functional consequence (e.g., from PolyPhen), sequence conservation (e.g., phyloP), splicing site alteration (e.g., MaxEntScan), etc. Since the contribution of each of these aspects to disease outcome may not be equal, different weights can be applied to the different scores in order to produce a single aggregated score to serve as an index of the variant's overall pathogenicity. VPOT, as an example, allows application of user defined weighting factors to each score to calculate a final, aggregated score for variant ranking and prioritization.

11.5 Classification of Variants Based on Pathogenicity

11.5.1 Classification of Germline Variants

In terms of the potential role they play in disease development, ranked variants can be anywhere in a continuous range from benign to pathogenic. To proceed to clinical reporting, they need to be classified based on existing evidence of their pathogenicity. In the United States, using the current joint guidelines from the American College of Medical Genetics and Genomics (ACMG) and the Association for Molecular Pathology (AMP), germline variants for Mendelian diseases are classified using a five-tier system. This system places variants into one of the following tiers: (1) Pathogenic: directly contributing to the development of the disease; (2) Likely Pathogenic: highly possible (with over 90% certainty) to cause the disease; (3) Uncertain Significance: currently available data and literature being inconclusive to declare the pathogenicity of the variant; (4) Likely Benign: high possible (with over 90% certainty) that the variant is not the cause of the disease; and (5) Benign: not disease causing.

To standardize classification of variants into these five tiers, specific evidence-based criteria have been established. For pathogenic variants, the evidence of pathogenicity is divided into four levels: (1) Very Strong (coded as PVS1) – these are null variants (nonsense, frameshift, canonical +/−1 or 2 splicing sites, initiation codon, mono- or multi-exon deletion, etc.) in a gene of which loss of function is established as a known mechanism of disease; (2) Strong (PS) – ranges from missense variants causing the same amino acid alteration as a pathogenic variant established earlier, to variants whose prevalence level is significantly higher in affected individuals than controls; (3) Moderate (PM) – encompasses variants from those located in a gene's hot

TABLE 11.1

Major Tools and Databases for Filtering Variants Using Various Criteria

Name	Full Name	Description	Reference
Variant Allele Frequency Databases			
1KGP	1000 Genomes Project	Contains genomes of thousands of individuals from dozens of populations, reconstructed with low-coverage WGS, deep WES, and dense array genotyping	[30]
gnomAD	Genome Aggregation Database	Aggregation of harmonized exome and genome sequencing data from a variety of large-scale sequencing projects	[29]
UK10K	UK10,000 Genomes/Exomes Project	Includes WGS data from ~4,000 healthy phenotyped individuals, and WES data from ~6,000 individuals with rare disease, severe obesity, and neurodevelopmental disorders	[32]
TOPMed	Trans-Omics for Precision Medicine (Supported by NIH/NHLBI)	WGS and other -omics data collected for heart, lung, blood, and sleep disorders from large and diverse cohorts, data available via dbGaP	[31]
ESP	Exome Sequencing Project (Supported by NIH/NHLBI)	WES data collected to discover variants associated with heart, lung, and blood diseases or traits, available via dbGaP	[33]
Predicted Functional Consequence (Missense Variants)			
PolyPhen	Polymorphism Phenotyping	Predicts possible effect of amino acid substitutions on protein function based on structural attributes and conservation profiles. Scores of 0.0–0.15 are thought as benign while 0.15–1.0 as damaging	[45]
SIFT	Sorting Intolerant From Tolerant	Classifies an amino acid coding change as tolerated or deleterious based on protein sequence conservation captured into a SIFT score (0–0.05: deleterious; 0.05–1.0: tolerated)	[47]
PROVEAN	Protein Variation Effect Analyzer	Computes a pairwise sequence alignment score to quantify the impact of a variant on protein function, with low score representing deleterious and high score neutral effect	[46]

(*continued*)

TABLE 11.1 (Continued)

Major Tools and Databases for Filtering Variants Using Various Criteria

Name	Full Name	Description	Reference
MutationTaster	NA	Deploys machine learning to predict whether a coding or non-coding variant is deleterious, in part based on evolutionary conservation, variant frequency, and patient phenotype	[48]
MutationAssessor	NA	Derives functional impact score for amino acid residue changes, based on sequence conservation at the residue positions determined by aligning families and subfamilies of sequence homologs within and between species	[49]
VEST	Variant Effect Scoring Tool	Applies a Random Forest classifier to 86 variant features to generate a pathogenicity score and an associated *p*-value to determine the confidence of calling a variant pathogenic	[50]
FATHMM	Functional Analysis Through Hidden Markov Models	Predicts functional effects of missense variants by integrating protein sequence conservation with pathogenicity weights, which represent the general tolerance to mutations of proteins and their domains	[51]
LRT	Likelihood Ratio Test	Identifies deleterious mutations using the likelihood ratio test, based on the null model that each codon evolves neutrally with no difference in the rate of missense to synonymous substitution	[53]
MutPred	NA	Quantifies pathogenicity of amino acid substitutions, and models their impacts on protein structure and function	[52]
Evolutionary Conservation			
phyloP	Phylogenetic P-value	Measures evolutionary conservation at an aligned position against a model of neutral drift. Positive *p*-values indicate conservation while negative *p*-values fast evolution	[54]
phastCons	NA	Identifies conserved sequences through fitting a phylogenetic hidden Markov model to multiple sequence alignment data	[54]
GERP	Genomic Evolutionary Rate Profiling	Uses maximum likelihood evolutionary rate estimation to identify evolutionarily constrained sequence elements through multiple alignments	[78]

SiPhy	SIte-specific PHYlogenetic analysis	Predicts constrained sequence elements using a hidden Markov model based on the pattern of base substitutions from sequence alignments	[56]
Ensemble Methods			
CADD	Combined Annotation Dependent Depletion	Integrates diverse annotations about a variant, including evolutionary constraint, functional prediction, epigenetic marking, etc., into a single score to measure deleteriousness	[57]
M-CAP	Mendelian Clinically Applicable Pathogenicity	Uses a gradient boosting tree classifier to separate pathogenic and benign variants using both existing pathogenicity scores (e.g., CADD, SIFT, and PolyPhen) and direct measures of sequence conservation	[58]
MetaRNN	NA	Employs a recurrent neural network model to integrate major existing variant annotation scores, sequence conservation, and allele frequency to generate a meta-score for variant pathogenicity	[59]
REVEL	Rare Exome Variant Ensemble Learner	Uses Random Forest to predict pathogenicity of missense variants based on features from an ensemble of existing methods	[60]
Eigen	NA	Estimates predictive accuracy of various existing functional annotations of a variant, and uses such estimates to derive an aggregate functional score	[61]
VAAST	Variant Annotation, Analysis and Search Tool	Incorporates information on amino acid substitution severity, variant frequency, and phylogenetic conservation for variant prioritization	[62]
Predicted Splicing Alterations			
SpliceAI	NA	Employs deep neural network to predict variants that cause cryptic splicing, through modeling of long-range genomic sequence information around the variant site	[38]
MaxEntScan	Maximum Entropy Scan	Generates splice prediction scores through modeling of short sequence motifs that accounts for dependencies between non-adjacent as well as adjacent positions, based on maximum entropy distribution	[39]
NNSplice	Splice Site Prediction by Neural Network	Uses neural network to predict splice site through exploiting pairwise correlations between adjacent nucleotides	[40]

(*continued*)

TABLE 11.1 (Continued)

Major Tools and Databases for Filtering Variants Using Various Criteria

Name	Full Name	Description	Reference
Databases of Variants with Known Disease Associations			
ClinVar	NA	Public database maintained by NCBI that contains both germline and somatic variants in the nuclear and mitochondrial genomes, and their relationships to human diseases and phenotypes	[79]
COSMIC	Catalogue Of Somatic Mutations In Cancer	Expert curated database on the effects of somatic mutations across human cancers	[66]
OMIM	Online Mendelian Inheritance in Man	Compendium of human genes and genetic disorders, with curated descriptions of genes, phenotypes, and their relationships in a structured free-text format	[65]
HGMD	Human Gene Mutation Database	Collates published germline mutations in nuclear genes and their relationships to inherited human diseases	[64]
DGV	Database of Genomic Variants	Provides a curated catalog of genomic structural variants in health controls	[34]
DECIPHER	Database of Chromosomal Imbalance and Phenotype in Humans Using Ensembl Resources	Repository of CNVs and other genomic variants associated with disease phenotypes, to aid in finding pathogenic variants in patients of rare genetic disorders	[80]
Phenotype Match			
Exomiser	NA	Starts with phenotypes presented in HPO terms and a VCF file from WES/WGS data, uses a Random Forest model to find likely causative variants	[81]
PHIVE	PHenotypic Interpretation of Variants in Exomes	Conducts cross-species phenotype matching and calculates phenotypic similarity between human diseases and animal models. Part of Exomiser	[69]
Phenolyzer	Phenotype Based Gene Analyzer	Prioritizes disease genes based on disease/phenotype information entered as free text	[82]

Phenomizer	NA	Provides a semantic similarity search algorithm to statistically match a patient's phenotypes to diseases. Also used by other tools such as Phen-Gen and PhenIX	[72]
Phevor	NA	Uses as input a list of variants generated from a variant ranking tool, reprioritizes the variants based on existing knowledge in multiple biomedical ontologies of the affected genes, patient phenotype, and human diseases	[73]
PhenIX	Phenotypic Interpretation of eXomes	Ranks variants in known disease genes based on similarity of patient phenotypes to phenotypes associated with the disease genes	[28]
eXtasy	NA	Uses a Random Forest classifier to rank missense variants based on patient phenotypes, predicted variant deleteriousness and haploinsufficiency	[70]
Phen-Gen	NA	Combines patient phenotypic and variant prediction data within a Bayesian framework to locate causative variants	[74]
DeepPVP	Deep PhenomeNET Variant Predictor	Identifies likely causative variants based on phenotype match using a deep neural network model	[76]

mutation spot affecting key functional domain of the coded protein, to those presumed to be *de novo*; and (4) Supporting (PP) – covers from those that show co-segregation with disease in multiple members of the affected family in a gene known to cause the disease, to those that are called pathogenic by a reputable source but with no available evidence.

For benign variants, the evidence of benign impact is divided into three levels: (1) Stand-Alone (coded as BA1) – this level contains variants with allele frequency above 5% in ESP, 1KGP, or gnomAD; (2) Strong (BS) – covers from those with allele frequency higher than expected for the disease to those lacking segregation in the affected family; and (3) Supporting (BP) – includes missense variants in a gene where truncation variants are the primary mechanism of disease to synonymous variants with no evolutionary conservation and no predicted effect on splicing.

For detailed definition and interpretation of the various levels of evidence, and the rules for classifying variants into one of the five tiers based on different combinations of these various evidence (summarized in Figure 11.2), readers can refer to the original ACMG/AMP publication ([89], particularly Tables 3, 4, and 5) and subsequent revisions. To follow the ACMG/AMP guidelines, all available evidence on the pathogenicity or benignity of a variant needs to be considered. Such evidence includes those gathered from the current case, data in public databases (such as those listed on Table 11.1) and scientific literature, and/or the clinal sequencing lab's internal data. Through reviewing the aggregated evidence and then applying the ACMG/AMP guidelines, the classification of variants can then be attained. Open-source or commercial tools, such as InterVar (Clinical Interpretation of Genetic Variants, open source) or its web version wInterVar [90], QIAGEN Clinical Insight (QCI) Interpret (commercial), may help classify variants by providing automated application of many of the criteria from the ACMG/AMP guidelines. CardioClassifier [91] and CardioVAI [92] are other examples of decision support tools specifically developed for particular diseases, as both aim to classify genes related to cardiac diseases. Prior to reporting, output from decision support tools needs to be examined, and modified if necessary, by testing lab personnel.

11.5.2 Classification of Somatic Variants

As the above ACMG/AMP five-tier system has been developed for germline mutations in Mendelian disorders to classify somatic variants in tumor samples, a different system has been proposed and recommended by AMP, American Society of Clinical Oncology (ASCO), and College of American Pathologists (CAP) [93]. This is a four-tier system with emphasis on variants' impact on clinical care including diagnosis, prognosis, therapy selection, and inclusion in clinical trials. The four tiers are: Tier I – variants of strong clinical significance, Tier II – variants of potential clinical significance, Tier III – those of unknown clinical significance, and Tier IV – benign or likely

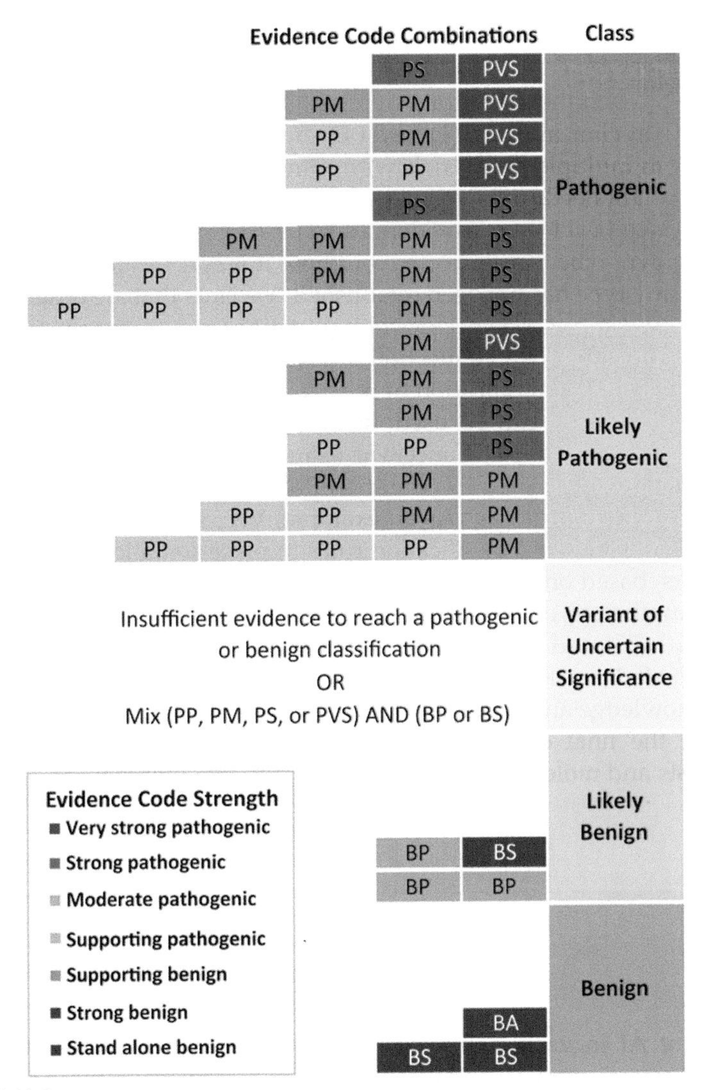

FIGURE 11.2
ACMG/AMP rules on how to classify variants based on combination of pathogenicity/benignity evidence at different levels. (From SE Brnich, EA Rivera-Munoz, JS Berg, Quantifying the potential of functional evidence to reclassify variants of uncertain significance in the categorical and Bayesian interpretation frameworks, *Human Mutation* 2018, 39(11):1531–1541. With permission.)

benign variants. To categorize into one of these four tiers, four levels of evidence are evaluated: (1) Level A – variants that can serve as biomarkers to predict response or resistance to approved therapies, or are included in professional guidelines for a specific cancer type; (2) Level B – those that can predict response or resistance to therapies based on well-powered studies with expert consensus; (3) Level C – those that can predict response or resistance to approved therapies for a different tumor type, or qualify patients to participate in clinical trials; (4) Level D – those that can predict response or resistance in multiple small trials with some consensus. To be classified into Tier I, a variant needs to have Levels A and B evidence. For Tier II, a variant needs to have Level C or D evidence. A Tier III variant is one that occurs in a gene known to be cancer related, but how it is specifically associated with cancer of any type has not been established. Variants that are observed in the general population (or a specific subpopulation) at a frequency that is high enough to preclude their connection with any cancer belong to Tier IV.

To help facilitate classification of oncogenic somatic variants, open-source or commercial tools can be used, such as VIC (Variant Interpretation for Cancer, open source) [94], CancerVar (Cancer Variants interpretation, open source) [95], Roche's NAVIFY Mutation Profiler (commercial), and Qiagen's QCI Interpret (commercial). As an example, VIC, a Java-based tool, applies the AMP/ASCO/CAP classification rules to place somatic variants into one of the tiers, based on assessment of their clinical impacts using information from a number of databases including CGI (Cancer Genome Interpreter) [96], CIViC (Clinical Interpretations of Variants in Cancer) [97], and PMKB (Precision Medicine Knowledge Base) [98]. Following this automated process, expert knowledge and patient-specific case information can be incorporated to adjust the final classification for reporting and utilization by clinical oncologists and molecular tumor board for decision making.

11.6 Clinical Review and Reporting

11.6.1 Use of Artificial Intelligence in Variant Reporting

The use of AI in variant reporting can be exemplified by an effort from Massachusetts General Hospital in which an AI system was built and validated using variants that had been prioritized and reported by expert pathologists from established patient cases [99]. To build machine learning models, this system uses a large number of features, including the many mentioned above such as variant frequency and prediction scores from CADD, SIFT, PolyPhen, LRT, MutationTaster, MutationAssessor, FATHMM, PROVEAN, VEST, phyloP, phastCons, GERP, SiPhy, etc. Machine learning algorithms used to build these models included logistic regression and random forests among others. This decision support tool generates a contiguous prediction score between 0 (not reportable) to 1 (reportable) to help decide whether a variant should be considered for reporting with human interpretable rationale. Other commercially available AI-based decision support tools, such as MOON [27], are also available to assist with the variant reporting process. While they cannot entirely replace clinical geneticists and pathologists, these AI-based decision support tools can help filter out most non-reportable variants to allow experts to focus on a more manageable number of potentially reportable variants, making this key analytic process more efficient.

11.6.2 Expert Review

While the variant reporting process can be automated with promising results using the tools introduced above, review of reportable pathogenic or likely pathogenic germline variants, or Tier I or II somatic variants, by molecular pathologists and oncologists is still needed to examine the entire evidence matrix for their pathogenic/oncogenic role prior to reporting. Current medical literature and database entries should be checked for their links with known diseases or implicated biological pathways. This step should also be contextualized in consideration of the patient's phenotype and family history. Such expert manual review is needed to avoid non-pathogenic/non-oncogenic variants from being reported as false positives.

11.6.3 Generation of Testing Report

After expert review, a final clinical report needs to be prepared by the testing lab to convey the results to clinicians for decision making. While there is no standard format for such a report, as a general rule the report should be designed toward helping with bedside decision making, and therefore needs to be concise, informative, actionable, and easy to understand to non-geneticists (Figure 11.3 shows an example). At the beginning of the report, an executive summary of major findings should be provided succinctly, followed by more detailed interpretations of the variants. Variants in the major findings may be limited only to those that are classified as pathogenic or likely pathogenic, but some labs may opt to also include variants of unknown significance. Typically variants that are benign or likely benign are not reported. It is possible to report a negative finding, i.e., no genetic cause of the patient's phenotype is identified. On the report the variants and the genes harboring them should be described using the nomenclature from Human Genome Variation Society (HGVS) and Human Genome Organization Gene Nomenclature Committee (HGNC) [100, 101]. Each variant should be presented with a reference sequence, a letter prefix (such as "c" for a DNA coding reference sequence, "p" for a protein reference, or "r" for an RNA transcript sequence), description of the sequence change at a DNA/RNA/protein sequence location, zygosity, medical condition, inheritance pattern, and classification (e.g., *ABCA4*, NM_000350.2, c.5882G>A, p.Gly1961Glu, heterozygous, age-related macular degeneration, autosomal dominant, likely pathogenic). A detailed description for each of the identified variants may include evidence level(s)

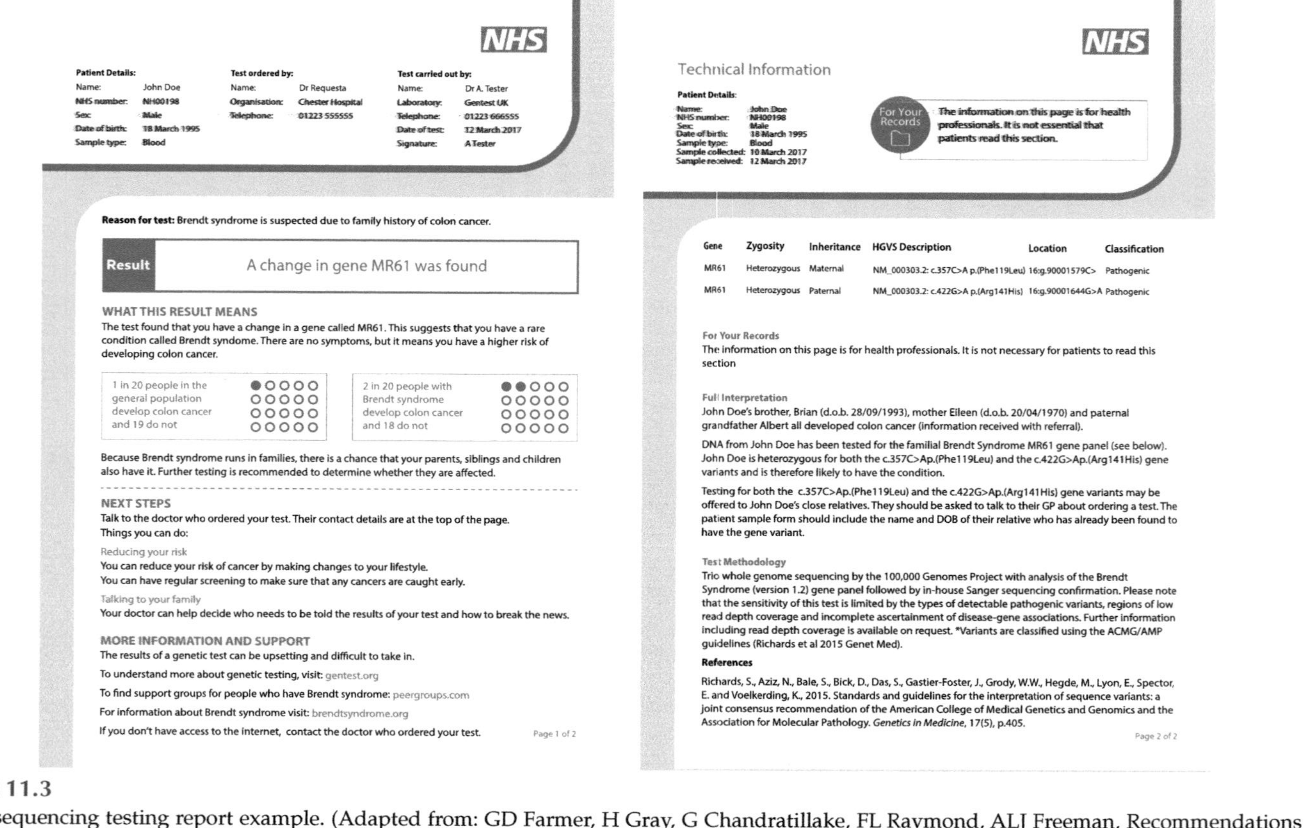

NHS

Patient Details:
Name: John Doe
NHS number: NH00198
Sex: Male
Date of birth: 18 March 1995
Sample type: Blood

Test ordered by:
Name: Dr Requesta
Organisation: Chester Hospital
Telephone: 01223 555555

Test carried out by:
Name: Dr A. Tester
Laboratory: Gentest UK
Telephone: 01223 666555
Date of test: 12 March 2017
Signature: A Tester

Reason for test: Brendt syndrome is suspected due to family history of colon cancer.

Result A change in gene MR61 was found

WHAT THIS RESULT MEANS
The test found that you have a change in a gene called MR61. This suggests that you have a rare condition called Brendt syndome. There are no symptoms, but it means you have a higher risk of developing colon cancer.

1 in 20 people in the general population develop colon cancer and 19 do not

2 in 20 people with Brendt syndrome develop colon cancer and 18 do not

Because Brendt syndrome runs in families, there is a chance that your parents, siblings and children also have it. Further testing is recommended to determine whether they are affected.

NEXT STEPS
Talk to the doctor who ordered your test. Their contact details are at the top of the page.
Things you can do:

Reducing your risk
You can reduce your risk of cancer by making changes to your lifestyle.
You can have regular screening to make sure that any cancers are caught early.

Talking to your family
Your doctor can help decide who needs to be told the results of your test and how to break the news.

MORE INFORMATION AND SUPPORT
The results of a genetic test can be upsetting and difficult to take in.

To understand more about genetic testing, visit: gentest.org

To find support groups for people who have Brendt syndrome: peergroups.com

For information about Brendt syndrome visit: brendtsyndrome.org

If you don't have access to the internet, contact the doctor who ordered your test.

Page 1 of 2

Technical Information

NHS

Patient Details:
Name: John Doe
NHS number: NH00198
Sex: Male
Date of birth: 18 March 1995
Sample type: Blood
Sample collected: 10 March 2017
Sample received: 12 March 2017

For Your Records

The information on this page is for health professionals. It is not essential that patients read this section.

Gene	Zygosity	Inheritance	HGVS Description	Location	Classification
MR61	Heterozygous	Maternal	NM_000303.2: c.357C>A p.(Phe119Leu)	16:g.90001579C>	Pathogenic
MR61	Heterozygous	Paternal	NM_000303.2: c.422G>A p.(Arg141His)	16:g.90001644G>A	Pathogenic

For Your Records
The information on this page is for health professionals. It is not necessary for patients to read this section

Full Interpretation
John Doe's brother, Brian (d.o.b. 28/09/1993), mother Eileen (d.o.b. 20/04/1970) and paternal grandfather Albert all developed colon cancer (information received with referral).

DNA from John Doe has been tested for the familial Brendt Syndrome MR61 gene panel (see below). John Doe is heterozygous for both the c.357C>Ap.(Phe119Leu) and the c.422G>Ap.(Arg141His) gene variants and is therefore likely to have the condition.

Testing for both the c.357C>Ap.(Phe119Leu) and the c.422G>Ap.(Arg141His) gene variants may be offered to John Doe's close relatives. They should be asked to talk to their GP about ordering a test. The patient sample form should include the name and DOB of their relative who has already been found to have the gene variant.

Test Methodology
Trio whole genome sequencing by the 100,000 Genomes Project with analysis of the Brendt Syndrome (version 1.2) gene panel followed by in-house Sanger sequencing confirmation. Please note that the sensitivity of this test is limited by the types of detectable pathogenic variants, regions of low read depth coverage and incomplete ascertainment of disease-gene associations. Further information including read depth coverage is available on request. *Variants are classified using the ACMG/AMP guidelines (Richards et al 2015 Genet Med).

References

Richards, S., Aziz, N., Bale, S., Bick, D., Das, S., Gastier-Foster, J., Grody, W.W., Hegde, M., Lyon, E., Spector, E. and Voelkerding, K., 2015. Standards and guidelines for the interpretation of sequence variants: a joint consensus recommendation of the American College of Medical Genetics and Genomics and the Association for Molecular Pathology. *Genetics in Medicine*, 17(5), p.405.

Page 2 of 2

FIGURE 11.3
Clinical sequencing testing report example. (Adapted from: GD Farmer, H Gray, G Chandratillake, FL Raymond, ALJ Freeman, Recommendations for designing genetic test reports to be understood by patients and non-specialists, *European Journal of Human Genetics* 2020, 28(7):885. Used under the terms of the Creative Commons Attribution 4.0 International License, https://creativecommons.org/licenses/by/4.0, ©2020 Farmer et al.)

for the classification of the variant, literature summary of disease/phenotype connection, and other key information supporting the variant's pathogenicity. This can be followed by recommendations for further actions, such as strategies for intervention and/or patient management including prognosis and diagnosis. Such recommendations need to be based on evidence, and considered in the context of the patient's specific condition. If a clinical trial is available, e.g., for a cancer patient carrying a Tier I somatic variant, this information should also be included in the report.

Other sections of the clinical report should also contain technical information about the test, including what gene or genome content was assayed, how the test was performed, as well as other technological factors such as the genome build used for analysis that may have an impact on test result generation and interpretation. Such factors include limit of detection, minimum sequencing coverage, actual test performance metrics, etc. Limitations of the test, such as low-coverage and non-targeted genomic regions, should also be made clear. To provide further reference, literature citations or links to database sites should be provided toward the end. Additional clinically relevant and actionable information may also be provided from resources such as GeneReviews, which provides a quick reference to clinicians on hereditary disorders. For patients and their family members, a section that cites easy-to-understand genetic health information from additional resources, such as MedlinePlus Genetics, may also be helpful.

11.6.4 Variant Validation

While traditionally germline SNVs or small indels on a clinical report need to be validated using an orthogonal technology, with Sanger sequencing being the golden standard, a number of studies have shown that this is not always necessary and NGS is often more accurate than Sanger sequencing [102, 103]. CNVs are typically validated using orthogonal techniques such as multiplex ligation-dependent probe amplification (MLPA) or comparative genomic hybridization arrays (aCGH), but validation using an NGS CNV pipeline has been reported [104]. Somatic variants detected in cancerous samples, on the other hand, should be validated using an orthogonal method, including Sanger sequencing, digital PCR, or pyrosequencing, with the latter two especially suited for those detected at low frequencies. It should be noted again that for both germline and somatic variants, the preliminary variant screening step mentioned earlier to filter out low-quality variants is an important step to minimize the rate of false-positive variants. In addition, to ensure call accuracy of the variants on the final report, their raw reads should be manually examined by visually checking pileup of reads that align to their genomic locations using a visualization tool such as Integrative Genomics Viewer (IGV). This simple step can help further reduce false positives caused by factors such as insufficient sequencing coverage. The use of these QC measures, as well as sound medical judgment and good clinical practice, are fundamental to generating a report that captures representative variants in the patient. How the reported variants are validated or confirmed should also be included in the report.

11.6.5 Incorporation into a Patient's Electronic Health Record

A patient clinical sequencing report is increasingly incorporated into the patient's EHR. While the report can be entered into EHR as an unstructured PDF, for better integration with the rest of EHR, new data standards and interfaces need to be developed. Currently, Health Level Seven International (HL7) and Global Alliance for Genomics and Health (GA4GH) are developing new data models and data exchange standards to facilitate collection, coding, and retrieval of clinical genomics data. HL7, for example, establishes standards for EHR data exchange, integration, and retrieval that are widely accepted and used by clinical computer systems in hospitals and other health care-related organizations. Fast Healthcare Interoperability Resources (FHIR) Genomics develops and provides implementation of the HL7 data standards for clinical genomics data. Incorporation of patient sequencing variant report into EHR using such standards provides future-proof clinical decision support and analytics.

11.6.6 Reporting of Secondary Findings

When performing WGS/WES, besides reporting variants related to the indication that leads to the ordering of clinical sequencing, ACMG recommends to also report Pathogenic or Likely Pathogenic variants in a short list of medically significant but actionable genes. The genes on this list are known to lead to disease phenotypes, such as cancer and cardiovascular phenotypes, that have severe medical implications and high lifetime penetrance. The goal of reporting such variants, termed secondary finding (previously called incidental finding, or unsolicited finding), is to help improve health by reducing morbidity and mortality through taking appropriate actions. The current list (ACMG SF v3.0) [105] contains 73 genes, but the number is bound to change with the addition of new genes or removal of existing genes. Based on the current ACMG recommendations [106], patients and family members should be consented on receiving secondary findings with the option to opt out.

11.6.7 Patient Counseling and Periodic Report Updates

To provide interpretation to the patient (and their family members) on primary as well as secondary findings, and help them understand the implications of such findings, genetic counseling is needed. On primary findings, the role of genetic counselors is to help patients and families better understand the testing results and treatment options, as well as limitations of the test. On secondary findings, the counseling includes providing interpretation on the probability of developing the indicated disorder(s), strategies to prevent or manage the indicated risk(s), as well as support on emotional concerns and referral to other health providers and support groups. In addition to post-test counseling, genetic counselors also provide pre-testing counseling for individuals who have symptoms of a particular genetic disorder, are likely to be affected by a genetic disorder because of family history, or are concerned about transmitting a condition to the next generation for the purpose of family planning.

Variant classification is a dynamic process. With accumulation of new evidence for gene and variant pathogenicity over time, it frequently becomes necessary to reclassify variants on the original report, for example, from Unknown Significance to Likely Pathogenic or Likely Benign, or less frequently from Likely Pathogenic to Benign, etc. In addition, new patient phenotypic data (e.g., new symptoms) and/or new family history information, or reports of other new cases that share similar symptoms or variants, may also prompt reanalysis of the original sequencing data. Furthermore, advancement in bioinformatics tool development provides another impetus for reanalysis. Besides improving variant classification, newer tools may also lead to identification of variants that were not found initially. Because of such advances in knowledge generation, patient phenotyping, and tool development, it is beneficial to periodically update the original report in order to provide new guidance that would impact patient clinical management. Performance of reclassification of variants on the original report and/or reanalysis of the original data to provide periodic updates, however, requires time and resources from testing labs, clinicians, and genetic counselors. Such longevity report updates may also have long-term implications, including elicitation of psychological distress to patients and their families, as a result some may prefer not to be recontacted. While the ethical and legal implications of reinterpretation/reanalysis/recontact are still being debated [107, 108], current guidelines from professional organizations such as ACMG focus on maximizing clinical impact of providing such updates while minimizing the burden on the test lab and health care system [109].

11.7 Bioinformatics Pipeline Validation

Besides wet lab workflow, a dry lab bioinformatics pipeline must also be rigorously tested and validated before deployment. This is to meet accreditation requirements by professional organizations such as CAP, and more importantly, to ensure that all variables that might affect specificity, sensitivity, and precision/reproducibility of a test are well defined and valid. On bioinformatics pipeline validation, AMP and CAP have published detailed standards and guidelines [110], which are summarized below. Validation is needed not only for labs that develop their own custom steps, but also for those that opt to use a commercially available platform. In the latter case, although the service provider may offer a pre-validated platform, the testing lab still needs to go through familiarization and optimization, and validate that the platform meets the designed analytical goals of the test. To help validate a new or existing pipeline and evaluate its performance, the reference standard samples used for validating the lab workflow (see Section 11.1.2), including the GIAB reference DNA and bio-engineered DNA that contains synthetic variants at predefined frequencies, are equally valuable here since they provide the ground truth. In addition, bioinformatically generated reference dataset using programs such as BAMSurgeon [111] may also be used.

For the validation, all analytic steps in the pipeline need to be clearly defined with required hardware and software specified for each step. Such specifics include hardware configuration, operating system, name and version number of software and their dependencies, data storage and transmission system, and network connection protocol. In addition, other analytic details such as parameters used in each software, reference genome used for alignment, and databases accessed for annotation and filtering, should also be specified. Any sequencing reads altering operations, such as trimming, should be fully evaluated to determine whether they are appropriate, or need to be revised or dropped. Quality metrics for each of the analytic steps, such as mean reads on-target coverage and percentage of target genomic regions with coverage over a threshold (for reads alignment), and depth of coverage for each called variant (variant calling), should be compared to pre-defined performance criteria. Applied variant filters should be evaluated carefully to make sure that true positives are not filtered out.

If the pipeline uses internally developed software tools and scripts, the computer code should be deposited in a source code repository, such as GitHub, BitBucket, or SourceForge. If using externally developed software, the source code should also be documented if accessible. Also as part of the validation process, the strategies established to back up data and maintain the integrity of raw and analyzed files during transfer should be evaluated. On the issue of legal compliance, the pipeline should follow all applicable laws at the national and local levels. In the United States, the applicable laws include the Health Insurance Portability and Accountability Act (HIPAA) and other national/state/local laws that pertain to clinical genetic testing. According to these laws, patient genetic information needs to be protected like other patient information including patient identity and other health records, and should be secured throughout the analytic process. When using a commercial system, it is the responsibility of the testing lab to verify such compliance issues. To avoid accidental mixing of patient data, based on the AMP/CAP guidelines the identity of a patient sample must be preserved throughout the analytic process using at least four unique identifiers to encompass sample, patient, run, and test location [110]. As all analytic pipelines have limitations, such limitations for a validated pipeline should be clearly documented and reported. After a pipeline is validated, if there is any update or change applied to any part of the pipeline, supplemental or new validation is needed depending on the scope of the update/change.

References

1. Patel LR, Nykter M, Chen K, Zhang W. Cancer genome sequencing: understanding malignancy as a disease of the genome, its conformation, and its evolution. *Cancer letters* 2013, 340(2):152–160.
2. Steuer CE, Ramalingam SS. Tumor mutation burden: leading immunotherapy to the era of precision medicine? *J Clin Oncol* 2018, 36(7):631–632.
3. Farnaes L, Hildreth A, Sweeney NM, Clark MM, Chowdhury S, Nahas S, Cakici JA, Benson W, Kaplan RH, Kronick R *et al.* Rapid whole-genome sequencing decreases infant morbidity and cost of hospitalization. *NPJ Genom Med* 2018, 3:10.
4. Jones S, Anagnostou V, Lytle K, Parpart-Li S, Nesselbush M, Riley DR, Shukla M, Chesnick B, Kadan M, Papp E *et al.* Personalized genomic analyses for cancer mutation discovery and interpretation. *Sci Transl Med* 2015, 7(283):283ra253.
5. Srinivasan M, Sedmak D, Jewell S. Effect of fixatives and tissue processing on the content and integrity of nucleic acids. *Am J Pathol* 2002, 161(6):1961–1971.
6. Hedegaard J, Thorsen K, Lund MK, Hein AM, Hamilton-Dutoit SJ, Vang S, Nordentoft I, Birkenkamp-Demtroder K, Kruhoffer M, Hager H *et al.* Next-generation sequencing of RNA and DNA isolated from paired fresh-frozen and formalin-fixed paraffin-embedded samples of human cancer and normal tissue. *PLoS One* 2014, 9(5):e98187.
7. Do H, Dobrovic A. Sequence artifacts in DNA from formalin-fixed tissues: causes and strategies for minimization. *Clin Chem* 2015, 61(1):64–71.
8. McDonough SJ, Bhagwate A, Sun ZF, Wang C, Zschunke M, Gorman JA, Kopp KJ, Cunningham JM. Use of FFPE-derived DNA in next generation sequencing: DNA extraction methods. *Plos One* 2019, 14(4).
9. Oreskovic A, Brault ND, Panpradist N, Lai JJ, Lutz BR. Analytical comparison of methods for extraction of short cell-free DNA from urine. *J Mol Diagn* 2019, 21(6):1067–1078.
10. Diefenbach RJ, Lee JH, Kefford RF, Rizos H. Evaluation of commercial kits for purification of circulating free DNA. *Cancer Genet* 2018, 228–229:21–27.
11. Alborelli I, Generali D, Jermann P, Cappelletti MR, Ferrero G, Scaggiante B, Bortul M, Zanconati F, Nicolet S, Haegele J *et al.* Cell-free DNA analysis in healthy individuals by next-generation sequencing: a proof of concept and technical validation study. *Cell Death Dis* 2019, 10(7):534.
12. Jiang P, Chan CW, Chan KC, Cheng SH, Wong J, Wong VW, Wong GL, Chan SL, Mok TS, Chan HL *et al.* Lengthening and shortening of plasma DNA in hepatocellular carcinoma patients. *Proc Natl Acad Sci U S A* 2015, 112(11):E1317–1325.

13. Wyatt AW, Annala M, Aggarwal R, Beja K, Feng F, Youngren J, Foye A, Lloyd P, Nykter M, Beer TM *et al.* Concordance of circulating tumor DNA and matched metastatic tissue biopsy in prostate cancer. *J Natl Cancer Inst* 2017, 109(12).
14. Chen M, Zhao H. Next-generation sequencing in liquid biopsy: cancer screening and early detection. *Hum Genomics* 2019, 13(1):34.
15. Jones AG, Small CM, Paczolt KA, Ratterman NL. A practical guide to methods of parentage analysis. *Mol Ecol Resour* 2010, 10(1):6–30.
16. Zhang L, Dong X, Lee M, Maslov AY, Wang T, Vijg J. Single-cell whole-genome sequencing reveals the functional landscape of somatic mutations in B lymphocytes across the human lifespan. *Proc Natl Acad Sci U S A* 2019, 116(18):9014–9019.
17. Petrackova A, Vasinek M, Sedlarikova L, Dyskova T, Schneiderova P, Novosad T, Papajik T, Kriegova E. Standardization of sequencing coverage depth in NGS: recommendation for detection of clonal and subclonal mutations in cancer diagnostics. *Front Oncol* 2019, 9:851.
18. Meggendorfer M, Jobanputra V, Wrzeszczynski KO, Roepman P, de Bruijn E, Cuppen E, Buttner R, Caldas C, Grimmond S, Mullighan CG *et al.* Analytical demands to use whole-genome sequencing in precision oncology. *Semin Cancer Biol* 2021, 84:16–22.
19. Salk JJ, Schmitt MW, Loeb LA. Enhancing the accuracy of next-generation sequencing for detecting rare and subclonal mutations. *Nat Rev Genet* 2018, 19(5):269–285.
20. Schmitt MW, Kennedy SR, Salk JJ, Fox EJ, Hiatt JB, Loeb LA. Detection of ultra-rare mutations by next-generation sequencing. *Proc Natl Acad Sci U S A* 2012, 109(36):14508–14513.
21. Zook JM, McDaniel J, Olson ND, Wagner J, Parikh H, Heaton H, Irvine SA, Trigg L, Truty R, McLean CY *et al.* An open resource for accurately benchmarking small variant and reference calls. *Nat Biotechnol* 2019, 37(5):561–566.
22. Miller NA, Farrow EG, Gibson M, Willig LK, Twist G, Yoo B, Marrs T, Corder S, Krivohlavek L, Walter A *et al.* A 26-hour system of highly sensitive whole genome sequencing for emergency management of genetic diseases. *Genome Med* 2015, 7(1):100.
23. Mestek-Boukhibar L, Clement E, Jones WD, Drury S, Ocaka L, Gagunashvili A, Le Quesne Stabej P, Bacchelli C, Jani N, Rahman S *et al.* Rapid Paediatric Sequencing (RaPS): comprehensive real-life workflow for rapid diagnosis of critically ill children. *J Med Genet* 2018, 55(11):721–728.
24. Kendig KI, Baheti S, Bockol MA, Drucker TM, Hart SN, Heldenbrand JR, Hernaez M, Hudson ME, Kalmbach MT, Klee EW *et al.* Sentieon DNASeq variant calling workflow demonstrates strong computational performance and accuracy. *Front Genet* 2019, 10:736.
25. Loka TP, Tausch SH, Renard BY. Reliable variant calling during runtime of Illumina sequencing. *Sci Rep* 2019, 9(1):16502.
26. Stranneheim H, Engvall M, Naess K, Lesko N, Larsson P, Dahlberg M, Andeer R, Wredenberg A, Freyer C, Barbaro M *et al.* Rapid pulsed whole genome sequencing for comprehensive acute diagnostics of inborn errors of metabolism. *BMC Genomics* 2014, 15:1090.
27. Clark MM, Hildreth A, Batalov S, Ding Y, Chowdhury S, Watkins K, Ellsworth K, Camp B, Kint CI, Yacoubian C *et al.* Diagnosis of genetic diseases in seriously

ill children by rapid whole-genome sequencing and automated phenotyping and interpretation. *Sci Transl Med* 2019, 11(489):eaat6177.

28. Zemojtel T, Kohler S, Mackenroth L, Jager M, Hecht J, Krawitz P, Graul-Neumann L, Doelken S, Ehmke N, Spielmann M *et al.* Effective diagnosis of genetic disease by computational phenotype analysis of the disease-associated genome. *Sci Transl Med* 2014, 6(252):252ra123.
29. Karczewski KJ, Francioli LC, Tiao G, Cummings BB, Alföldi J, Wang Q, Collins RL, Laricchia KM, Ganna A, Birnbaum DP *et al.* The mutational constraint spectrum quantified from variation in 141,456 humans. *Nature* 2020, 581(7809):434–443.
30. Siva N. 1000 Genomes project. *Nat Biotechnol* 2008, 26(3):256.
31. Taliun D, Harris DN, Kessler MD, Carlson J, Szpiech ZA, Torres R, Taliun SAG, Corvelo A, Gogarten SM, Kang HM *et al.* Sequencing of 53,831 diverse genomes from the NHLBI TOPMed Program. *Nature* 2021, 590(7845):290–299.
32. Consortium UK, Walter K, Min JL, Huang J, Crooks L, Memari Y, McCarthy S, Perry JR, Xu C, Futema M *et al.* The UK10K project identifies rare variants in health and disease. *Nature* 2015, 526(7571):82–90.
33. Exome Variant Server, NHLBI GO Exome Sequencing Project (ESP) (http://evs.gs.washington.edu/EVS/)
34. MacDonald JR, Ziman R, Yuen RK, Feuk L, Scherer SW. The Database of Genomic Variants: a curated collection of structural variation in the human genome. *Nucleic Acids Res* 2014, 42(Database issue):D986–992.
35. Firth HV, Richards SM, Bevan AP, Clayton S, Corpas M, Rajan D, Van Vooren S, Moreau Y, Pettett RM, Carter NP. DECIPHER: Database of Chromosomal Imbalance and Phenotype in Humans Using Ensembl Resources. *Am J Hum Genet* 2009, 84(4):524–533.
36. Lappalainen I, Lopez J, Skipper L, Hefferon T, Spalding JD, Garner J, Chen C, Maguire M, Corbett M, Zhou G *et al.* DbVar and DGVa: public archives for genomic structural variation. *Nucleic Acids Res* 2013, 41(Database issue):D936–941.
37. Obenchain V, Lawrence M, Carey V, Gogarten S, Shannon P, Morgan M. VariantAnnotation: a Bioconductor package for exploration and annotation of genetic variants. *Bioinformatics* 2014, 30(14):2076–2078.
38. Jaganathan K, Kyriazopoulou Panagiotopoulou S, McRae JF, Darbandi SF, Knowles D, Li YI, Kosmicki JA, Arbelaez J, Cui W, Schwartz GB *et al.* Predicting splicing from primary sequence with deep learning. *Cell* 2019, 176(3):535–548 e524.
39. Yeo G, Burge CB. Maximum entropy modeling of short sequence motifs with applications to RNA splicing signals. *J Comput Biol* 2004, 11(2–3):377–394.
40. Reese MG, Eeckman FH, Kulp D, Haussler D. Improved splice site detection in Genie. *J Comput Biol* 1997, 4(3):311–323.
41. Jian X, Boerwinkle E, Liu X. In silico prediction of splice-altering single nucleotide variants in the human genome. *Nucleic Acids Res* 2014, 42(22):13534–13544.
42. Wai HA, Lord J, Lyon M, Gunning A, Kelly H, Cibin P, Seaby EG, Spiers-Fitzgerald K, Lye J, Ellard S *et al.* Blood RNA analysis can increase clinical diagnostic rate and resolve variants of uncertain significance. *Genet Med* 2020, 22(6):1005–1014.
43. Riepe TV, Khan M, Roosing S, Cremers FPM, t Hoen PAC. Benchmarking deep learning splice prediction tools using functional splice assays. *Hum Mutat* 2021, 42(7):799–810.

44. Rentzsch P, Schubach M, Shendure J, Kircher M. CADD-Splice-improving genome-wide variant effect prediction using deep learning-derived splice scores. *Genome Med* 2021, 13(1):31.
45. Adzhubei IA, Schmidt S, Peshkin L, Ramensky VE, Gerasimova A, Bork P, Kondrashov AS, Sunyaev SR. A method and server for predicting damaging missense mutations. *Nat Methods* 2010, 7(4):248–249.
46. Choi Y, Chan AP. PROVEAN web server: a tool to predict the functional effect of amino acid substitutions and indels. *Bioinformatics* 2015, 31(16):2745–2747.
47. Vaser R, Adusumalli S, Leng SN, Sikic M, Ng PC. SIFT missense predictions for genomes. *Nat Protoc* 2016, 11(1):1–9.
48. Steinhaus R, Proft S, Schuelke M, Cooper DN, Schwarz JM, Seelow D. MutationTaster2021. *Nucleic Acids Res* 2021, 49(W1):W446–W451.
49. Reva B, Antipin Y, Sander C. Predicting the functional impact of protein mutations: application to cancer genomics. *Nucleic Acids Res* 2011, 39(17):e118.
50. Carter H, Douville C, Stenson PD, Cooper DN, Karchin R. Identifying Mendelian disease genes with the variant effect scoring tool. *BMC Genomics* 2013, 14 Suppl 3:S3.
51. Shihab HA, Gough J, Cooper DN, Stenson PD, Barker GL, Edwards KJ, Day IN, Gaunt TR. Predicting the functional, molecular, and phenotypic consequences of amino acid substitutions using hidden Markov models. *Hum Mutat* 2013, 34(1):57–65.
52. Pejaver V, Urresti J, Lugo-Martinez J, Pagel KA, Lin GN, Nam HJ, Mort M, Cooper DN, Sebat J, Iakoucheva LM *et al.* Inferring the molecular and phenotypic impact of amino acid variants with MutPred2. *Nat Commun* 2020, 11(1):5918.
53. Chun S, Fay JC. Identification of deleterious mutations within three human genomes. *Genome Res* 2009, 19(9):1553–1561.
54. Siepel A, Bejerano G, Pedersen JS, Hinrichs AS, Hou M, Rosenbloom K, Clawson H, Spieth J, Hillier LW, Richards S *et al.* Evolutionarily conserved elements in vertebrate, insect, worm, and yeast genomes. *Genome Res* 2005, 15(8):1034–1050.
55. Cooper GM, Stone EA, Asimenos G, Program NCS, Green ED, Batzoglou S, Sidow A. Distribution and intensity of constraint in mammalian genomic sequence. *Genome Res* 2005, 15(7):901–913.
56. Garber M, Guttman M, Clamp M, Zody MC, Friedman N, Xie X. Identifying novel constrained elements by exploiting biased substitution patterns. *Bioinformatics* 2009, 25(12):i54–62.
57. Rentzsch P, Witten D, Cooper GM, Shendure J, Kircher M. CADD: predicting the deleteriousness of variants throughout the human genome. *Nucleic Acids Res* 2019, 47(D1):D886–D894.
58. Jagadeesh KA, Wenger AM, Berger MJ, Guturu H, Stenson PD, Cooper DN, Bernstein JA, Bejerano G. M-CAP eliminates a majority of variants of uncertain significance in clinical exomes at high sensitivity. *Nat Genet* 2016, 48(12):1581–1586.
59. Li C, Zhi D, Wang K, Liu X. MetaRNN: Differentiating Rare Pathogenic and Rare Benign Missense SNVs and InDels using deep learning. *bioRxiv* 2021.
60. Ioannidis NM, Rothstein JH, Pejaver V, Middha S, McDonnell SK, Baheti S, Musolf A, Li Q, Holzinger E, Karyadi D *et al.* REVEL: an ensemble method for

predicting the pathogenicity of rare missense variants. *Am J Hum Genet* 2016, 99(4):877–885.
61. Ionita-Laza I, McCallum K, Xu B, Buxbaum JD. A spectral approach integrating functional genomic annotations for coding and noncoding variants. *Nat Genet* 2016, 48(2):214–220.
62. Hu H, Huff CD, Moore B, Flygare S, Reese MG, Yandell M. VAAST 2.0: improved variant classification and disease-gene identification using a conservation-controlled amino acid substitution matrix. *Genet Epidemiol* 2013, 37(6):622–634.
63. Landrum MJ, Lee JM, Benson M, Brown GR, Chao C, Chitipiralla S, Gu B, Hart J, Hoffman D, Jang W *et al.* ClinVar: improving access to variant interpretations and supporting evidence. *Nucleic Acids Res* 2018, 46(D1):D1062–D1067.
64. Stenson PD, Ball EV, Mort M, Phillips AD, Shaw K, Cooper DN. The Human Gene Mutation Database (HGMD) and its exploitation in the fields of personalized genomics and molecular evolution. *Curr Protoc Bioinformatics* 2012, Chapter 1:1.13.1-1.13.20.
65. Online Mendelian Inheritance in Man, OMIM (https://omim.org/)
66. Tate JG, Bamford S, Jubb HC, Sondka Z, Beare DM, Bindal N, Boutselakis H, Cole CG, Creatore C, Dawson E *et al.* COSMIC: the Catalogue Of Somatic Mutations In Cancer. *Nucleic Acids Res* 2019, 47(D1):D941–D947.
67. Grossman RL, Heath AP, Ferretti V, Varmus HE, Lowy DR, Kibbe WA, Staudt LM. Toward a Shared Vision for Cancer Genomic Data. *N Engl J Med* 2016, 375(12):1109–1112.
68. Consortium APG. AACR Project GENIE: powering precision medicine through an International Consortium. *Cancer Discov* 2017, 7(8):818–831.
69. Robinson PN, Kohler S, Oellrich A, Sanger Mouse Genetics P, Wang K, Mungall CJ, Lewis SE, Washington N, Bauer S, Seelow D *et al.* Improved exome prioritization of disease genes through cross-species phenotype comparison. *Genome Res* 2014, 24(2):340–348.
70. Sifrim A, Popovic D, Tranchevent LC, Ardeshirdavani A, Sakai R, Konings P, Vermeesch JR, Aerts J, De Moor B, Moreau Y. eXtasy: variant prioritization by genomic data fusion. *Nat Methods* 2013, 10(11):1083–1084.
71. Smedley D, Schubach M, Jacobsen JOB, Kohler S, Zemojtel T, Spielmann M, Jager M, Hochheiser H, Washington NL, McMurry JA *et al.* A whole-genome analysis framework for effective identification of pathogenic regulatory variants in Mendelian disease. *Am J Hum Genet* 2016, 99(3):595–606.
72. Kohler S, Schulz MH, Krawitz P, Bauer S, Dolken S, Ott CE, Mundlos C, Horn D, Mundlos S, Robinson PN. Clinical diagnostics in human genetics with semantic similarity searches in ontologies. *Am J Hum Genet* 2009, 85(4):457–464.
73. Singleton MV, Guthery SL, Voelkerding KV, Chen K, Kennedy B, Margraf RL, Durtschi J, Eilbeck K, Reese MG, Jorde LB *et al.* Phevor combines multiple biomedical ontologies for accurate identification of disease-causing alleles in single individuals and small nuclear families. *Am J Hum Genet* 2014, 94(4):599–610.
74. Javed A, Agrawal S, Ng PC. Phen-Gen: combining phenotype and genotype to analyze rare disorders. *Nat Methods* 2014, 11(9):935–937.
75. Stelzer G, Plaschkes I, Oz-Levi D, Alkelai A, Olender T, Zimmerman S, Twik M, Belinky F, Fishilevich S, Nudel R *et al.* VarElect: the phenotype-based variation prioritizer of the GeneCards Suite. *BMC Genomics* 2016, 17 Suppl 2:444.

76. Boudellioua I, Kulmanov M, Schofield PN, Gkoutos GV, Hoehndorf R. DeepPVP: phenotype-based prioritization of causative variants using deep learning. *BMC Bioinformatics* 2019, 20(1):65.
77. Rodriguez-Garcia MA, Gkoutos GV, Schofield PN, Hoehndorf R. Integrating phenotype ontologies with PhenomeNET. *J Biomed Semantics* 2017, 8(1):58.
78. Davydov EV, Goode DL, Sirota M, Cooper GM, Sidow A, Batzoglou S. Identifying a high fraction of the human genome to be under selective constraint using GERP++. *PLoS Comput Biol* 2010, 6(12):e1001025.
79. Landrum MJ, Chitipiralla S, Brown GR, Chen C, Gu B, Hart J, Hoffman D, Jang W, Kaur K, Liu C *et al.* ClinVar: improvements to accessing data. *Nucleic Acids Res* 2020, 48(D1):D835–D844.
80. Bragin E, Chatzimichali EA, Wright CF, Hurles ME, Firth HV, Bevan AP, Swaminathan GJ. DECIPHER: database for the interpretation of phenotype-linked plausibly pathogenic sequence and copy-number variation. *Nucleic Acids Res* 2014, 42(Database issue):D993–D1000.
81. Smedley D, Jacobsen JO, Jager M, Kohler S, Holtgrewe M, Schubach M, Siragusa E, Zemojtel T, Buske OJ, Washington NL *et al.* Next-generation diagnostics and disease-gene discovery with the Exomiser. *Nat Protoc* 2015, 10(12):2004–2015.
82. Yang H, Robinson PN, Wang K. Phenolyzer: phenotype-based prioritization of candidate genes for human diseases. *Nat Methods* 2015, 12(9):841–843.
83. Holtgrewe M, Stolpe O, Nieminen M, Mundlos S, Knaus A, Kornak U, Seelow D, Segebrecht L, Spielmann M, Fischer-Zirnsak B *et al.* VarFish: comprehensive DNA variant analysis for diagnostics and research. *Nucleic Acids Res* 2020, 48(W1):W162–W169.
84. Li MX, Gui HS, Kwan JS, Bao SY, Sham PC. A comprehensive framework for prioritizing variants in exome sequencing studies of Mendelian diseases. *Nucleic Acids Res* 2012, 40(7):e53.
85. Flygare S, Hernandez EJ, Phan L, Moore B, Li M, Fejes A, Hu H, Eilbeck K, Huff C, Jorde L *et al.* The VAAST Variant Prioritizer (VVP): ultrafast, easy to use whole genome variant prioritization tool. *BMC Bioinformatics* 2018, 19(1):57.
86. Geoffroy V, Pizot C, Redin C, Piton A, Vasli N, Stoetzel C, Blavier A, Laporte J, Muller J. VaRank: a simple and powerful tool for ranking genetic variants. *PeerJ* 2015, 3:e796.
87. Alexander J, Mantzaris D, Georgitsi M, Drineas P, Paschou P. Variant Ranker: a web-tool to rank genomic data according to functional significance. *BMC Bioinformatics* 2017, 18(1):341.
88. Ip E, Chapman G, Winlaw D, Dunwoodie SL, Giannoulatou E. VPOT: A Customizable Variant Prioritization Ordering Tool for Annotated Variants. *Genomics Proteomics Bioinformatics* 2019, 17(5):540–545.
89. Richards S, Aziz N, Bale S, Bick D, Das S, Gastier-Foster J, Grody WW, Hegde M, Lyon E, Spector E *et al.* Standards and guidelines for the interpretation of sequence variants: a joint consensus recommendation of the American College of Medical Genetics and Genomics and the Association for Molecular Pathology. *Genet Med* 2015, 17(5):405–424.
90. Li Q, Wang K. InterVar: Clinical Interpretation of Genetic Variants by the 2015 ACMG-AMP Guidelines. *Am J Hum Genet* 2017, 100(2):267–280.

91. Whiffin N, Walsh R, Govind R, Edwards M, Ahmad M, Zhang X, Tayal U, Buchan R, Midwinter W, Wilk AE *et al.* CardioClassifier: disease- and gene-specific computational decision support for clinical genome interpretation. *Genet Med* 2018, 20(10):1246–1254.
92. Nicora G, Limongelli I, Gambelli P, Memmi M, Malovini A, Mazzanti A, Napolitano C, Priori S, Bellazzi R. CardioVAI: An automatic implementation of ACMG-AMP variant interpretation guidelines in the diagnosis of cardiovascular diseases. *Hum Mutat* 2018, 39(12):1835–1846.
93. Li MM, Datto M, Duncavage EJ, Kulkarni S, Lindeman NI, Roy S, Tsimberidou AM, Vnencak-Jones CL, Wolff DJ, Younes A *et al.* Standards and Guidelines for the Interpretation and Reporting of Sequence Variants in Cancer: A Joint Consensus Recommendation of the Association for Molecular Pathology, American Society of Clinical Oncology, and College of American Pathologists. *J Mol Diagn* 2017, 19(1):4–23.
94. He MM, Li Q, Yan M, Cao H, Hu Y, He KY, Cao K, Li MM, Wang K. Variant Interpretation for Cancer (VIC): a computational tool for assessing clinical impacts of somatic variants. *Genome Med* 2019, 11(1):53.
95. Li Q, Ren Z, Cao K, Li MM, Wang K, Zhou Y. CancerVar: an artificial intelligence-empowered platform for clinical interpretation of somatic mutations in cancer. *Sci Adv* 2022, 8(18):eabj1624.
96. Tamborero D, Rubio-Perez C, Deu-Pons J, Schroeder MP, Vivancos A, Rovira A, Tusquets I, Albanell J, Rodon J, Tabernero J *et al.* Cancer Genome Interpreter annotates the biological and clinical relevance of tumor alterations. *Genome Med* 2018, 10(1):25.
97. Griffith M, Spies NC, Krysiak K, McMichael JF, Coffman AC, Danos AM, Ainscough BJ, Ramirez CA, Rieke DT, Kujan L *et al.* CIViC is a community knowledgebase for expert crowdsourcing the clinical interpretation of variants in cancer. *Nat Genet* 2017, 49(2):170–174.
98. Huang L, Fernandes H, Zia H, Tavassoli P, Rennert H, Pisapia D, Imielinski M, Sboner A, Rubin MA, Kluk M *et al.* The cancer precision medicine knowledge base for structured clinical-grade mutations and interpretations. *J Am Med Inform Assoc* 2017, 24(3):513–519.
99. Zomnir MG, Lipkin L, Pacula M, Meneses ED, MacLeay A, Duraisamy S, Nadhamuni N, Al Turki SH, Zheng Z, Rivera M *et al.* Artificial intelligence approach for variant reporting. *JCO Clin Cancer Inform* 2018, 2.
100. den Dunnen JT, Dalgleish R, Maglott DR, Hart RK, Greenblatt MS, McGowan-Jordan J, Roux AF, Smith T, Antonarakis SE, Taschner PE. HGVS Recommendations for the Description of Sequence Variants: 2016 update. *Hum Mutat* 2016, 37(6):564–569.
101. Braschi B, Denny P, Gray K, Jones T, Seal R, Tweedie S, Yates B, Bruford E. Genenames.org: the HGNC and VGNC resources in 2019. *Nucleic Acids Res* 2019, 47(D1):D786–D792.
102. Arteche-Lopez A, Avila-Fernandez A, Romero R, Riveiro-Alvarez R, Lopez-Martinez MA, Gimenez-Pardo A, Velez-Monsalve C, Gallego-Merlo J, Garcia-Vara I, Almoguera B *et al.* Sanger sequencing is no longer always necessary based on a single-center validation of 1109 NGS variants in 825 clinical exomes. *Sci Rep* 2021, 11(1):5697.

103. Beck TF, Mullikin JC, Program NCS, Biesecker LG. Systematic evaluation of sanger validation of next-generation sequencing variants. *Clin Chem* 2016, 62(4):647–654.
104. Kerkhof J, Schenkel LC, Reilly J, McRobbie S, Aref-Eshghi E, Stuart A, Rupar CA, Adams P, Hegele RA, Lin H *et al.* Clinical validation of copy number variant detection from targeted next-generation sequencing panels. *J Mol Diagn* 2017, 19(6):905–920.
105. Miller DT, Lee K, Chung WK, Gordon AS, Herman GE, Klein TE, Stewart DR, Amendola LM, Adelman K, Bale SJ *et al.* ACMG SF v3.0 list for reporting of secondary findings in clinical exome and genome sequencing: a policy statement of the American College of Medical Genetics and Genomics (ACMG). *Genet Med* 2021, 23(8):1381–1390.
106. Miller DT, Lee K, Gordon AS, Amendola LM, Adelman K, Bale SJ, Chung WK, Gollob MH, Harrison SM, Herman GE *et al.* Recommendations for reporting of secondary findings in clinical exome and genome sequencing, 2021 update: a policy statement of the American College of Medical Genetics and Genomics (ACMG). *Genet Med* 2021, 23(8):1391–1398.
107. Appelbaum PS, Parens E, Berger SM, Chung WK, Burke W. Is there a duty to reinterpret genetic data? The ethical dimensions. *Genet Med* 2020, 22(3):633–639.
108. Clayton EW, Appelbaum PS, Chung WK, Marchant GE, Roberts JL, Evans BJ. Does the law require reinterpretation and return of revised genomic results? *Genet Med* 2021, 23(5):833–836.
109. Deignan JL, Chung WK, Kearney HM, Monaghan KG, Rehder CW, Chao EC, Committee ALQA. Points to consider in the reevaluation and reanalysis of genomic test results: a statement of the American College of Medical Genetics and Genomics (ACMG). *Genet Med* 2019, 21(6):1267–1270.
110. Roy S, Coldren C, Karunamurthy A, Kip NS, Klee EW, Lincoln SE, Leon A, Pullambhatla M, Temple-Smolkin RL, Voelkerding KV *et al.* Standards and Guidelines for Validating Next-Generation Sequencing Bioinformatics Pipelines: A Joint Recommendation of the Association for Molecular Pathology and the College of American Pathologists. *J Mol Diagn* 2018, 20(1):4–27.
111. Ewing AD, Houlahan KE, Hu Y, Ellrott K, Caloian C, Yamaguchi TN, Bare JC, P'ng C, Waggott D, Sabelnykova VY *et al.* Combining tumor genome simulation with crowdsourcing to benchmark somatic single-nucleotide-variant detection. *Nat Methods* 2015, 12(7):623–630.

12

De Novo Genome Assembly with Long and/or Short Reads

De novo genome assembly is a fundamental endeavor for genomics research, as it creates the reference genome against which most other NGS applications rely on for alignment. Accurate representation of the reference genome is of utmost importance, as the assembly quality directly impacts the results of other NGS applications that depend on it. Not very long ago Sanger sequencing was considered the golden standard for *de novo* genome assembly. However, it is prohibitively expensive and time-consuming to assemble a genome using this first-generation technology, as it took $3 billion and 13 years to generate the human genome draft assembly. The demand for low-cost and fast genome sequencing provides the very impetus for the development of NGS technologies. The dramatically reduced cost of NGS makes whole-genome sequencing much more affordable, much faster, and readily accessible to individual labs for deciphering any species. While increasingly more and more organisms' reference genomes are assembled, *de novo* assembly continues to be an important application of NGS for not only novel unsequenced genomes, but also completed genomes in terms of resolving challenging regions (such as highly repetitive regions), removing biases, and searching for novel components or structural variations. *De novo* genome assembly from the relatively short and enormous number of reads generated from most NGS platforms, however, poses serious challenges to assembling algorithms that were designed for Sanger sequences. The usually shorter length of NGS reads means that they carry less information and as a result lead to more uncertainties in the assembling process. To remedy this situation, higher coverage is required, which significantly increases the number of reads required and therefore computational complexity. For example, using Sanger sequences with length up to 800 bp, assembling the human genome used approximately 8× coverage; for NGS reads of 35–100 bp, the same task needs 50–100× coverage [1].

Since Sanger sequence assemblers cannot effectively deal with these challenges, new *de novo* genome assemblers have been developed for NGS data. The development of Velvet [2] and ABySS [3] in 2008–2009 showed that *de novo* high-quality genome assembly can be achieved, even for large genomes, using massive numbers of ultra-short (as short as 30 bp) reads. The first *de novo* assembly of a human genome with the use of only short

DOI: 10.1201/9780429329180-15

NGS reads was accomplished in 2010 with the development SOAPdenovo [4]. Subsequently, to increase assembly contiguity, a hybrid strategy is often used to integrate long but error-prone sequencing reads from PacBio (or Nanopore) with short reads [5]. Now with significantly improved accuracy of long-read sequencing, assembling a genome from long-read sequencing data alone has become a reality [6].

12.1 Genomic Factors and Sequencing Strategies for *De Novo* Assembly

12.1.1 Genomic Factors That Affect *De Novo* Assembly

The size of a target genome to a large degree determines the difficulty of assembling it. All NGS *de novo* assemblers (to be detailed next) can handle small genomes (<10 Mb), such as those of bacteria, without difficulty. For genomes of medium size (10 Mb – 1 Gb), such as those of lower plants and insects, most of the assemblers should still work without much problem. For large genomes (>1 Gb), while some assemblers, such as the aforementioned SOAPdenovo, have been shown to have the capability to assemble the human or other mammalian genomes, in general it is not an easy task to put them together with only short reads. In addition, assembling a large genome *de novo* is the most computationally demanding among all NGS applications.

The amount of repetitive sequences in a genome is another major factor that affects *de novo* genome assembly. Some genomes are inherently more difficult to assemble than others because they contain more repetitive sequences. Because they produce reads that are not unique due to their repetitive nature, repetitive regions create serious challenges in the genome assembly process. The challenges come from the inability to assemble reads from these regions into contiguous segments (contigs) or scaffolds, and determine the locations of these reads in relation to contigs or scaffolds assembled from reads from non-repetitive regions. As a result, these regions become gaps in a draft assembly. Besides repetitive elements, genomic heterozygosity is another factor that may affect *de novo* assembly. Genomic heterozygosity is a measure of allelic differences in a genome, and allelic differences in a diploid or polyploid genome lead to uncertainty in assembling their reads together. In addition, other genome features, such as local GC content, may also affect *de novo* genome assembly.

12.1.2 Sequencing Strategies for *De Novo* Assembly

Filling the gaps caused by repetitive regions is important for most *de novo* genome assembly projects, and how to fill them should be a major consideration when devising an appropriate sequencing strategy. The basic approach to connect contigs or scaffolds across the gaps is to use read pairs, or long reads, that span a distance longer than the gaps. For paired reads they have to be generated from paired-end and/or mate-pair sequencing, and the known distances between the read pairs provide guidance to align the contigs or scaffolds over the gaps. Mate-pair sequencing differs from paired-end sequencing (see Chapter 4) in that the mate-pair approach is designed to "jump" sequence two ends of a larger DNA fragment. To conduct mate-pair sequencing, a DNA fragment is first circularized to have the two ends joined. This circular DNA is then fragmented, and the segment that contains the junction of the two ends is selected and sequenced with paired-end sequencing. To span repetitive regions in different sizes, sequencing reads generated from mate-pair libraries of varying insert sizes, e.g., from 2 to 40 Kb, as well as regular paired-end reads, are often used [7, 8].

The combined use of paired-end and mate-pair libraries of different insert sizes is a key strategy in assembling a genome from NGS reads. The paired-end sequencing generates reads at the shorter size range (e.g., 300–350 bp) for assembling of non-repeat sequences as well as resolving short repeat sequences, while the mate-pair "jump" sequencing produces reads at the larger size range for resolving intermediate and long range repeat regions and fill the corresponding gaps. Gaps of substantial sizes that are beyond the covering range of mate pair libraries cannot be filled.

Besides the use of paired-end and mate-pair sequencing, read length is also a key parameter for *de novo* genome assembly. While mammalian genomes have been assembled from reads shorter than 75 bp (e.g., [4, 8]), longer reads are always better. With long-read sequencing platforms such as PacBio and ONT, obtaining long reads is becoming more routine. As detailed in Chapter 4, CCS long reads from PacBio can reach 25 Kb, and 200 Kb can be reached with CLR sequencing. On ONT flow cells, maximum read length can reach over 4 Mb. In addition to native long reads, synthetic long reads produced from approaches such as single-tube Long Fragment Read (or stLFR) from MGI [9] and Transposase Enzyme Linked Long-read Sequencing (TELL-seq) from Universal Sequencing Technology [10] can reach 20–300 Kb and over 100 Kb, respectively. While synthetic long reads are based on short-read sequencing, they differ from regular short reads in that instead of directly breaking down genomic DNA into fragments of hundreds of base pairs for sequencing, these approaches first shear DNA into rather large fragments (up to hundreds of Kb). These large fragments are then used to produce short fragments in such a way that short fragments derived from the same large fragment all carry the same barcode, based on which short read sequences are then linked together for synthesis of the original large fragment sequence. 10× Genomics introduced one of the first synthetic long-read strategies called linked-reads sequencing [11], but it was discontinued in 2020. Synthetic long read approaches have greatly aided *de novo* genome assembly.

Sequencing depth is another important factor to consider for a *de novo* assembly project. While it varies by project and is dependent on the other factors (including the number of repeats and level of heterozygosity in the genome as well as read length and error rate), a coverage that is too low will undoubtedly result in a highly fragmented assembly. As a rough guide, in the combined use of paired-end and mate-pair libraries of various insert sizes, 45–50x coverage is needed for the short-insert-size paired-end and intermediate-size (3–10 Kb) mate-pair libraries, and 1–5x coverage for the long-insert (10–40 Kb) mate-pair libraries [12, 13]. For *de novo* assembly of a human genome using long reads alone, 30x or 35x coverage can be sufficient for PacBio CCS or ONT reads respectively with appropriately chosen tools [14, 15]. It should also be noted that while higher coverage may lead to improvement in the final assembly quality, additional increase in coverage also means increased data volume, computational complexity, and processing time. There are also studies showing that beyond certain level of coverage further increase in sequencing depth does not necessarily lead to increase in assembly quality in terms of the size of assembled contigs [4].

12.2 Assembly of Contigs

12.2.1 Sequence Data Preprocessing, Error Correction, and Assessment of Genome Characteristics

The *de novo* assembly of a genome from NGS reads is a multi-step process (Figure 12.1). As the first step, sequence data quality needs to be inspected. Data QC steps described in Chapter 5 can be performed here to examine per-base error rate, quality score distribution, read size distribution, contamination of adaptor sequences, etc. Low-quality reads need to be filtered out, and portions of reads that contain low-quality basecalls (usually the 3′ end), ambiguities (reported as Ns), or adaptor sequences should be trimmed off. As part of data preprocessing, paired-end reads with part of their sequences overlapped need to be merged to generate longer reads. The read merging can also correct errors if discrepancy at some base positions are observed, in which case the higher quality basecall is used. The merging process can be handled by tools such as FLASH2 [16], PEAR [17], fastq-join [18], PANDAseq [19], and VSEARCH [20].

Sequencing error correction is an important step for *de novo* read assembly, more so than for most other NGS applications due to the fact that the assembly process is much more sensitive to these errors. The data QC measures mentioned above cannot totally remove sequencing errors, as high basecall quality scores alone cannot guarantee a read is free of sequencing errors. If left uncorrected, the errors will lead to prolonged computational time,

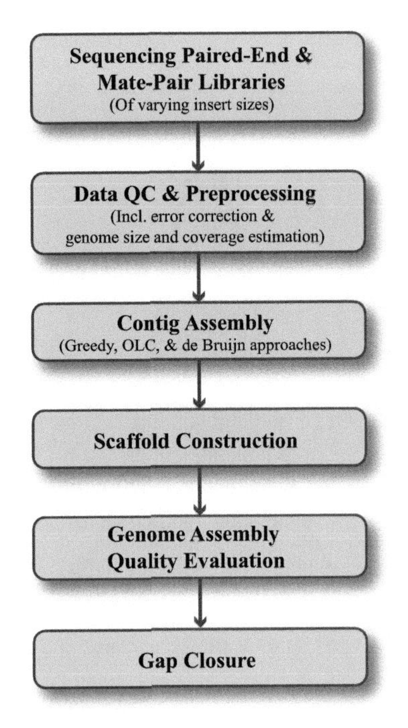

FIGURE 12.1
General workflow for *de novo* genome assembly.

erroneous contigs, and low-quality genome assembly. While it can be time consuming, an additional error correction step can improve final assembly quality. There are multiple options to carry out this step. For example, BFC [21], BLESS [22], Lighter [23], Musket [24], Fiona [25], and Coral [26] can be used as stand-alone tools, while some assemblers (see next) have their own error correction modules, such as ALLPATHS-LG [27] and SGA [28]. Most error correction methods, including BFC, BLESS, Lighter, Musket, the ALLPATHS-LG error correction module (can be used stand-alone), and the SGA default correction method, are based on *k*-mer filtering [29]. *K*-mer refers to all the possible subsequences of length *k* in a read, and breaking reads to *k*-mers makes the complicated task of genome assembly more tractable. When all reads are converted to *k*-mers, most *k*-mers in the pool are represented multiple times. Having a *k*-mer that appears only once or twice is an indication of sequencing error (Figure 12.2). The general error correction approach is to find the smallest number of base changes to make all *k*-mers contained in a read "strong," i.e., with the frequency of these *k*-mers from all reads above a threshold level. To determine the appropriate threshold level for error correction, the distribution of the frequency of *k*-mers can be plotted using data from a *k*-mer counting software such as Jellyfish [30] or Meryl [31]. Besides error correction, the *k*-mer frequency information from Jellyfish or

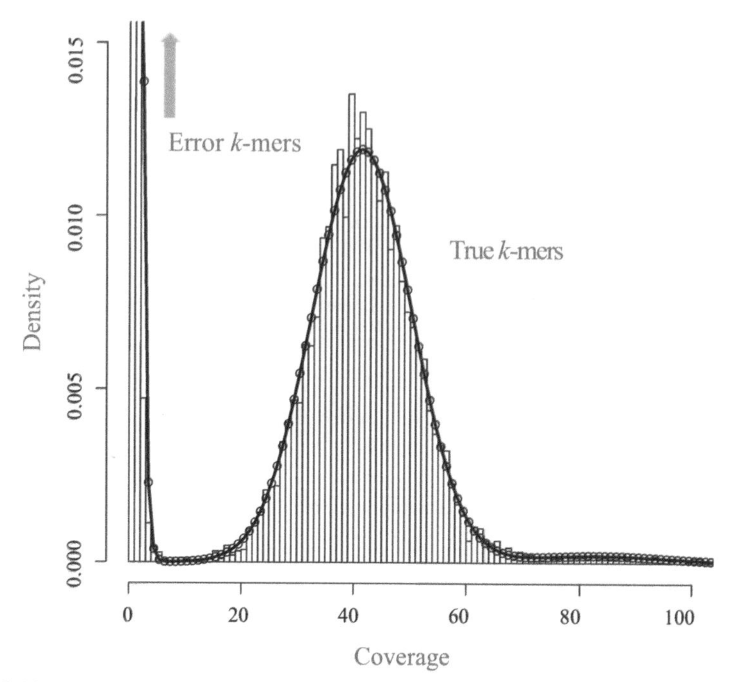

FIGURE 12.2
The coverage profile of true k-mers and those with sequencing errors. (Adapted from Kelley D.R. et al. (2010) Quake: quality-aware detection and correction of sequencing errors, *Genome Biology*, 11:R116. Used under the terms of the Creative Commons Attribution License (http://creativecommons.org/licenses/by/2.0). © 2010 Kelley et al.)

Meryl can also be input into GenomeScope [32] to estimate genome size, level of duplication, and heterozygosity. Besides using k-mer filtering, suffix tree/array and multiple sequence alignment (MSA) are also often used to correct sequencing errors. For example, Fiona is an example of the suffix tree/array-based approach. Coral, on the other hand, is based on the MSA approach. The approach aligns reads that share common k-mers, and error corrections are made based on alignment results and consensus sequences.

For long reads from PacBio or ONT that have higher error rates, sequencing error correction is even more needed and can be performed with tools such as FMLRC [33], PBcR [34], LoRDEC [35], LSC [36], Nanocorr [37], proovread [38], and DeepConsensus [39]. Many long-read assemblers (to be detailed next) contain their own error correction modules, such as Canu [40], Hifiasm [41], FALCON [42], MARVEL [43], MECAT [44], and NECAT [14]. In general, long-read error correction methods can be divided into two groups, with one using redundant information in long reads alone for error correction, while the other using a hybrid approach to leverage more accurate short reads to help correct errors in long reads. Long-read-only correction methods belonging to the first group include LoRMA [45] and FLAS [46]. The error correction methods used by Canu, MARVEL, and MECAT, are also in this group since they are based on consensus correction using sampling redundancy (or overlaps) within long reads. As an example, to make corrections Canu builds overlaps between long reads using a probabilistic sequence overlapping algorithm called the MinHash alignment process (or MHAP) [47]. The overlapping results are then used to identify regions that need correction based on sequence consensus, followed by estimation of corrected read length and generation of corrected reads. Besides error correction, MARVEL also has a "patching" module, which aims to repair apparent large-scale errors (e.g., regions that contain a lot of errors) based on comparisons between reads. Examples of the hybrid approach to use more accurate short reads to correct error-prone long reads include FMLRC, PBcR, LoRDEC, LSC, Nanocorr, and proovread. Based on how they work, these methods can be further divided into those based on assembly and those on alignment. Assembly-based correction methods, such as FMLRC and LoRDEC, first use short reads to perform assembly, and then align long reads to the assemblies for making corrections. Alignment-based methods, including PBcR, LSC, Nanocorr, and proovread, align short reads to long reads and sequencing errors in the long reads are corrected based on the alignment results.

12.2.2 Contig Assembly Algorithms

Fundamentally different from the reference-based alignment process, which is used by most of the other NGS applications in this book, *de novo* genome assembly attempts to construct a superstring (or superstrings) of DNA letters based on the overlapping of sequence reads. This assembly process was previously modeled by Lander and Waterman with the use of ideal (error- and repeat-free) sequence data [48]. In this model, if two reads overlap and the overlap is above a cutoff level, the two reads are merged into a contig and this process reiterates until the contig cannot be extended further. Although this guiding model is straightforward, finding all possible overlaps between millions of short reads and assembling them into contigs are computationally intensive and challenging. Added to this challenge are other complicating factors such as sequencing errors, sequencing bias, heterozygosity, and repetitive sequences. To deal with these challenges, a number of assemblers that employ different methodologies have been developed.

The currently available *de novo* genome assemblers can be classified into three major categories – those using (1) the Greedy approach; (2) the Overlap-Layout-Consensus (OLC) approach; and (3) the de Bruijn graph. Although all of them are based on graphs, the Greedy approach is the one that is based on the maximization of local sequence similarity. It was used by Sanger sequence assemblers, such as phrap and the TIGR assembler, and early NGS reads assemblers, such as SSAKE [49], SHARCGS [50], and VCAKE [51]. Since it is a local approach, the Greedy approach does not consider the global relationship between reads. Therefore, more recent NGS-based assemblers

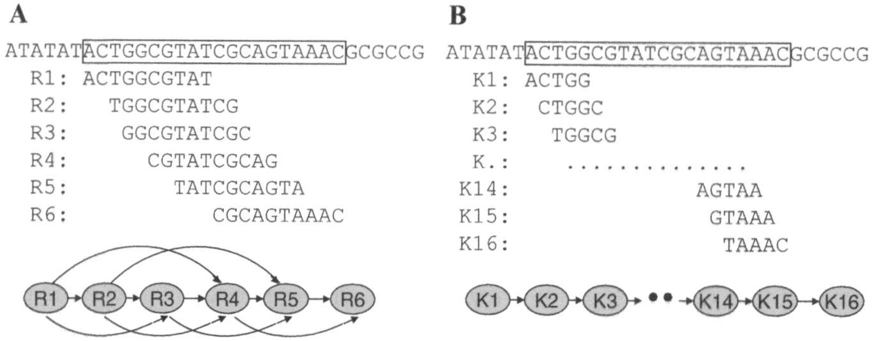

```
A
ATATATACTGGCGTATCGCAGTAAACGCGCCG
  R1: ACTGGCGTAT
  R2:   TGGCGTATCG
  R3:    GGCGTATCGC
  R4:      CGTATCGCAG
  R5:        TATCGCAGTA
  R6:           CGCAGTAAAC

R1 → R2 → R3 → R4 → R5 → R6

B
ATATATACTGGCGTATCGCAGTAAACGCGCCG
  K1: ACTGG
  K2:  CTGGC
  K3:   TGGCG
  K.:    ..............
 K14:              AGTAA
 K15:               GTAAA
 K16:                TAAAC

K1 → K2 → K3 → •• → K14 → K15 → K16
```

FIGURE 12.3
Comparison of the OLC (A) and the de Bruijn graph (B) approaches for global *de novo* genome assembly. In the OLC example, six sequence reads (R1–R6) are shown for the illustrated genomic region, with each read being 10 bp in length and the overlap between them set at ≥ 5 bp. The reads are laid out in order based on how they overlap. The OLC graph is shown at the bottom, with many nodes having more than one incoming or outgoing connections. In the de Bruijn graph example, the reads are cut into a series of k-mers (k=5). In total there are 16 such k-mers, many of which occur in more than one read. The k-mers are arranged sequentially based on how they overlap, and the de Bruijn graph built from this approach is shown at the bottom. Different from those in the OLC graph, the majority of the nodes in this graph have only one incoming and one outgoing connection. (From Li Z. et al. Comparison of the two major classes of assembly algorithms: overlap–layout–consensus and de-bruijn-graph. *Briefings in Functional Genomics*, 2012, 11(1): 25–37, by permission of Oxford University Press.)

no longer use this approach, as it cannot take advantage of the global relationship offered by paired-end and mate-pair reads.

The OLC and the de Bruijn graph approaches are global by design and both assemble reads into contigs using reads overlapping information based on the Lander-Waterman model. They approach the task, however, in different ways (Figure 12.3). The OLC approach, as the name suggests, involves three steps: (1) detecting potential overlaps between all reads; (2) laying out all reads with their overlaps in a graph; and (3) constructing consensus sequence superstring. The first step is computationally intensive and the run time increases quadratically with the increase in the total number of reads. The graph created in the second step consists of vertices (or nodes) representing reads, and edges between them representing their overlaps. The construction of a consensus sequence superstring equals to finding a path in the graph that visits each node exactly once, which is known as a Hamiltonian path in graph theory. While there are a small number of OLC-based short-read assemblers available such as Edena [52] and Fermi [53], the OLC approach is more suitable and mostly used to assemble long reads generated from PacBio and ONT sequencers. In fact, most long read *de novo* assembly methods are based on OLC, including Canu, FALCON, Hifiasm,

MECAT, NECAT, miniasm [54], HINGE [55], Peregrine [56], Shasta [57], Raven [58], NextDenovo [59], and SMARTdenovo [60]. In the case of Canu, after the aforementioned error correction process, unsupported sequences are trimmed off to prepare corrected reads for assembly. In the assembly stage, reads are scanned one last time for errors, and then used to construct overlap graph, before output of consensus contig sequences and an assembly graph. HiCanu is a modified version of Canu developed for using the PacBio HiFi (CCS) reads [61].

Solving the Hamiltonian path problem in the OLC approach is NP-hard. To reduce the high computing demand imposed by the OLC approach, a simplified version called the String graph has been employed to merge and reduce redundant vertices and edges, along with identification and removal of false vertices and edges [62]. The implementation of a string indexing data structure (FM-index) has improved the performance of assemblers such as SGA and ReadJoiner [63]. FALCON also first builds a string graph that contains bubbles representing structural variation between haplotypes through the use of a process called HGAP (or hierarchical genome assembly process). The subsequent "Unzip" process creates haplotype-resolved assembly graph, making FALCON a diploid-aware assembler. Other long-read assemblers, such as Miniasm, Hifiasm, NECAT, and NextDenovo, also use string graph in their assembly processes. Hifiasm, for example, produces fully phased assembly for each haplotype from phased string graph created from PacBio HiFi reads.

Compared to the OLC approach, the de Bruijn graph-based approach takes an alternative, computationally more tractable route. This approach does not involve a step to find all possible overlaps between reads. Instead, the reads are first cut into *k*-mers. For instance, the sequence read ATTACGTCGA can be cut into a series of *k*-mers, e.g., ATT, TTA, TAC, ACG, CGT, GTC, TCG, and CGA, when $k = 3$. These *k*-mers are then used as vertices in the de Bruijn graph. An edge that connects two nodes represents a convergence of the two nodes, e.g., the edge that connects ATT and TTA is ATTA. Using the de Bruijn graph, the assembly process is equivalent to finding a shortest path that visits each node at least once, which is known as the Chinese Postman Problem in graph theory. An Eulerian path, if it exists, represents the solution to this problem. Computationally, finding an Eulerian path is much easier than finding a Hamiltonian path for the OLC approach. The major drawback of this approach, however, is that it is highly sensitive to sequencing errors. Therefore, to use assemblers in this category, error correction is mandatory. Assemblers that use this approach include AbySS [3], ALLPATHS-LG, Euler-SR [29], SOAPdenovo/SOAPdenovo2 [4, 64], SPAdes [65], and Velvet [2]. Because of higher sequencing error rates, long reads are not usually assembled with this approach. There are, however, a small number of long read assemblers that employ variants of de Bruijn Graph. For example, Flye [66] uses the A-Bruijn variant, which tolerates the higher level of sequencing errors in long reads as it uses approximate sequence matches instead of exact *k*-mer matches. Using PacBio or ONT long reads as input, it first creates a repeat graph, and then through resolving repetitions by identifying variations in repeat copies in the graph a final assembly is reconstructed. Wtdbg2 [67] is another example based on another de Bruijn variant called fuzzy Bruijn graph (or FBG), which permits mismatches and gaps. This method assembles a genome through first cutting long reads into smaller segments, and then merging the segments based on their adjacency to create contigs.

Hybrid assemblers integrate both short and long reads toward achieving improved accuracy and contiguity. Some of the assemblers, such as MaSuRCA [68] and DBG2OLC [69], use a combination of the de Bruijn graph and OLC approaches for short and long reads, respectively. Others take alternative approaches. For example, WENGAN uses de Bruijn graph to assemble short reads, but avoids the all-versus-all read comparison used in the OLC process commonly used for long reads to increase efficiency [70].

12.2.3 Polishing

Besides error correction prior to genome assembly, assembly quality can be further improved after draft assembly is created through a process called polishing. In general terms, the polishing process uses information from alignment of reads to the draft assembly as input, examines how reads map to the draft assembly at each location, and then decides whether assembly sequences at certain locations need to be modified. To perform this process, there are a selection of tools available, including Pilon [71], Medaka [72], Racon [73], Nanopolish [74], MarginPolish & HELEN [57], NextPolish [75], POLCA [76], and NeuralPolish [77]. These polishers typically use either short or long reads to polish a draft assembly. For example, Pilon takes as input an assembled genome in FASTA format, and alignment of short reads to the assembly in BAM format. After searching the alignment for inconsistencies, assembly sequences are modified to provide improvements to the input assembly through reduction of mismatches and indels, as well as gap filling and misassembly identification. Medaka, the first neural network-based polisher, on the other hand, uses ONT long reads for polishing through creation of consensus sequences. As input it requires a draft assembly in FASTA format and basecalls in either FASTA or FASTQ format. Prior to creating consensus sequences, the alignment of the reads to the input assembly is carried out by minimap2. Nanopolish uses a third approach. Instead of using called bases, it takes raw sequencing signals recorded from an ONT sequencer as input, and applies an HMM-based signal-to-assembly analysis to generate improved consensus sequences for the draft assembly.

12.3 Scaffolding and Gap Closure

After assembly of contigs, the next step is to organize the contigs into a "scaffold" structure to improve continuity rather than leave them disjointed. This scaffolding process orders and orients the contigs, and estimates the lengths of the gaps between them (Figure 12.4). To establish positional relationship between contigs, scaffolding algorithms use long reads or mate-pair reads that span different contigs.

For input, scaffolding algorithms typically take pre-assembled contigs, and long reads or sometimes mate-pair reads. The first and also an important step in the scaffolding process is to map the input long reads or read pairs to the contigs. To improve mapping results, sequencing errors in the reads should be corrected prior to mapping as covered above. To assemble the contigs into scaffolds using the guiding information in the long or mate-pair reads, scaffolders usually take a graph-based approach similar to the contig assembly process, but here with contigs as nodes and connecting long reads (or read pairs) as edges. The quality of the assembled scaffolds is dependent on the quality of input contigs, the complexity of the genome, and the quality of long or mate-pair read libraries. The sizes of the scaffolds are limited by the length of long reads or the insert size of mate-pair libraries, as the scaffolds cannot span repetitive regions larger than the insert size or read length. Besides long reads or mate-pair reads, scaffolding information may also come from experimental data generated from other techniques including Hi-C, synthetic long-read sequencing, and optical mapping.

Currently available stand-alone scaffolders include SSPACE-LongRead [78], LINKS [79], OPERA-LG [80], LRScaf [81], SMIS [82], and npGraph [83]. To

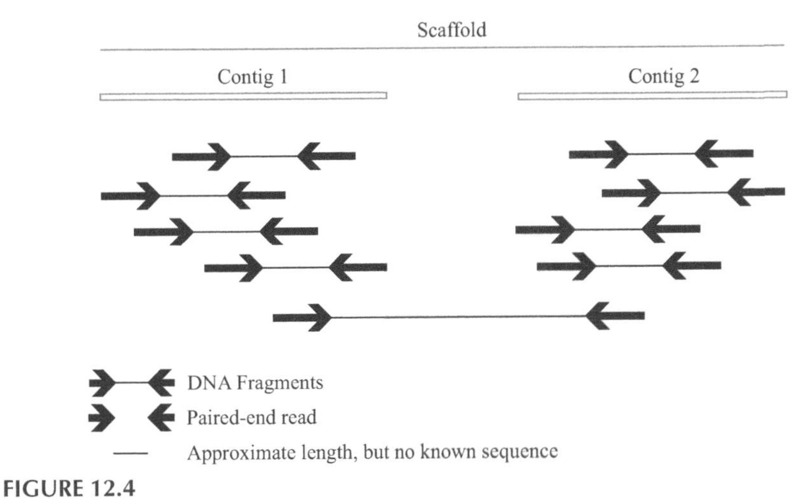

FIGURE 12.4
Assembling contigs into a scaffold.

provide an example of how scaffolding can be performed, LRScaf first maps long reads using minimap2 to identify those that overlap the ends of contigs, and then connects contigs based on reads that map to multiple contigs. From the connections, a scaffold graph similar to a string graph is constructed, and scaffolding of contigs is achieved through traversal of the graph to expose all linear paths. Besides these stand-alone methods, many contig assemblers, including ABySS, SGA, SOAPdenovo2, SPAdes, and MaSuRCA, DBG2OLC, also have built-in scaffolding modules. The performance of different scaffolders varies with data sets and analysis parameters. Therefore, before deciding on an appropriate scaffolder for a project, it is helpful to first try out different scaffolders using different parameters and then evaluate the results (see Section 12.4).

Among the other techniques that generate data for scaffolding, Hi-C is an NGS-based approach that uses proximity ligation to interrogate chromosomal interactions within the nucleus. The principle behind the use of Hi-C for scaffolding is that the probability of intrachromosomal interactions is much higher than that of interchromosomal contacts. Therefore, Hi-C reads provide long-range information on linear arrangement of contig sequences along individual chromosomes. Scaffolders that utilize Hi-C data include HiRise [84], 3D-DNA [85], and SALSA [86]. Similar to native long reads generated from ONT and PacBio sequencers, synthetic long reads also provide long-range sequence information required for scaffolding of contigs. Currently available scaffolding methods that use synthetic long reads include ARCS [87], Architect [88], ARKS [87], and SLR-superscaffolder [89].

The final stage of finishing a *de novo* genome assembly is to close the gaps between scaffolded contigs. Gap closure has greatly benefited from the constant advancements in long reads sequencing. There are currently a number of gap filling algorithms available, including PBjelly [5], GMcloser [90], Sealer [91], LR_Gapcloser [92], TGS-GapCloser [93], SAMBA [94], and gapless [95]. Among these algorithms SAMBA and gapless can perform both scaffolding and gap filling. Some assemblers, such as ABySS, ALLPATHS-LG, and SOAPdenovo2, also contain gap filling modules. Figure 12.5 shows an example of how long reads are used to finish gaps by TGS-GapCloser.

12.4 Assembly Quality Evaluation

Contiguity, completeness, and accuracy are key indices of the quality of an assembly. Contiguity is reflected by the total number of assembled contigs or scaffolds and their size distribution, i.e., whether the assembly is composed of a small number of large fragments or a large number of small fragments. It can be measured by statistics such as mean or median

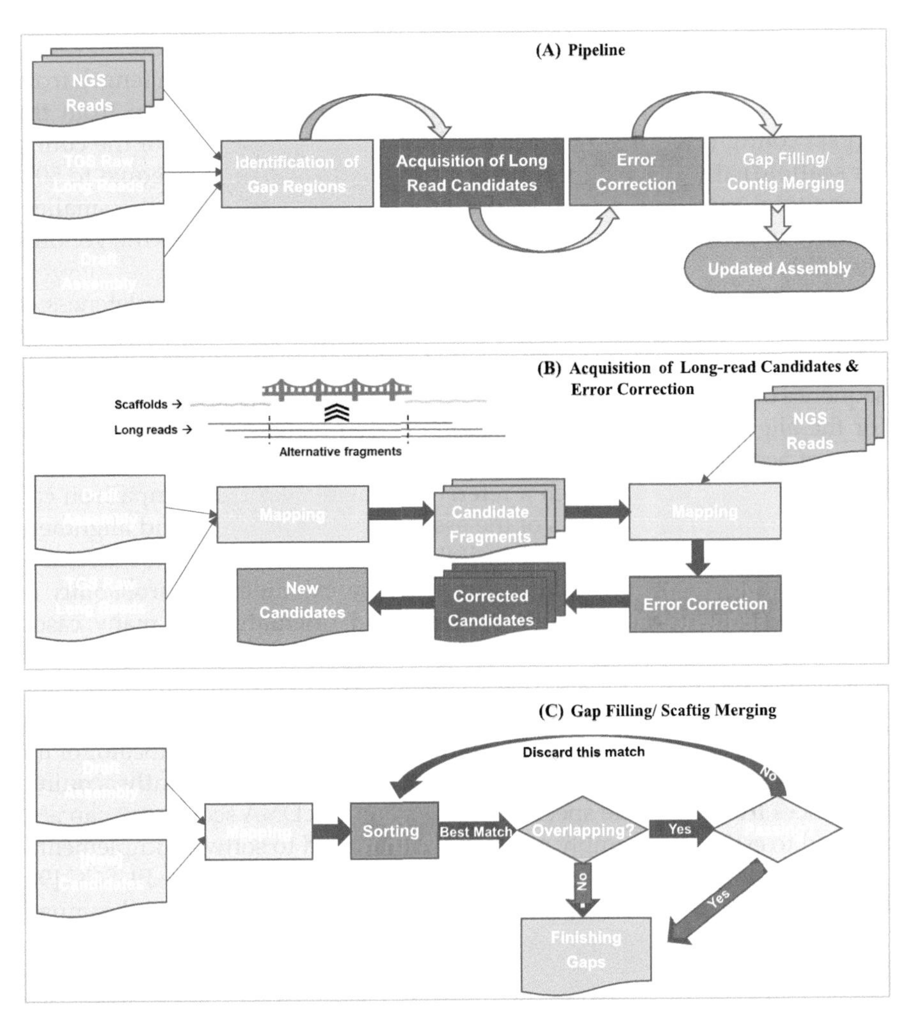

FIGURE 12.5
Gap closing with the TGS-GapCloser workflow. Panel (A) displays the overall pipeline. Panel (B) shows the steps of identification of gap regions and long-read candidates that map to the gap regions, and error correction of long-read candidates using short reads. Panel (C) details how the gaps are filled with error-corrected long reads. As input, TGS-GapCloser can accept long reads generated from any platform or other preassembled contigs to fill gaps in a draft assembly. The unknown nucleotides marked as N's between two neighboring contigs in the scaffolds are identified as gap regions. Long reads are then mapped to the gap regions using minimap2 to acquire candidate fragments. The subsequent error correction on the long-read candidates is carried out with Pilon or Racon. The corrected new long read candidates are then realigned to the gaps, and the read with the best match for a gap is used to fill the gap. To increase computational efficiency, overlapped candidates are clipped and merged. (Adapted from Xu M. et al. (2020) TGS-GapCloser: A fast and accurate gap closer for large genomes with low coverage of error-prone long reads. *GigaScience*, 9, 1–11. Used under the terms of the Creative Commons Attribution License (http://creativecommons.org/licenses/by/4.0). © 2020 Xu et al.)

length, but the most commonly used statistics are N50 and L50. To calculate the N50, all contigs (or scaffolds) are first ranked based on length from the largest to the smallest. Their lengths are then summed up from the largest contig (or scaffold) downward. N50 refers to the size of the contig (or scaffold) at which the summed length becomes greater or equal to 50% of the total assembly size. L50, on the other hand, refers to the smallest number of contigs (or scaffolds) that add up to at least 50% of the genome assembly.

The total assembly size, however, does not measure the completeness of the assembly. To determine completeness, the original DNA reads are aligned to the assembly and the percentage of reads aligned is calculated. Other sequence data from the same species, such as RNA-seq data, may also be used for the alignment and rough estimation of completeness. On the measurement of accuracy, the assembly can be compared to a high-quality reference genome of the species, if such a reference is available. This comparison can be carried out on two aspects of the assembly: base accuracy and alignment accuracy. Base accuracy determines if the right base is called in the assembly at a given position, while alignment accuracy examines the probability of placing a sequence at the right position and orientation. In many cases, however, a reference map is not available and instead is the very goal of the assembling process. For these cases, a measurement on internal consistency, through aligning original reads to the assembly and checking for evenness and congruence in coverage across the assembly, provides an indicator of the assembly quality. Comparison of the assembly with independently acquired sequences from the same species, such as gene or cDNA sequences, can also be used to estimate assembly accuracy. With regard to software implementation on evaluating assembly quality, the often used tools include BUSCO [96] and QUAST [97], which help perform the above measurements and compare different contig and scaffold assembly algorithms and settings.

12.5 Limitations and Future Development

Among all NGS applications covered in this book, *de novo* genome assembly remains one of the most challenging due to limitations of current sequencing technologies and the complexity of genomic landscape. Due to their cost effectiveness, short read sequencing remains to be the most accessible; however, the short read length poses a limit on *de novo* genome assembly. This, combined with other factors including sequencing errors, repetitive elements, and uneven regional coverage, leads to ambiguities, false-positive and branched paths in the assembly graph, and early terminations in contig extension, limiting the completeness of assembled sequences. As a result, the assembled sequences are usually fragmented and exist in the suboptimal form of large numbers of contigs. Among the contigs, there are also certain (sometimes high) levels of falsely assembled contigs, due to chimeric joining. In addition, the gapped regions between the assembled contigs may not be filled completely. To overcome some of these limitations and increase assembly quality, the use of a reference genome, even from a remotely related species, can be very helpful. This reference-assisted assembly approach works especially well when scaffolding information from paired reads is not available or exhausted. With the quickly increasing number of sequenced genomes, improving assembly quality with this reference-assisted approach becomes more feasible. Some tools have recently been developed to provide this functionality, either as dedicated packages such as RaGOO [98], Chromosomer [99], and RACA [100], or components of existing assemblers including ALLPATHS-LG, IDBA-Hybrid, and Velvet.

With the constant advancements achieved in long read sequencing, the landscape of *de novo* genome assembly has been shifting. The publication of a gapless human genome assembly by the Telomere-to-Telomere (T2T) Consortium has demonstrated the utility of long reads in *de novo* genome assembly [101]. Built on this success, more and better designed assemblers are bound to be continuously developed to take full advantage of what new sequencing technologies have to offer. Undoubtedly such future developments will further overcome the limitations and challenges facing the community today.

References

1. Schatz MC, Delcher AL, Salzberg SL. Assembly of large genomes using second-generation sequencing. *Genome Res* 2010, 20(9):1165–1173.
2. Zerbino DR, Birney E. Velvet: algorithms for de novo short read assembly using de Bruijn graphs. *Genome Res* 2008, 18(5):821–829.
3. Simpson JT, Wong K, Jackman SD, Schein JE, Jones SJ, Birol I. ABySS: a parallel assembler for short read sequence data. *Genome Res* 2009, 19(6):1117–1123.
4. Li R, Zhu H, Ruan J, Qian W, Fang X, Shi Z, Li Y, Li S, Shan G, Kristiansen K *et al.* De novo assembly of human genomes with massively parallel short read sequencing. *Genome Res* 2010, 20(2):265–272.
5. English AC, Richards S, Han Y, Wang M, Vee V, Qu J, Qin X, Muzny DM, Reid JG, Worley KC *et al.* Mind the gap: upgrading genomes with Pacific Biosciences RS long-read sequencing technology. *PLoS One* 2012, 7(11):e47768.
6. Rayamajhi N, Cheng CC, Catchen JM. Evaluating Illumina-, Nanopore-, and PacBio-based genome assembly strategies with the bald notothen, Trematomus borchgrevinki. *G3* 2022, 12(11):jkac192.
7. van Heesch S, Kloosterman WP, Lansu N, Ruzius FP, Levandowsky E, Lee CC, Zhou S, Goldstein S, Schwartz DC, Harkins TT *et al.* Improving mammalian

genome scaffolding using large insert mate-pair next-generation sequencing. *BMC Genomics* 2013, 14:257.

8. Li R, Fan W, Tian G, Zhu H, He L, Cai J, Huang Q, Cai Q, Li B, Bai Y *et al.* The sequence and de novo assembly of the giant panda genome. *Nature* 2010, 463(7279):311–317.
9. Wang O, Chin R, Cheng X, Wu MKY, Mao Q, Tang J, Sun Y, Anderson E, Lam HK, Chen D *et al.* Efficient and unique cobarcoding of second-generation sequencing reads from long DNA molecules enabling cost-effective and accurate sequencing, haplotyping, and de novo assembly. *Genome Res* 2019, 29(5):798–808.
10. Chen Z, Pham L, Wu TC, Mo G, Xia Y, Chang PL, Porter D, Phan T, Che H, Tran H *et al.* Ultralow-input single-tube linked-read library method enables short-read second-generation sequencing systems to routinely generate highly accurate and economical long-range sequencing information. *Genome Res* 2020, 30(6):898–909.
11. Zheng GX, Lau BT, Schnall-Levin M, Jarosz M, Bell JM, Hindson CM, Kyriazopoulou-Panagiotopoulou S, Masquelier DA, Merrill L, Terry JM *et al.* Haplotyping germline and cancer genomes with high-throughput linked-read sequencing. *Nat Biotechnol* 2016, 34(3):303–311.
12. Nagarajan N, Pop M. Sequence assembly demystified. *Nature Rev Genet* 2013, 14(3):157–167.
13. Desai A, Marwah VS, Yadav A, Jha V, Dhaygude K, Bangar U, Kulkarni V, Jere A. Identification of optimum sequencing depth especially for de novo genome assembly of small genomes using next generation sequencing data. *PLoS One* 2013, 8(4):e60204.
14. Chen Y, Nie F, Xie SQ, Zheng YF, Dai Q, Bray T, Wang YX, Xing JF, Huang ZJ, Wang DP *et al.* Efficient assembly of nanopore reads via highly accurate and intact error correction. *Nature Commun* 2021, 12(1):60.
15. Chen Y, Zhang Y, Wang AY, Gao M, Chong Z. Accurate long-read de novo assembly evaluation with Inspector. *Genome Biol* 2021, 22(1):312.
16. Magoc T, Salzberg SL. FLASH: fast length adjustment of short reads to improve genome assemblies. *Bioinformatics* 2011, 27(21):2957–2963.
17. Zhang J, Kobert K, Flouri T, Stamatakis A. PEAR: a fast and accurate Illumina Paired-End reAd mergeR. *Bioinformatics* 2014, 30(5):614–620.
18. Aronesty E. Comparison of sequencing utility programs. *Open Bioinformatics J* 2013, 7(1):1–8.
19. Masella AP, Bartram AK, Truszkowski JM, Brown DG, Neufeld JD. PANDAseq: paired-end assembler for Illumina sequences. *BMC Bioinformatics* 2012, 13:31.
20. Rognes T, Flouri T, Nichols B, Quince C, Mahe F. VSEARCH: a versatile open source tool for metagenomics. *PeerJ* 2016, 4:e2584.
21. Li H. BFC: correcting Illumina sequencing errors. *Bioinformatics* 2015, 31(17):2885–2887.
22. Heo Y, Wu XL, Chen D, Ma J, Hwu WM. BLESS: bloom filter-based error correction solution for high-throughput sequencing reads. *Bioinformatics* 2014, 30(10):1354–1362.
23. Song L, Florea L, Langmead B. Lighter: fast and memory-efficient sequencing error correction without counting. *Genome Biol* 2014, 15(11):509.

24. Liu Y, Schroder J, Schmidt B. Musket: a multistage k-mer spectrum-based error corrector for Illumina sequence data. *Bioinformatics* 2013, 29(3):308–315.
25. Schulz MH, Weese D, Holtgrewe M, Dimitrova V, Niu S, Reinert K, Richard H. Fiona: a parallel and automatic strategy for read error correction. *Bioinformatics* 2014, 30(17):i356–363.
26. Salmela L, Schroder J. Correcting errors in short reads by multiple alignments. *Bioinformatics* 2011, 27(11):1455–1461.
27. Gnerre S, Maccallum I, Przybylski D, Ribeiro FJ, Burton JN, Walker BJ, Sharpe T, Hall G, Shea TP, Sykes S *et al.* High-quality draft assemblies of mammalian genomes from massively parallel sequence data. *Proc Natl Acad Sci U S A* 2011, 108(4):1513–1518.
28. Simpson JT, Durbin R. Efficient de novo assembly of large genomes using compressed data structures. *Genome Res* 2012, 22(3):549–556.
29. Pevzner PA, Tang H, Waterman MS. An Eulerian path approach to DNA fragment assembly. *Proc Natl Acad Sci U S A* 2001, 98(17):9748–9753.
30. Marcais G, Kingsford C. A fast, lock-free approach for efficient parallel counting of occurrences of k-mers. *Bioinformatics* 2011, 27(6):764–770.
31. Rhie A, Walenz BP, Koren S, Phillippy AM. Merqury: reference-free quality, completeness, and phasing assessment for genome assemblies. *Genome Biol* 2020, 21(1):245.
32. Ranallo-Benavidez TR, Jaron KS, Schatz MC. GenomeScope 2.0 and Smudgeplot for reference-free profiling of polyploid genomes. *Nat Commun* 2020, 11(1):1432.
33. Wang JR, Holt J, McMillan L, Jones CD. FMLRC: Hybrid long read error correction using an FM-index. *BMC Bioinformatics* 2018, 19(1):50.
34. Koren S, Schatz MC, Walenz BP, Martin J, Howard JT, Ganapathy G, Wang Z, Rasko DA, McCombie WR, Jarvis ED *et al.* Hybrid error correction and de novo assembly of single-molecule sequencing reads. *Nat Biotechnol* 2012, 30(7):693–700.
35. Salmela L, Rivals E. LoRDEC: accurate and efficient long read error correction. *Bioinformatics* 2014, 30(24):3506–3514.
36. Au KF, Underwood JG, Lee L, Wong WH. Improving PacBio long read accuracy by short read alignment. *PLoS One* 2012, 7(10):e46679.
37. Goodwin S, Gurtowski J, Ethe-Sayers S, Deshpande P, Schatz MC, McCombie WR. Oxford Nanopore sequencing, hybrid error correction, and de novo assembly of a eukaryotic genome. *Genome Res* 2015, 25(11):1750–1756.
38. Hackl T, Hedrich R, Schultz J, Forster F. proovread: large-scale high-accuracy PacBio correction through iterative short read consensus. *Bioinformatics* 2014, 30(21):3004–3011.
39. Baid G, Cook DE, Shafin K, Yun T, Llinares-López F, Berthet Q, Wenger AM, Rowell WJ, Nattestad M, Yang H *et al.* DeepConsensus improves the accuracy of sequences with a gap-aware sequence transformer. *Nat Biotechnol* 2023, 41(2):232–238.
40. Koren S, Walenz BP, Berlin K, Miller JR, Bergman NH, Phillippy AM. Canu: scalable and accurate long-read assembly via adaptive k-mer weighting and repeat separation. *Genome Res* 2017, 27(5):722–736.
41. Cheng H, Concepcion GT, Feng X, Zhang H, Li H. Haplotype-resolved de novo assembly using phased assembly graphs with Hifiasm. *Nat Methods* 2021, 18(2):170–175.

42. Chin CS, Peluso P, Sedlazeck FJ, Nattestad M, Concepcion GT, Clum A, Dunn C, O'Malley R, Figueroa-Balderas R, Morales-Cruz A *et al.* Phased diploid genome assembly with single-molecule real-time sequencing. *Nat Methods* 2016, 13(12):1050–1054.
43. Nowoshilow S, Schloissnig S, Fei JF, Dahl A, Pang AWC, Pippel M, Winkler S, Hastie AR, Young G, Roscito JG *et al.* The axolotl genome and the evolution of key tissue formation regulators. *Nature* 2018, 554(7690):50–55.
44. Xiao CL, Chen Y, Xie SQ, Chen KN, Wang Y, Han Y, Luo F, Xie Z. MECAT: fast mapping, error correction, and de novo assembly for single-molecule sequencing reads. *Nat Methods* 2017, 14(11):1072–1074.
45. Salmela L, Walve R, Rivals E, Ukkonen E. Accurate self-correction of errors in long reads using de Bruijn graphs. *Bioinformatics* 2017, 33(6):799–806.
46. Bao E, Xie F, Song C, Song D. FLAS: fast and high-throughput algorithm for PacBio long-read self-correction. *Bioinformatics* 2019, 35(20):3953–3960.
47. Berlin K, Koren S, Chin CS, Drake JP, Landolin JM, Phillippy AM. Assembling large genomes with single-molecule sequencing and locality-sensitive hashing. *Nat Biotechnol* 2015, 33(6):623–630.
48. Lander ES, Waterman MS. Genomic mapping by fingerprinting random clones: a mathematical analysis. *Genomics* 1988, 2(3):231–239.
49. Warren RL, Sutton GG, Jones SJ, Holt RA. Assembling millions of short DNA sequences using SSAKE. *Bioinformatics* 2007, 23(4):500–501.
50. Dohm JC, Lottaz C, Borodina T, Himmelbauer H. SHARCGS, a fast and highly accurate short-read assembly algorithm for de novo genomic sequencing. *Genome Res* 2007, 17(11):1697–1706.
51. Jeck WR, Reinhardt JA, Baltrus DA, Hickenbotham MT, Magrini V, Mardis ER, Dangl JL, Jones CD. Extending assembly of short DNA sequences to handle error. *Bioinformatics* 2007, 23(21):2942–2944.
52. Hernandez D, Francois P, Farinelli L, Osteras M, Schrenzel J. De novo bacterial genome sequencing: millions of very short reads assembled on a desktop computer. *Genome Res* 2008, 18(5):802–809.
53. Li H. Exploring single-sample SNP and INDEL calling with whole-genome de novo assembly. *Bioinformatics* 2012, 28(14):1838–1844.
54. Li H. Minimap and miniasm: fast mapping and de novo assembly for noisy long sequences. *Bioinformatics* 2016, 32(14):2103–2110.
55. Kamath GM, Shomorony I, Xia F, Courtade TA, Tse DN. HINGE: long-read assembly achieves optimal repeat resolution. *Genome Res* 2017, 27(5):747–756.
56. Chin C-S, Khalak A. Human Genome Assembly in 100 Minutes. *bioRxiv* 2019, doi: https://doi.org/10.1101/705616
57. Shafin K, Pesout T, Lorig-Roach R, Haukness M, Olsen HE, Bosworth C, Armstrong J, Tigyi K, Maurer N, Koren S *et al.* Nanopore sequencing and the Shasta toolkit enable efficient de novo assembly of eleven human genomes. *Nat Biotechnol* 2020, 38(9):1044–1053.
58. Vaser R, Šikić M. Time-and memory-efficient genome assembly with Raven. *Nat Comput Sci* 2021, 1(5):332–336.
59. NextDenovo (https://github.com/Nextomics/NextDenovo)
60. Liu H, Wu S, Li A, Ruan J. SMARTdenovo: a de novo assembler using long noisy reads. *Gigabyte* 2021:1–9.

61. Nurk S, Walenz BP, Rhie A, Vollger MR, Logsdon GA, Grothe R, Miga KH, Eichler EE, Phillippy AM, Koren S. HiCanu: accurate assembly of segmental duplications, satellites, and allelic variants from high-fidelity long reads. *Genome Res* 2020, 30(9):1291–1305.
62. Myers EW. Toward simplifying and accurately formulating fragment assembly. *J Comput Biol* 1995, 2(2):275–290.
63. Gonnella G, Kurtz S. Readjoiner: a fast and memory efficient string graph-based sequence assembler. *BMC Bioinformatics* 2012, 13:82.
64. Luo R, Liu B, Xie Y, Li Z, Huang W, Yuan J, He G, Chen Y, Pan Q, Liu Y *et al.* SOAPdenovo2: an empirically improved memory-efficient short-read de novo assembler. *GigaScience* 2012, 1(1):18.
65. Nurk S, Bankevich A, Antipov D, Gurevich AA, Korobeynikov A, Lapidus A, Prjibelski AD, Pyshkin A, Sirotkin A, Sirotkin Y *et al.* Assembling single-cell genomes and mini-metagenomes from chimeric MDA products. *J Comput Biol* 2013, 20(10):714–737.
66. Kolmogorov M, Yuan J, Lin Y, Pevzner PA. Assembly of long, error-prone reads using repeat graphs. *Nat Biotechnol* 2019, 37(5):540–546.
67. Ruan J, Li H. Fast and accurate long-read assembly with wtdbg2. *Nat Methods* 2020, 17(2):155–158.
68. Zimin AV, Marcais G, Puiu D, Roberts M, Salzberg SL, Yorke JA. The MaSuRCA genome assembler. *Bioinformatics* 2013, 29(21):2669–2677.
69. Ye C, Hill CM, Wu S, Ruan J, Ma ZS. DBG2OLC: Efficient Assembly of Large Genomes Using Long Erroneous Reads of the Third Generation Sequencing Technologies. *Sci Rep* 2016, 6:31900.
70. Di Genova A, Buena-Atienza E, Ossowski S, Sagot MF. Efficient hybrid de novo assembly of human genomes with WENGAN. *Nat Biotechnol* 2021, 39(4):422–430.
71. Walker BJ, Abeel T, Shea T, Priest M, Abouelliel A, Sakthikumar S, Cuomo CA, Zeng Q, Wortman J, Young SK *et al.* Pilon: an integrated tool for comprehensive microbial variant detection and genome assembly improvement. *PLoS One* 2014, 9(11):e112963.
72. Medaka (https://github.com/nanoporetech/medaka)
73. Vaser R, Sovic I, Nagarajan N, Sikic M. Fast and accurate de novo genome assembly from long uncorrected reads. *Genome Res* 2017, 27(5):737–746.
74. Loman NJ, Quick J, Simpson JT. A complete bacterial genome assembled de novo using only nanopore sequencing data. *Nat Methods* 2015, 12(8):733–735.
75. Hu J, Fan J, Sun Z, Liu S. NextPolish: a fast and efficient genome polishing tool for long-read assembly. *Bioinformatics* 2020, 36(7):2253–2255.
76. Zimin AV, Salzberg SL. The genome polishing tool POLCA makes fast and accurate corrections in genome assemblies. *PLoS Comput Biol* 2020, 16(6):e1007981.
77. Huang N, Nie F, Ni P, Luo F, Gao X, Wang J. NeuralPolish: a novel Nanopore polishing method based on alignment matrix construction and orthogonal Bi-GRU Networks. *Bioinformatics* 2021, 37(19):3120–3127.
78. Boetzer M, Pirovano W. SSPACE-LongRead: scaffolding bacterial draft genomes using long read sequence information. *BMC Bioinformatics* 2014, 15:211.

79. Warren RL, Yang C, Vandervalk BP, Behsaz B, Lagman A, Jones SJ, Birol I. LINKS: Scalable, alignment-free scaffolding of draft genomes with long reads. *GigaScience* 2015, 4:35.
80. Gao S, Bertrand D, Chia BK, Nagarajan N. OPERA-LG: efficient and exact scaffolding of large, repeat-rich eukaryotic genomes with performance guarantees. *Genome Biol* 2016, 17:102.
81. Qin M, Wu S, Li A, Zhao F, Feng H, Ding L, Ruan J. LRScaf: improving draft genomes using long noisy reads. *BMC Genomics* 2019, 20(1):955.
82. SMIS (Single Molecular Integrative Scaffolding) (www.sanger.ac.uk/tool/smis/)
83. Nguyen SH, Cao MD, Coin LJM. Real-time resolution of short-read assembly graph using ONT long reads. *PLoS Comput Biol* 2021, 17(1):e1008586.
84. Putnam NH, O'Connell BL, Stites JC, Rice BJ, Blanchette M, Calef R, Troll CJ, Fields A, Hartley PD, Sugnet CW *et al.* Chromosome-scale shotgun assembly using an in vitro method for long-range linkage. *Genome Res* 2016, 26(3):342–350.
85. Dudchenko O, Batra SS, Omer AD, Nyquist SK, Hoeger M, Durand NC, Shamim MS, Machol I, Lander ES, Aiden AP *et al.* De novo assembly of the Aedes aegypti genome using Hi-C yields chromosome-length scaffolds. *Science* 2017, 356(6333):92–95.
86. Ghurye J, Rhie A, Walenz BP, Schmitt A, Selvaraj S, Pop M, Phillippy AM, Koren S. Integrating Hi-C links with assembly graphs for chromosome-scale assembly. *PLoS Comput Biol* 2019, 15(8):e1007273.
87. Yeo S, Coombe L, Warren RL, Chu J, Birol I. ARCS: scaffolding genome drafts with linked reads. *Bioinformatics* 2018, 34(5):725–731.
88. Kuleshov V, Snyder MP, Batzoglou S. Genome assembly from synthetic long read clouds. *Bioinformatics* 2016, 32(12):i216–i224.
89. Guo L, Xu M, Wang W, Gu S, Zhao X, Chen F, Wang O, Xu X, Seim I, Fan G *et al.* SLR-superscaffolder: a de novo scaffolding tool for synthetic long reads using a top-to-bottom scheme. *BMC Bioinformatics* 2021, 22(1):158.
90. Kosugi S, Hirakawa H, Tabata S. GMcloser: closing gaps in assemblies accurately with a likelihood-based selection of contig or long-read alignments. *Bioinformatics* 2015, 31(23):3733–3741.
91. Paulino D, Warren RL, Vandervalk BP, Raymond A, Jackman SD, Birol I. Sealer: a scalable gap-closing application for finishing draft genomes. *BMC Bioinformatics* 2015, 16:230.
92. Xu GC, Xu TJ, Zhu R, Zhang Y, Li SQ, Wang HW, Li JT. LR_Gapcloser: a tiling path-based gap closer that uses long reads to complete genome assembly. *GigaScience* 2019, 8(1):giy157.
93. Xu M, Guo L, Gu S, Wang O, Zhang R, Peters BA, Fan G, Liu X, Xu X, Deng L *et al.* TGS-GapCloser: A fast and accurate gap closer for large genomes with low coverage of error-prone long reads. *GigaScience* 2020, 9(9):giaa094.
94. Zimin AV, Salzberg SL. The SAMBA tool uses long reads to improve the contiguity of genome assemblies. *PLoS Comput Biol* 2022, 18(2):e1009860.
95. Schmeing S, Robinson MD. Gapless provides combined scaffolding, gap filling and assembly correction with long reads. *bioRxiv* 2022, doi: https://doi.org/10.1101/2022.03.08.483466

96. Simao FA, Waterhouse RM, Ioannidis P, Kriventseva EV, Zdobnov EM. BUSCO: assessing genome assembly and annotation completeness with single-copy orthologs. *Bioinformatics* 2015, 31(19):3210–3212.
97. Gurevich A, Saveliev V, Vyahhi N, Tesler G. QUAST: quality assessment tool for genome assemblies. *Bioinformatics* 2013, 29(8):1072–1075.
98. Alonge M, Soyk S, Ramakrishnan S, Wang X, Goodwin S, Sedlazeck FJ, Lippman ZB, Schatz MC. RaGOO: fast and accurate reference-guided scaffolding of draft genomes. *Genome Biol* 2019, 20(1):224.
99. Tamazian G, Dobrynin P, Krasheninnikova K, Komissarov A, Koepfli KP, O'Brien SJ. Chromosomer: a reference-based genome arrangement tool for producing draft chromosome sequences. *GigaScience* 2016, 5(1):38.
100. Kim J, Larkin DM, Cai Q, Asan, Zhang Y, Ge RL, Auvil L, Capitanu B, Zhang G, Lewin HA *et al.* Reference-assisted chromosome assembly. *Proc Natl Acad Sci U S A* 2013, 110(5):1785–1790.
101. Nurk S, Koren S, Rhie A, Rautiainen M, Bzikadze AV, Mikheenko A, Vollger MR, Altemose N, Uralsky L, Gershman A *et al.* The complete sequence of a human genome. *Science* 2022, 376(6588):44–53.

13

Mapping Protein-DNA Interactions with ChIP-Seq

13.1 Principle of ChIP-Seq

Without the involvement of DNA-interacting proteins, the information coded in DNA cannot be accessed, transcribed, and maintained. Besides a large number of transcription factors and coactivators, key DNA-interacting proteins include histones, DNA and RNA polymerases, and enzymes for DNA repair and modification (e.g., methylation). Through their DNA-interacting domains, such as helix-turn-helix, zinc finger, and leucine zipper domains, these proteins interact with their DNA targets by hydrogen bonding, hydrophobic interactions, or base stacking. Because the intimate relationship between DNA and these proteins plays an important role in the functioning of the genome, studying how proteins and DNA interact and where DNA-interacting proteins bind across the genome provides key insights into the many roles these proteins play in various aspects of genomic function, including information exposure, transcription, and maintenance.

ChIP-seq is an NGS-based technology to locate binding sites of a DNA-interacting protein in the genome. An exemplary scenario in using ChIP-seq is to study transcription factor binding profiles in the genome under different conditions, such as development stages or pathological conditions. To achieve this, the protein of interest is first cross-linked covalently to DNA in cells with a chemical agent, usually formaldehyde (Figure 13.1). Then the cells are disrupted, and subsequently sonicated or enzymatically digested to shear chromatin into fragments that contain 100–300 bp DNA, followed by enrichment of the target protein with its bound DNA by immunoprecipitation using an antibody specific for the protein. Subsequently, the enriched protein-DNA complex is dissociated by reversing the cross-links previously formed between them, and the released DNA fragments are subjected to NGS. One key experimental factor in the ChIP-seq process is the quality of the antibody used in the enrichment step, as the use of a poor-quality antibody can lead to high experimental noise due to non-specific precipitation of DNA fragments.

Based on the size the region that they bind, DNA-interacting proteins can be divided into three groups: (1) punctate-binding: these proteins, usually

DOI: 10.1201/9780429329180-16

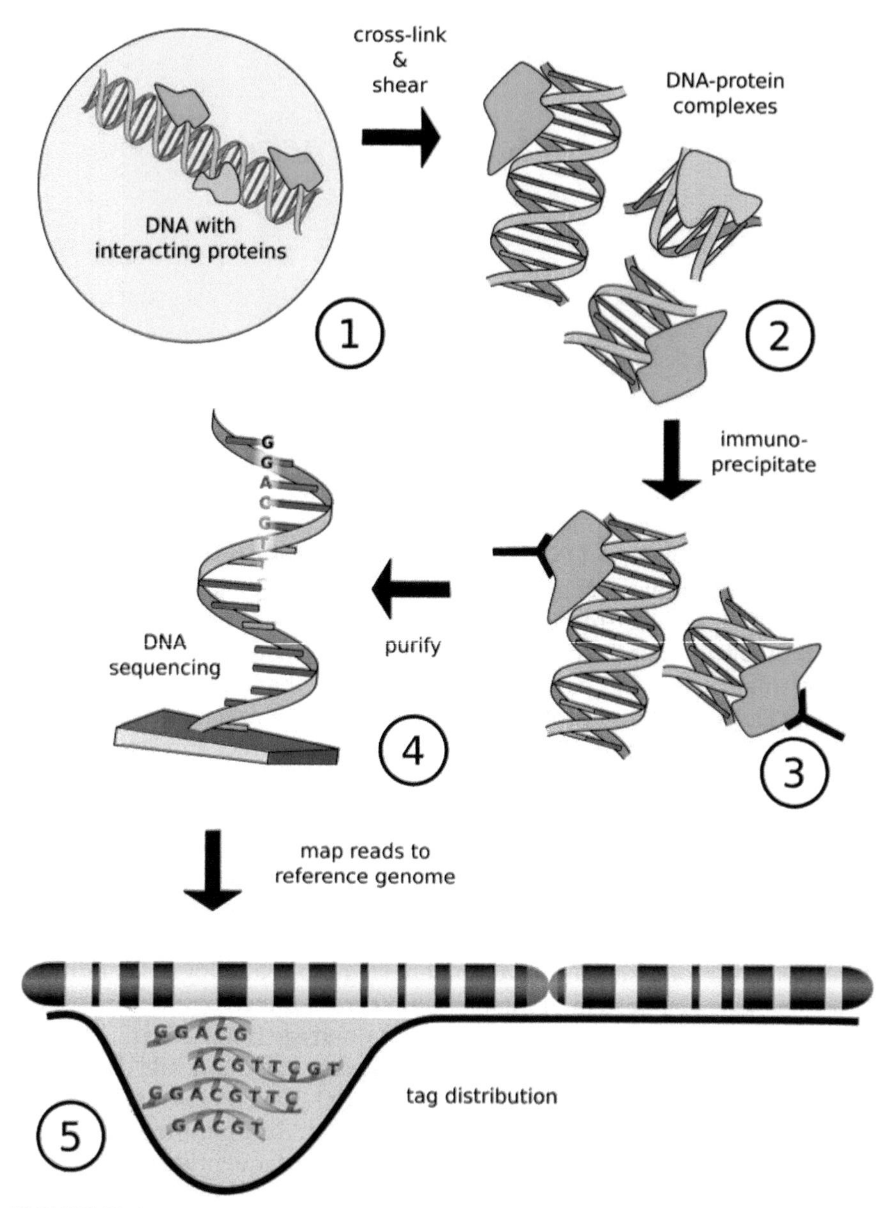

FIGURE 13.1
The basic steps of ChIP-seq. (From AM Szalkowski, CD Schmid, Rapid innovation in ChIP-seq peak-calling algorithms is outdistancing benchmarking efforts, *Briefings in Bioinformatics* 2011, 12(6):626–33. With permission.)

transcription factors, bind to a genomic region that is a few hundred bp or less in size; (2) broad-binding: chemically modified histones (such as histone H3 lysine 27 trimethylation or H3K27me3, and H3 lysine 79 dimethylation or H3K79me2), or other proteins associated with chromatin domains, bind to a much larger area of the genome up to several hundred thousand bp; and (3) mixed or intermediate-binding: these include some histone modifications (such as H3 lysine 27 acetylation or H3K27ac, and H3 lysine 4 trimethylation or H3K4me3), and proteins such as RNA polymerase II, which bind to regions of the genome that are a few thousand bp in size.

13.2 Experimental Design

13.2.1 Experimental Control

Appropriate control for a ChIP-seq experiment is key to accounting for artifacts or biases that might be introduced into the experimental process. These artifacts and biases may include potential antibody cross-reactivity with non-specific protein factors, higher signal from open chromatin regions (since they are easier to be fragmented than closed regions [1]), and uneven sequencing of captured genomic regions due to variations in base composition. Two major types of controls are usually set up for ChIP-seq signal adjustment. One is input control, i.e., chromatins extracted from cells or tissues, which are subjected to the same cross-linking and fragmentation procedure but without the immunoprecipitation process. The other is "mock" control, which is processed by the same procedure including immunoprecipitation; the immunoprecipitation, however, is carried out using an irrelevant antibody (e.g., IgG). While it may seem to serve as the better control between the two, the "mock" control often produces much less DNA for sequencing than real experimental ChIP samples. Although sequencing can be carried out on amplified DNA in this circumstance, the amplification process adds additional artifact and bias to the sequencing data, which justifies the use of input DNA as experimental control in many cases.

13.2.2 Library Preparation

To prepare ChIP DNA for library prep, 1–10 million cells are typically needed [2]. Within this range, studies of broad-binding protein factors require less cells than those of punctate-binding proteins. In terms of the amount of immunoprecipitated DNA required for library prep, while it is often suggested to start with 5–10 ng ChIP DNA, it is common to obtain less DNA, which often still generates high-quality libraries for sequencing. For library prep, the steps involved include end repair, A-tailing, incorporation of 3′ and 5′ adapter sequences, size selection, and PCR amplification for final library generation. The number of cycles used in the PCR amplification step is important, and overamplification should be avoided as it can affect fragment representation and library complexity. If obtaining large numbers of cells is challenging or not feasible, alternative methods such as CUT&RUN (or Cleavage Under Targets and Release Using Nuclease), which can generate high-quality data from as low as 100 cells due to its low background [3], can be employed.

13.2.3 Sequencing Length and Depth

Single-end 50–75 base reads are typically used for ChIP-seq libraries [4]. Paired-end sequencing can also be used to help increase mapping, especially for proteins that may bind to repetitive regions of the genome, but is not required in most cases. How many reads to obtain for a ChIP-seq experiment depends on the size of the genome and how many binding sites the protein of interest has in the genome. A good indication of having reached sufficient sequence depth is when the number of protein binding sites reaches plateau with increasing numbers of reads. As a practical guide, for analyzing a transcription factor that has thousands of binding sites in the mammalian genome, 20 million reads may be sufficient. Fewer reads may suffice for a smaller genome, while more reads are required for proteins that bind to the genome at a higher frequency or with larger "footprint." To locate binding regions of these proteins, including histone marks, 30–50 million reads might be needed for a genome at the scale of the human genome [5]. Higher sequencing depth is required for control samples in order to obtain background signals from most regions of the genome. If using CUT&RUN when starting from low numbers of cells, paired-end sequencing is suggested, but short read length (e.g., 25 x 25 bases) at a depth of 5 million read pairs per sample is usually sufficient [3].

13.2.4 Replication

To examine the reproducibility of a ChIP-seq experiment, and also reduce FDR, replicate samples should be used. If a protein of interest binds to regions of the genome with high affinity, the bound regions should be identified in replicate samples. Regions that are not identified in replicates are most possibly due to experimental noise. Pearson correlation coefficient (PCC) between biological replicates serves as a measurement of experimental reproducibility, while irreproducible discovery rate (IDR) is another such metric. The calculation and usage of PCC and IDR are detailed later in this chapter.

13.3 Read Mapping, Normalization, and Peak Calling

13.3.1 Data Quality Control and Read Mapping

The first step in ChIP-seq data analysis (Figure 13.2) is to evaluate reads quality. The quality control (QC) metrics detailed in Chapter 5 need to be examined. If necessary, low-quality reads should be filtered out and low-quality bases trimmed off. Other aspects of determining ChIP-seq data quality include assessing library complexity and experimental reproducibility between replicates. Assessment of library complexity is important, as low-complexity libraries, caused by limited starting material, over-crosslinking, low antibody quality, or PCR over-amplification, can lead to skewed reads distribution. Library complexity, defined as the proportion of non-redundant reads, can be examined with tools such as Preseq [6], or using PCR bottleneck coefficient (PBC), which is the ratio of N1/Nd, with N1 being the number

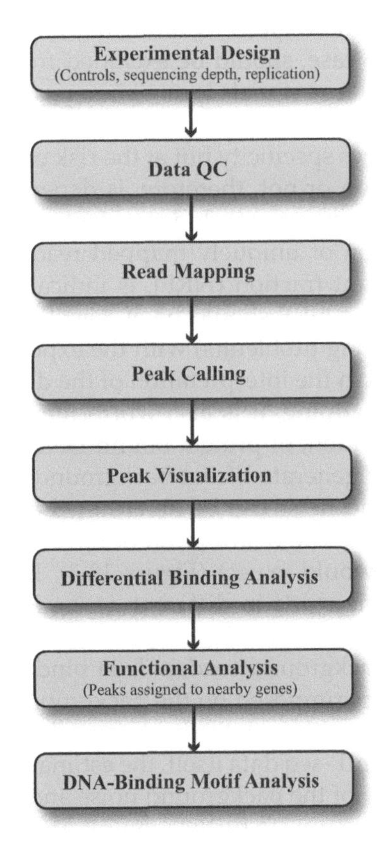

FIGURE 13.2
Basic ChIP-seq data analysis workflow.

of non-redundant, uniquely mapped reads, and Nd the number of uniquely mapped reads. PBC is calculated by a component of ENCODE Software Tools [7] called phantompeakqualtools, which, besides PBC, also calculates other quality metrics, i.e., normalized strand cross-correlation (NSC) and relative strand cross-correlation coefficients (RSC), as measures of sequence enrichment (NSC and RSC will be discussed more in the subsequent section on "Peak Calling"). The assessment of experimental reproducibility is usually performed by analyzing IDR, which can be calculated using another component of ENCODE Software Tools called "Irreproducible Discovery Rate (IDR)" [8].

The assessment of library complexity and experimental reproducibility by the ENCODE Software Tools, or the use of other ChIP-seq QC tools such as CHANCE [9], deepTools2 [10], and SSP [11], requires mapping the filtered/trimmed reads to a reference genome. For this mapping, the mappers introduced in Chapter 5, including Bowtie2 or BWA, can be used. One mapping parameter that directly affects subsequent binding site detection sensitivity and specificity is whether to use multireads, i.e., reads that map to multiple genomic regions. Multireads may represent background noise and, if this is the case, should be excluded from further analysis, but they may also represent true signals located in repeats or duplicated regions. Including them increases sensitivity but at the expense of higher FDR, while excluding them improves specificity but at the risk of losing true signals. The choice for their inclusion or not, therefore, is dependent on whether sensitivity or specificity is given priority. Independent of whether multireads are used or not, the fraction of uniquely mapped reads among all reads, also called the non-redundant fraction (NRF), is indicative of data quality. Per ENCODE guidelines, the NRF should be at least 0.8 [12]. If it drops below 0.5, it indicates concerning problem(s) with the experimental procedure and caution should be used in the interpretation of the data.

For ChIP-seq read mapping, it is also worth mentioning that ChIP is an enrichment, not purification, of protein-bound DNA sequences. As a result, more reads are usually generated from background noise than from bound regions. The background noise can be determined empirically with the use of input control samples. The distribution of observed background noise is not random as many would expect (Figure 13.3). Instead, it is affected by the density of mappable reads in different genomic regions, and the local chromatin structure (e.g., as mentioned previously, open chromatin structure generates more background reads). True binding signals in ChIP-seq samples are usually superimposed on the background noise. At the absence of input control samples, although the background noise could be estimated from modeling of the ChIP-seq data itself, the estimation cannot fully capture the inherent complexity of the background noise and therefore experimental controls should always be run. To further complicate the situation, the degree of protein-binding sequence enrichment may also vary from location to

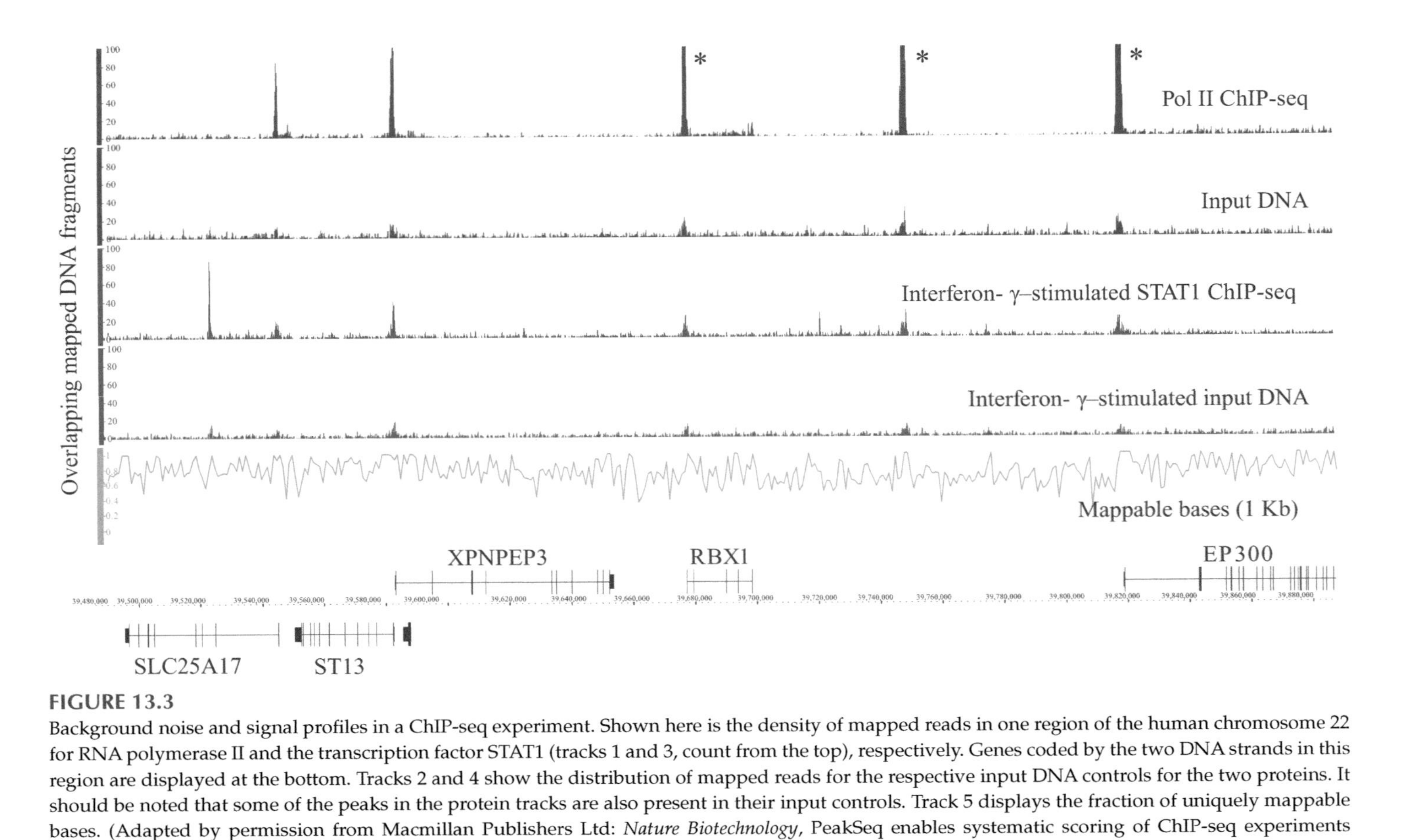

FIGURE 13.3
Background noise and signal profiles in a ChIP-seq experiment. Shown here is the density of mapped reads in one region of the human chromosome 22 for RNA polymerase II and the transcription factor STAT1 (tracks 1 and 3, count from the top), respectively. Genes coded by the two DNA strands in this region are displayed at the bottom. Tracks 2 and 4 show the distribution of mapped reads for the respective input DNA controls for the two proteins. It should be noted that some of the peaks in the protein tracks are also present in their input controls. Track 5 displays the fraction of uniquely mappable bases. (Adapted by permission from Macmillan Publishers Ltd: *Nature Biotechnology*, PeakSeq enables systematic scoring of ChIP-seq experiments relative to controls, J Rozowsky, G Euskirchen, RK Auerbach, ZD Zhang, T Gibson, R Bjornson, N Carriero, et al., copyright 2009.)

location, with some having strong signals while others more modest signals. The degree of enrichment at each location is not necessarily a reflection of their biological importance, as those with more modest enrichment may be equally important as those at the top of the enrichment list.

After mapping, reproducibility between replicate samples and overall similarity between different samples can be examined using established tools. For example, the multiBamSummary tool in deepTools2 can be used to check reads coverage across the entire genome in "bins" mode from two or more input BAM files, and another tool called plotCorrelation can take the output to compute and visualize sample correlations. The plotFingerprint tool in the same toolset can be used to visualize cumulative reads enrichment as an indication of target DNA enrichment efficiency. This tool is most informative for punctate-binding proteins such as transcription factors. Besides sample correlations and the other aforementioned QC measures such as PBC, additional QC analyses can also be performed. For example, visualization of the distribution of mapped reads in the genome, using the visualization tools introduced in Chapter 5, can offer further clues on data quality. This is especially true when some specific binding regions have already been known for the protein of interest. In comparison to those from input control samples, sequence reads from ChIP samples should show strong clustering in these regions.

13.3.2 Peak Calling

Peak calling, the process of finding regions of the genome to which the protein of interest binds, is a key step in ChIP-seq data analysis. It is basically achieved through locating regions where reads are mapped at levels significantly above the background. As this process is also applicable to ATAC-seq (assay for transposase-accessible chromatin using sequencing) and DNase-seq (DNase I hypersensitive sites sequencing), both of which aim to locate transcriptionally accessible regions of the genome, many of the methods introduced below can also be used for analysis of ATAC-seq and DNase-seq data. Among currently available peak calling methods, the simplest is to count the total number of reads mapped along the genome, and call each location with the number of mapped reads over a threshold as a peak. Due to the inherent complexity in signal generation, including uneven background noise and other confounding experimental factors, this approach is overly simplistic. Among the experimental factors, the way the immunoprecipitated DNA fragments are sequenced on most platforms has a direct influence on how peaks are called. Since the reads are usually short, only one end or both ends of a fragment, depending on whether single-end or paired-end reads are produced, are sequenced. To locate a target protein's binding regions, which are represented by the immunoprecipitated DNA fragments and not just the generated reads, peak calling algorithms need to either shift or extend the reads to cover the actual binding areas. For example, MACS2 shifts reads mapped to the two opposite strands toward the 3′ ends, based on the average DNA fragment length, to cover the most likely protein binding location [13]. SPP uses a similar strategy to shift each read mapped to either strand relative to each other until strand cross-correlation coefficient reaches the highest level at the shift that equals to the average length of the DNA fragments (Figure 13.4) [14]. PeakSeq uses an alternative approach to extend the reads in the 3′ direction to reach the average length of DNA fragments in the library [15].

The reads shift approach and the strand cross-correlation profile shown in Figure 13.4 can also be used to evaluate ChIP-seq data quality. When using short reads (usually less than 100 bases) to analyze large target genomes, which usually results in a significant number of reads being mapped to multiple genomic locations, a "phantom" peak also exists at a shift that equals to the read length (Figure 13.5). If a run is successful, the fragment-length ChIP peak should be significantly higher than the read-length "phantom" peak, as well as the background signal. The aforementioned ENCODE software phantompeakqualtools provides two indices for the examination of strand cross-correlation, i.e., NSC, being the ratio of the cross-correlation coefficient at the fragment-length peak over that of the background, and RSC, as the ratio of background-adjusted cross-correlation coefficient at the fragment-length peak over that at the "phantom" peak.

Shifting reads mapped to the positive and negative strands toward the center, or extending reads to reach the average fragment length, in order to count the number of aggregated reads at each base pair position is the first sub-step to peak calling. As illustrated in Figure 13.6, peak calling involves multiple substeps. First, a signal profile is created through smoothing of aggregated read count across each chromosome. Subsequently, background noise needs to be defined and the signals along the genome need be adjusted for the background. One simple approach is to subtract read counts in control sample, if available, from the signal across the genome, or use the signal/noise ratio. Subsequently, to call peaks from the background-adjusted ChIP-seq signal, often-used approaches include using absolute signal strength, signal enrichment in relation to background noise, or a combination of both. To facilitate determination of signal enrichment, statistical significance is often computed using Poisson or negative binomial distributions. Empirical estimation of false discovery rate (FDR) can be carried out by first calling peaks using control data (i.e., false positives), and subsequently calculating the ratio of peaks called from the control to those called from the ChIP sample. Alternatively, the Benjamini–Hochberg approach introduced in Chapter 7 can also be applied to correct for multiple comparisons. After peak calling, artifactual peaks need to be filtered out, including those that contain only one or a few reads that are most possibly due to PCR artifacts, or those that involve significantly imbalanced numbers of reads on the two strands (see Figure 13.6).

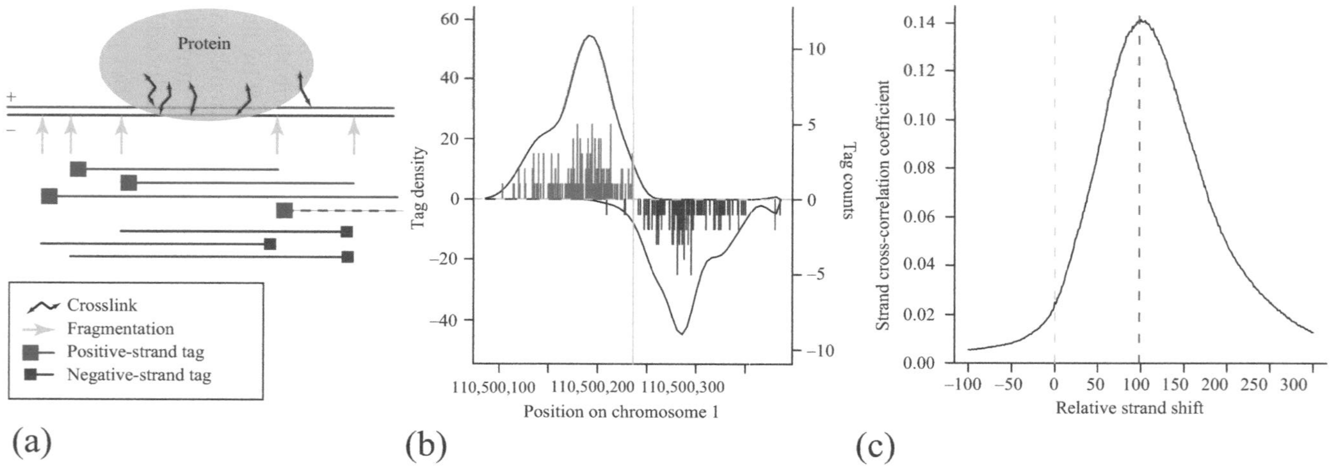

FIGURE 13.4
The distribution of ChIP-seq reads around the actual binding region and their positional shift on the two DNA strands. (a) How ChIP-seq reads are produced from cross-linked and fragmented DNA. The cross-linking between the protein and the bound DNA can occur at different sites, as is the fragmentation of the DNA. Each fragment is read at its 5′ end (indicated by the squares). These reads, serving as sequence tags of each fragment, are clustered around the actual binding region from the two sides depending on which strand they come from. The dashed red line depicts a fragment from a long cross-link. (b) The distribution of sequence tag signal around the binding region. Vertical lines represent counts of sequence tags whose 5′ end maps to each nucleotide position on the positive (red) and negative (blue) strands. The solid curves represent smoothed tag density along each strand. Since the two curves approach the binding site from the two sides, there is a gap between their peaks. (c) Strand cross-correlation associated with shifting the strands across the gap. Before shifting the strands, Pearson correlation coefficient is calculated between tag density of the two strands. When sequence tags mapped to the two strands are shifted relative to each other (shown on the x-axis), the Pearson correlation coefficient gradually changes (y-axis). The dashed gray line at x=0 corresponds to the strand cross-correlation before the shift, while the dashed red line at the peak to the highest cross-correlation coefficient at the strand shift that equals to the average length of the DNA fragments. (Adapted by permission from Macmillan Publishers Ltd: *Nature Biotechnology*, Design and analysis of ChIP-seq experiments for DNA-binding proteins, PV Kharchenko, MY Tolstorukov, PJ Park, copyright 2008.)

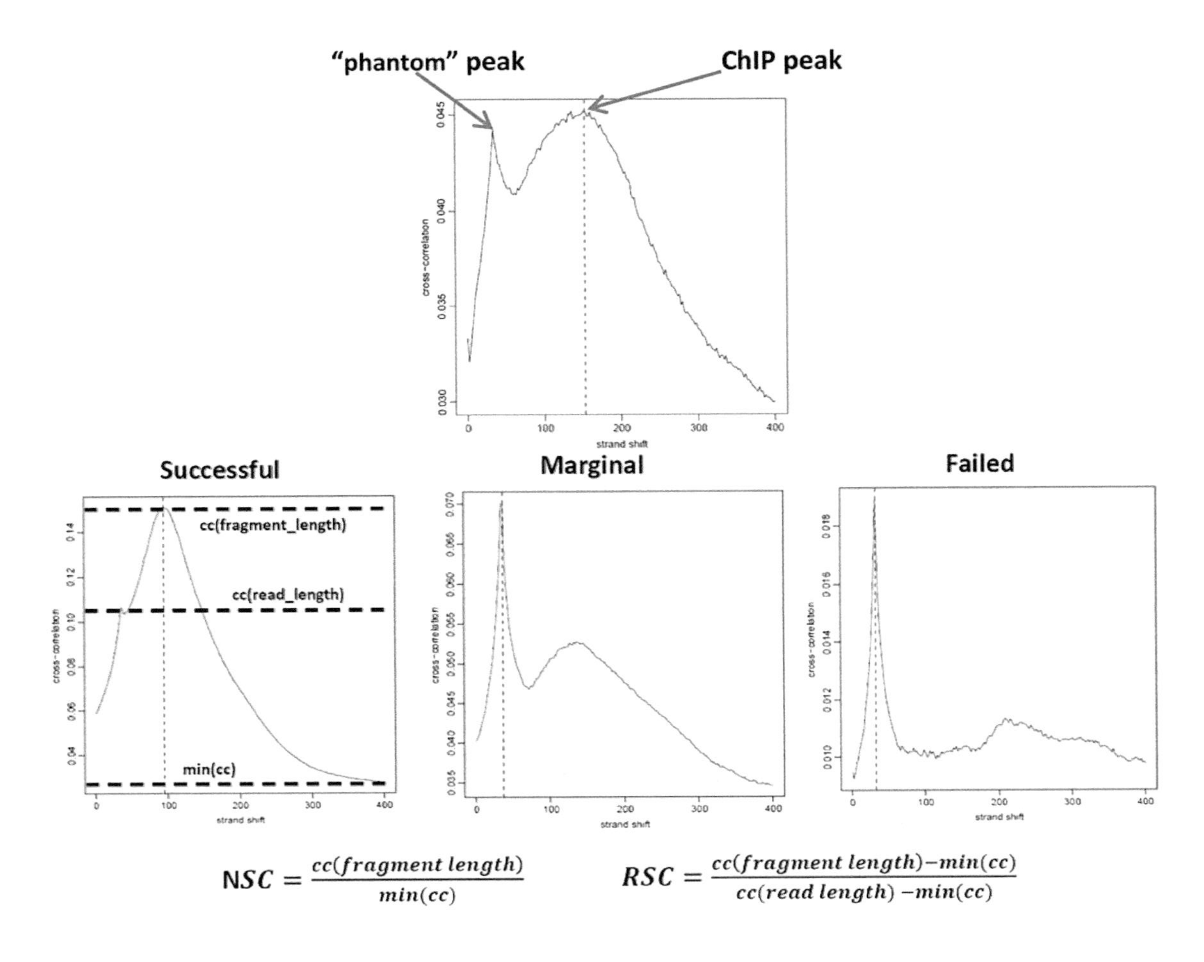
"phantom" peak
ChIP peak
cross-correlation
strand shift
Successful
Marginal
Failed
cc(fragment_length)
cc(read_length)
min(cc)
$NSC = \frac{cc(fragment\ length)}{min(cc)}$
$RSC = \frac{cc(fragment\ length) - min(cc)}{cc(read\ length) - min(cc)}$

FIGURE 13.5
The "phantom" peak and its use in determining ChIP-seq data quality. The "phantom" peak corresponds to the cross-correlation at the strand shift that equals to the read length, while the ChIP peak corresponds to the cross-correlation at the shift of the average DNA fragment length. A successful run is characterized by the existence of a predominant ChIP peak and a much weaker "phantom" peak. In marginally passed or failed runs, the former diminishes while the latter relatively becomes much stronger. (Adapted from SG Landt, GK Marinov, A Kundaje, P Kheradpour, F Pauli, S Batzoglou, BE Bernstein et al., ChIP-seq guidelines and practices of the ENCODE and modENCODE consortia, *Genome Research* 2012, 22(9):1813–1831. Used under the terms of the Creative Commons License (Attribution-NonCommercial 3.0 Unported License) as described at http://creativecommons.org/licenses/by-nc/3.0/. ©2012 Cold Spring Harbor Laboratory Press.)

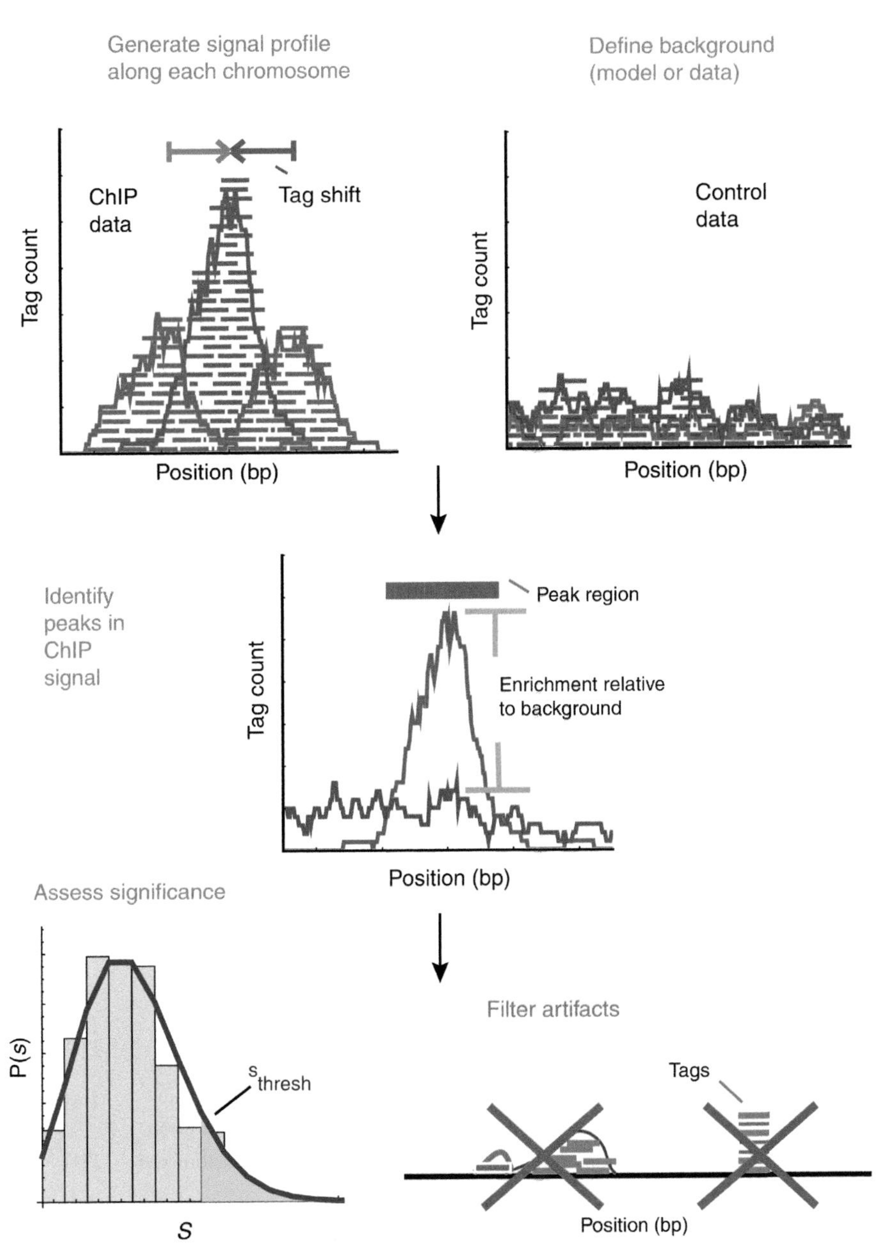

FIGURE 13.6

Basic substeps of calling peaks from ChIP-seq data. The P(s) at bottom left signifies the probability of observing a location covered by S mapped reads, and the s_{thresh} marks the threshold for calling a peak significant. Bottom right shows two types of possible artifactual peaks: single strand peaks and those based on mostly duplicate reads. (Adapted by permission from Macmillan Publishers Ltd: *Nature Methods*, Computation for ChIP-seq and RNA-seq studies, S Pepke, B Wold, A Mortazavi, copyright 2009.)

For implementation of this peak calling process, different peak callers use different methods, which can lead to differences in final outcomes. Among currently available peak callers, MACS2, HOMER (findPeaks module) [16], SICER2 [17], JAMM [18], SPP, and PeakSeq are among some of the popular ones (see Table 13.1). Among these methods, MACS2 uses Poisson distribution to model reads distribution across the genome. To achieve robust peak prediction, this modeling process considers the dynamic nature and effects of local background noise as caused by biological factors such as local chromatin structure and genome copy number variations, and technical biases introduced during library prep, sequencing, and mapping processes. Peaks are called from enrichment of reads in a genomic region with statistical significance calculated based on Poisson distribution. On FDR estimation, while the original MACS uses an empirical approach through exchanging ChIP and control samples, MACS2 applies the Benjamini–Hochberg method to adjust p-values. The findPeaks module in HOMER also uses Poisson distribution to identify putative peaks. To arrive at a list of high-quality peaks, different filters are then applied to these putative peaks. These filters are either based on the use of (1) control samples, i.e., peaks need to pass fold-change and cumulative Poisson p-value thresholds in comparison to control samples; (2) local read counts, i.e., the density of reads at a peak need to be

TABLE 13.1

A Short List of ChIP-Seq Peak Calling Algorithms

Name	Description	Reference
MACS/MACS2	Empirically models ChIP-seq read length to improve peak prediction, uses a dynamic Poisson distribution	[13]
HOMER (findPeaks module)	Identifies peaks based on the principle that more sequencing reads are found in these regions than expected by chance	[16]
PeakSeq	Based on a two-pass strategy to compensate for open chromatin signal	[15]
SPP	Includes binding profile normalization, peak detection, and estimation of read depth to achieve peak saturation	[14]
CisGenome	Features multifaceted interactive analysis and customized batch-mode computation	[21]
SICER2	Uses a clustering approach to identify enriched domains from histone modification ChIP-seq data	[17]
SiSSRs	Uses the direction and density of reads and the average DNA fragment length to identify binding sites	[22]
PeakRanger	Applies a staged algorithm to discover enriched regions and the summits within them	[23]
ZINBA	Models and accounts for factors co-varying with background or true signals	[24]

significantly higher than that in the surrounding region; or (3) clonal signal, to remove peaks with high number of reads that map to only a small number of unique positions. SPP, an R package, calculates smoothed read enrichment profile along the genome and identifies significantly enriched sites compared to input control using methods such as WTD (Window Tag Density). The WTD method considers sequence tag patterns immediately upstream and downstream of the center of the binding location, thereby increasing prediction accuracy. The peak calling employed in PeakSeq is a two-pass process. In its first pass, PeakSeq uses background modeling to identify initial potential binding regions. To adjust for background using control data, the fraction of reads located in the initially identified potential binding regions are excluded and the reads in the remainder of the genome in the ChIP-seq sample are normalized to the control data by linear regression. In the second pass, target peaks are called by scoring reads-enriched target regions based on calculation of fold enrichment in ChIP-seq sample vs. control, and statistical significance associated with each enriched target region is calculated from binomial distribution. More recently, newer algorithms based on the application of deep learning approaches have begun to emerge, such as CNN-Peaks [19] and LanceOtron [20].

To ensure the robustness of analysis results, it is recommended to use more than one method for peak calling. While IDR is usually used to measure the rate of irreproducible discoveries, i.e., peaks that are called in one replicate sample but not in another, it can also be used to compare peak calling results generated from different methods. The original use of IDR in assessing replicate reproducibility is based on the rationale that peaks of high significance are more consistently ranked across replicates and therefore have better reproducibility than those of low significance. As shown in Figure 13.7, to compare a pair of ranked lists of peaks identified in two replicates, IDR are plotted against the total numbers of ranked peaks. Since IDR computation relies on both high significance (more reproducible) and low significance (less reproducible) peaks, peak calling stringency needs to be relaxed to allow generation of both high and low confidence calls. The transition in this plot from reliable signal to noise is an index of overall experimental reproducibility. Because IDR is independent of any particular peak-calling method, it can be applied to compare the performance of different peak calling methods on a particular dataset and therefore help pick the most appropriate method(s) (Figure 13.8). IDR can also be used to evaluate reproducibility across experiments and labs.

13.3.3 Post-Peak Calling Quality Control

After peaks are called, prior to conducting further downstream analyses, it is a good practice to perform another QC step to determine the quality of peaks called. Tools such as deepTools2 and ChIPQC [25] can be used for this

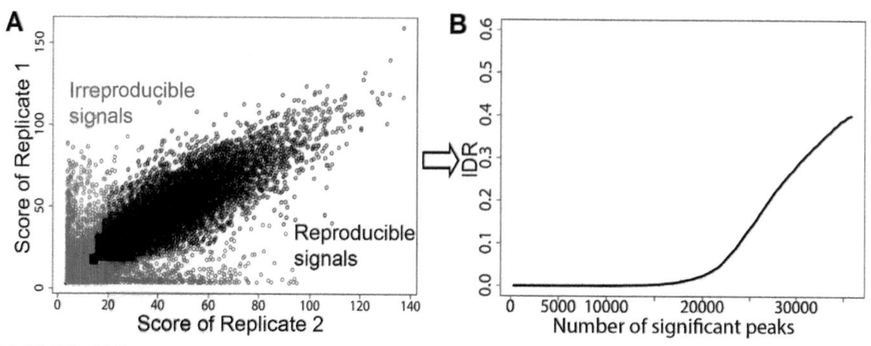

FIGURE 13.7

Use of irreproducible discovery rate (IDR) in assessing replicate reproducibility. Panel (A) shows the distribution of the significance scores of the peaks identified two replicate experiments. The IDR method computes the probability of being irreproducible for each peak, and classifies them as being reproducible (black) or irreproducible (red). Panel (B) displays the IDR at different rank thresholds when the peaks are sorted by the original significance score. (From T Bailey, P Krajewski, I Ladunga, C Lefebvre, Q Li, T Liu, P Madrigal, C Taslim, J Zhang, Practical guidelines for the comprehensive analysis of ChIP-seq data, *PLoS Computational Biology* 2013, 9:e1003326. Used under the terms of the Creative Commons Attribution License (http://creativecommons.org/licenses/by/3.0/). © 2013 Bailey et al.)

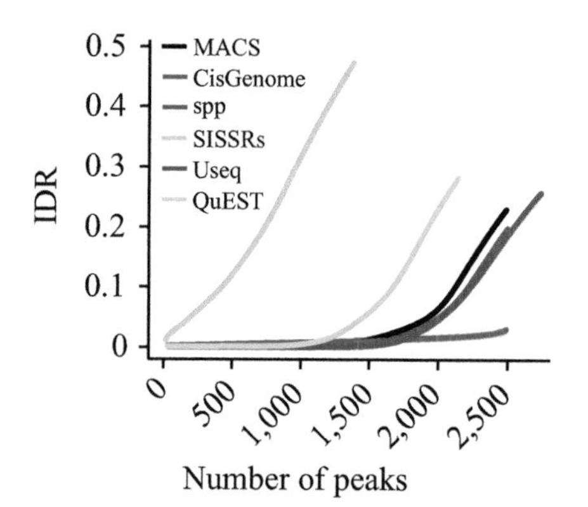

FIGURE 13.8

Evaluation of the performance of six peak callers using IDR. (Adapted by permission from Macmillan Publishers Ltd: *Nature Methods*, Systematic evaluation of factors influencing ChIP-seq fidelity, Y Chen, N Negre, Q Li, JO Mieczkowska, M Slattery, T Liu, Y Zhang et al., copyright 2012.)

step. For example, to check for consistency of called peaks between replicate samples, the aforementioned multiBamSummary tool in deepTools2 can be used in "BED-file" mode using BED files from peak callers as input, and the output passed on to plotCorrelation to generate visualization of samples based on their correlation coefficients.

ChIPQC reports a number of quality metrics related to called peaks, including peak signal strength, enrichment of reads in peaks, and relative enrichment of peaks in various genomic features. The metrics related to peak signal strength are relative strand cross-correlation coefficient, i.e., RSC, and estimated fragment length, which should be very close to the DNA fragment length produced during library prep. Enrichment of reads in peaks is represented by metrics including the percentage of reads in peaks, standardized standard deviation (SSD) of genomic coverage, and the percentage of reads in ENCODE blacklisted regions. Among these metrics, the percentage of reads in peaks, also often termed FRiP (or fraction of reads in peaks), is an overall indicator of immunoenrichment and ChIP-seq data quality. Usually only a small percentage of reads map to peak regions, and the majority of reads only represent background. As a general guideline, the ENCODE Consortium sets 1% as the minimum for an acceptable FRiP with MACS as the peak caller using default parameters. As they can vary with the use of different peak callers and parameters, FRiP values must be derived from the same algorithm using same parameters in order for them to be comparable across samples or experiments. The SSD metric is a measure of variability of reads genomic coverage. If reads are highly enriched in peak regions, as expected from a well-executed ChIP-seq experiment, the SSD values associated with ChIP samples should be much higher than those with control samples. The percentage of reads in blacklisted regions is another metric that affects peak calling outcome. Since some regions of the genome, especially highly repetitive regions, can produce artifactually high enrichment signals independent of experiment, the ENCODE Consortium has compiled a blacklist for such regions [26]. Since reads mapped to the blacklisted regions often confound peak calling and fragment length estimation, they should be monitored and filtered out. Lastly, on relative enrichment of peaks in various genomic features, the distribution of reads across major features including promoters, 5′ UTRs, coding regions, introns, and 3′ UTRs, provides another evaluation of ChIP-seq signal distribution. For instance, transcription factors are expected to have the highest enrichment levels in promoters and 5′ UTRs.

13.3.4 Peak Visualization

Visualizing peaks in their genomic context allows identification of overlapping or nearby functional elements, and thereby facilitates peak annotation and data interpretation. Many peak callers generate BED files containing peak chromosomal locations, along with WIG and bedGraph track files, all of which can be uploaded to a genome browser for peak visualization. Examination of peak regions in a genome browser and comparison with other data/annotation tracks allow identification of associated genomic features, such as promoters, enhancers, and other regulatory regions. BEDTools can also be applied to explore relationships between peaks and other genomic landmarks such as nearby protein-coding or non-coding genes.

13.4 Differential Binding Analysis

Binding of DNA-interacting proteins to their target genomic regions is a quantitative process, that is, they occupy these regions at different rates under different conditions. This is due to regional accessibility, presence/absence of other protein partners, and/or other factors that regulate their binding. Differential binding analysis answers the question of how a target protein changes its DNA-binding pattern under different conditions. There are two different approaches for this analysis, with one qualitative and the other quantitative. The qualitative approach compares peaks called in different conditions, and divides them into "shared" and "unique" [27]. This approach is simple, but it does not use the quantitative information generated in the peak calling process, so it is best used to produce a rough initial estimation of differential binding. The quantitative approach, through analysis of read counts or read densities in peak regions or on a sliding window basis across the genome, generates statistical assessment of the degree of differential binding between conditions. As this is similar to the RNA-seq-based differential expression analysis despite the higher level of background noise, data normalization is required to adjust for systematic biases that are unrelated to biological factors. For the comparison of two or more ChIP-seq samples, such biases include immunoprecipitation efficiency and sequencing depth.

Similar to normalizing RNA-seq data, adjusting for sequencing depth is the simplest approach. In this approach, the total numbers of reads in different samples are adjusted, by multiplying a scaling factor to each sample, to a same target level, e.g., the median or lowest total read count among the samples. The basic assumption for this approach is that the overall number of binding sites for a target protein does not change across different experimental conditions. Although this approach is simple and straightforward, it does not take into consideration the difference in signal-to-noise ratio that is often observed in different samples. If one sample library is noisier and contains more background reads, these reads, while not representing true signals, are still counted into the total read number. This situation will therefore lead to bias in the normalized data.

There are several currently available normalization approaches that consider this issue of signal-to-noise ratio variation among samples. For example, the normalization procedure used in diffReps first identifies and removes regions with low read count (mostly background noise) [28]. The subsequent normalization is then based on the remaining regions, using a linear procedure similar to that used by DESeq2. Another similar approach uses only reads mapped to peaks. In this modified sequencing depth-based normalization approach, the total number of reads mapped to the peak regions are used as the basis for calculating scaling factor for each sample [29]. Using this approach, the normalized peak signal is computed as the original peak sequence read count being scaled by the sum of read counts of all peaks, i.e.,

$$Z_{i,j} = \frac{X_{i,j}}{\sum_{j=1}^{N} X_{i,j}}$$

where $Z_{i,j}$ and $X_{i,j}$ are the normalized and original peak signal for sample i and peak j, and N is the total number of called peaks.

Normalization methods that were previously developed for microarray data have also been adapted for ChIP-seq data. MAnorm2, as an example, uses a hierarchical normalization process that is similar to the MA plot approach used for microarray data. Based on the assumption that signals from common peaks shared among all samples remain unchanged globally, MAnorm2 applies a linear transformation process to the raw read counts in order to make both the arithmetic mean of M values (or differences in signals between samples), and the covariance between M and A (average signals between two samples) values, to be zero [30]. ChIPnorm uses a modified version of quantile normalization [31]. A locally weighted regression (LOESS) normalization approach for ChIP-seq data [32] is similar to the LOESS procedure applied to cDNA microarray data normalization. All these approaches assume that the overall binding profile of the target protein does not vary across different conditions. This assumption does not hold, however, under conditions when the overall level or activity of the protein under study changes due to experimental perturbation. Under such conditions, normalization approaches based on spike-in of an exogenous reference epigenome at a constant amount [33], similar to the use of spike-in RNA controls in bulk RNA-seq, can be used.

Besides all the normalization approaches introduced above, good experimental design and consistent experimental procedure can minimize data variability in different samples and groups, thereby alleviating the burden on posterior normalization. For example, processing all samples side by side using the same experimental procedure and parameters, such as the same antibody, by the sample operator, will minimize sample-to-sample variability. When conducting an experiment in this way, the simpler normalization approach based on total library read count can be sufficient.

Since the ChIP-seq-based quantitative analysis of differential binding is similar to the RNA-seq-based differential expression analysis, packages such as edgeR and DESeq2 can be applied here. Table 13.2 lists some of the packages that are designed for ChIP-seq differential binding analysis. As listed these packages can be divided into two categories, with one composed of methods that are dependent on peak calling from an external application, while the other of those that handle peak calling internally or do away with peak calling altogether. For methods that require internally or externally

TABLE 13.2

Packages Developed for ChIP-Seq Differential Binding Analysis

Name	Description	Reference
Methods that require externally called peaks		
DiffBind	Uses statistical tests used in RNA-seq packages edgeR and DESeq to process peak sets and identify differentially bound regions	[38]
ChIPComp	Differential binding analysis taking into considerations of controls, signal-to-noise ratios, replicates, and multi-factor experimental design	[42]
MAnorm2	Conducts a hierarchical MA-plot-based normalization prior to differential analysis	[30]
DBChIP	Identifies differentially bound punctate binding sites in multiple conditions using RNA-seq DE approaches and accommodates controls	[39]
DIME	Differential binding analysis using a finite exponential-normal mixture model	[43]
MMDiff2	Takes a multivariate non-parametric approach to test differential binding	[44]
Methods that do not rely on pre-called peaks		
bdgdiff/ MACS2	A module of MACS2 that detects differential binding based on paired four bedGraph files	[13]
diffReps	Detects and annotates differential chromatin modification hotspots using a sliding window approach	[28]
ChIPDiff	Differential histone mark analysis based on HMM	[36]
RSEG	Uses HMM to locate differentially bound broad genomic domains associated with diffusive histone modification marks	[45]
csaw	Uses the sliding window approach for detection of differential binding regions, with different window sizes set for transcription factors and histone marks	[34]
PePr	Another method that uses sliding windows, and a negative binomial distribution model to detect differential peaks	[35]
THOR	Differential peak calling based on an HMM with a three-state topology	[37]
ChIPnorm	Carries out quantile normalization for differential binding sites identification	[31]

called peaks, robust peak calling is essential to produce quality results. For those that do not rely on peak calling, differential binding analysis aims to find significant ChIP-seq signal changes between conditions throughout the entire genome. These latter methods can be further divided into two subcategories. One subcategory, exemplified by diffReps [28], csaw [34], and PePr [35], uses sliding windows, the size of which is typically selected based on the footprint of the target protein. The other subcategory uses complex segmentation techniques, such as hidden Markov model (HMM), to first segment genome into bins and then infer the hidden state of each bin in order to detect differential protein-binding sites. ChIPDiff [36] and THOR [37] are examples of this approach.

To test for differential binding, statistical models based on Poisson or negative binomial distribution are often used, again similar to RNA-seq DE analysis. In fact, methods such as DiffBind [38] and DBChip [39] directly inherit statistical models from edgeR and DESeq2. In terms of detection targets, while some methods are specifically designed for punctate-binding protein factors (such as DBChip) and some others for more broad marks (such as SICER2 and RSEG), most of the methods can be used for binding regions of different sizes. In addition, in terms of handling replicate samples, some can work with experiments that do not use replicate samples, while others require replicates. As mentioned earlier, use of replication is suggested, which usually leads to increased detection precision with a much reduced number of differential binding regions, but in the meantime at the expense of reduced sensitivity. To help select appropriate methods for a particular application, existing benchmarking studies provide decision trees based on their comparative testing results [40, 41]. It should also be noted that like those devised for RNA-seq-based differential expression analysis, these packages are designed based on certain assumptions and therefore the user needs to be aware of these assumptions and ensure they are fulfilled prior to using them. For example, MAnorm2 is based on the assumption that there is no global change in binding at peak regions between conditions.

13.5 Functional Analysis

Often the data gathered from a ChIP-seq study is used to understand gene expression regulation and associated biological functions. To conduct functional analysis, peaks are first assigned to nearby genes using tools such as ChIPseeker [46], GREAT (Genomic Regions Enrichment of Annotations Tool) [47], and ChIPpeakAnno [48]. While it is debatable on what genes a peak should be assigned to, a straightforward approach is to assign it to the closest gene transcription start site. Once peaks are assigned to target genes, an integrated analysis of ChIP-seq and gene expression data (more on this in Section 13.7) can be carried out. Furthermore, Gene Ontology, biological pathway, gene network, or gene set enrichment analyses can be conducted using similar approaches as described in Chapter 7. Prior to carrying out these gene functional analyses, one should also bear in mind that the peak-to-gene assignment process is biased by gene size, as the presence of peak(s) has a positive correlation with the length of a gene. In addition, the distribution of gene size in different functional annotations such as GO categories is not uniform, with some categories having excess number of long genes while others having more short genes. To solve the problems caused by different gene sizes, methods that adjust for the effects of gene size should be used, such as ChIP-Enrich [49].

13.6 Motif Analysis

One of the goals of ChIP-seq data analysis is to identify DNA-binding motifs for the protein of interest. A DNA-binding motif is usually represented by a consensus sequence, or more accurately, a position specific frequency matrix. Figure 13.9 A shows an example of such a DNA-binding motif, the one bound by a previously introduced transcription factor NRF2 (see Chapter 2). To identify motifs from ChIP-seq data, all peak sequences need to be assembled and fed into multiple motif discovery tools. Some of the commonly used motif discovery tools are Cistrome [50], Gibbs motif sampler (part of CisGenome), HOMER (findMotifs module), MEME-ChIP [51] as part of the MEME suite [52], rGADEM [53], and RSAT peak-motifs [54]. The motif discovery phase usually ends up with one or more motifs, with one being the binding site of

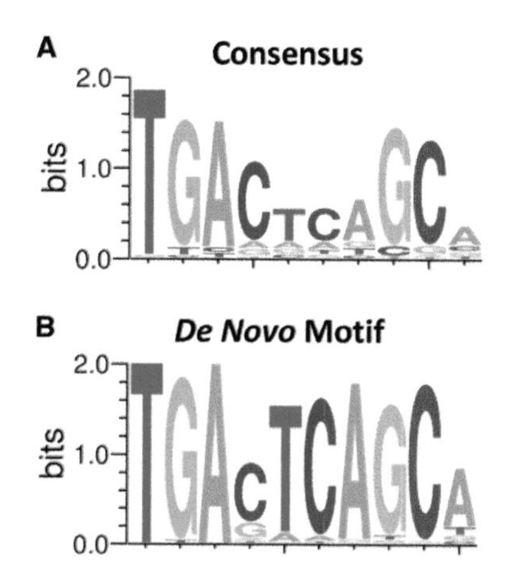

FIGURE 13.9
The consensus DNA-binding motif of the transcription factor NRF2. Panel (A) shows the currently known NRF2-binding motif, while panel (B) displays the result of a *de novo* motif analysis using NRF2 ChIP-seq data. (From BN Chorley, MR Campbell, X Wang, M Karaca, D Sambandan, F Bangura, P Xue, J Pi, SR Kleeberger, DA Bell, Identification of novel NRF2-regulated genes by ChIP-seq: influence on retinoid X receptor alpha, *Nucleic Acids Research* 2012, 40(15):7416–7429, With permission.)

the target protein and others being that of its partners. The discovered motif(s) can be compared with currently known motifs catalogued in databases such as JASPAR [55] to detect similarities, find relationships with other motifs, or locate other proteins that might bind at or near the peak region as part of a protein complex. Tools for motif comparison include STAMP [56] and Tomtom [57]. Motif enrichment analysis can also be carried out to find out if known motifs are enriched in the peak regions using tools such as AME [58], CentriMo [59], and SEA [60]. Finally motif scanning and mapping by tools like FIMO [61] allows visualization of the discovered motif(s) in the ChIP-seq peak areas. Some of these tools have been integrated into motif analysis pipelines, such as the MEME Suite [62], which includes MEME-ChIP, Tomtom, CentriMo, SEA, AME, and FIMO.

13.7 Integrated ChIP-Seq Data Analysis

As genomic functions are to a large degree controlled by concerted binding of a wide array of DNA-interacting proteins, integrated analysis of ChIP-seq data sets generated for a multitude of these proteins affords new opportunities to gain a comprehensive overview of the functional states of a genome and the host cell. As a good example, such an integrated analysis has led to the discovery of a large number of chromatin states, each of which display distinct sequence motifs and functional characteristics [63]. The discovery of these chromatin states was achieved with the use of a multivariate hidden Markov model on a large collection of ChIP-seq data, generated for 38 different histone methylation and acetylation marks, H2AZ (a variant of histone H2A), RNA polymerase II, and CTCF (a transcriptional repressor).

Besides meta-analysis of multiple ChIP-seq data sets, integrated analysis of ChIP-seq with other genomics data, such as RNA-seq data, offers further information on genome function and regulation. The majority of protein factors used in various ChIP-seq studies are transcription factors and histones that carry a large array of modified marks, all of which are key regulators of genome transcription. Coupled analysis of matched ChIP-seq and RNA-seq data augments the utility of both data types, and provides new insights that cannot be obtained from analyzing either data type alone. To carry out integrative analysis of ChIP-seq and RNA-seq data, Bayesian mixed and hierarchical models [64, 65] have been used. In addition, tools such as BETA [66], CEAS [67], and ChIPpeakAnno can also be used to help investigate the correlation between the DNA-binding profile and regulation of nearby gene transcription.

References

1. Chen Y, Negre N, Li Q, Mieczkowska JO, Slattery M, Liu T, Zhang Y, Kim TK, He HH, Zieba J *et al.* Systematic evaluation of factors influencing ChIP-seq fidelity. *Nat Methods* 2012, 9(6):609–614.
2. Visa N, Jordan-Pla A. ChIP and ChIP-Related Techniques: Expanding the fields of application and improving ChIP performance. *Methods Mol Biol* 2018, 1689:1–7.
3. Skene PJ, Henikoff JG, Henikoff S. Targeted in situ genome-wide profiling with high efficiency for low cell numbers. *Nat Protoc* 2018, 13(5):1006–1019.
4. Meyer CA, Liu XS. Identifying and mitigating bias in next-generation sequencing methods for chromatin biology. *Nat Rev Genet* 2014, 15(11):709–721.
5. Jordán-Pla A, Visa N. Considerations on experimental design and data analysis of chromatin immunoprecipitation experiments. *Methods Mol Biol* 2018, 1689:9–28.
6. Daley T, Smith AD. Predicting the molecular complexity of sequencing libraries. *Nat Methods* 2013, 10(4):325–327.
7. ENCODE Software Tools (www.encodeproject.org/software/)
8. Irreproducible Discovery Rate (IDR) (www.encodeproject.org/software/idr/)
9. Diaz A, Nellore A, Song JS. CHANCE: comprehensive software for quality control and validation of ChIP-seq data. *Genome Biol* 2012, 13(10):R98.
10. Ramirez F, Ryan DP, Gruning B, Bhardwaj V, Kilpert F, Richter AS, Heyne S, Dundar F, Manke T. deepTools2: a next generation web server for deep-sequencing data analysis. *Nucleic Acids Res* 2016, 44(W1):W160–165.
11. Nakato R, Shirahige K. Sensitive and robust assessment of ChIP-seq read distribution using a strand-shift profile. *Bioinformatics* 2018, 34(14):2356–2363.
12. Landt SG, Marinov GK, Kundaje A, Kheradpour P, Pauli F, Batzoglou S, Bernstein BE, Bickel P, Brown JB, Cayting P *et al.* ChIP-seq guidelines and practices of the ENCODE and modENCODE consortia. *Genome Res* 2012, 22(9):1813–1831.
13. Zhang Y, Liu T, Meyer CA, Eeckhoute J, Johnson DS, Bernstein BE, Nusbaum C, Myers RM, Brown M, Li W *et al.* Model-based analysis of ChIP-Seq (MACS). *Genome Biol* 2008, 9(9):R137.
14. Kharchenko PV, Tolstorukov MY, Park PJ. Design and analysis of ChIP-seq experiments for DNA-binding proteins. *Nat Biotechnol* 2008, 26(12):1351–1359.
15. Rozowsky J, Euskirchen G, Auerbach RK, Zhang ZD, Gibson T, Bjornson R, Carriero N, Snyder M, Gerstein MB. PeakSeq enables systematic scoring of ChIP-seq experiments relative to controls. *Nat Biotechnol* 2009, 27(1):66–75.
16. Heinz S, Benner C, Spann N, Bertolino E, Lin YC, Laslo P, Cheng JX, Murre C, Singh H, Glass CK. Simple combinations of lineage-determining transcription factors prime cis-regulatory elements required for macrophage and B cell identities. *Mol Cell* 2010, 38(4):576–589.
17. Zang C, Schones DE, Zeng C, Cui K, Zhao K, Peng W. A clustering approach for identification of enriched domains from histone modification ChIP-Seq data. *Bioinformatics* 2009, 25(15):1952–1958.

18. Ibrahim MM, Lacadie SA, Ohler U. JAMM: a peak finder for joint analysis of NGS replicates. *Bioinformatics* 2015, 31(1):48–55.
19. Oh D, Strattan JS, Hur JK, Bento J, Urban AE, Song G, Cherry JM. CNN-Peaks: ChIP-Seq peak detection pipeline using convolutional neural networks that imitate human visual inspection. *Sci Rep* 2020, 10(1):7933.
20. Hentges LD, Sergeant MJ, Cole CB, Downes DJ, Hughes JR, Taylor S. LanceOtron: a deep learning peak caller for genome sequencing experiments. *Bioinformatics* 2022, 38(18):4255–4263.
21. Ji H, Jiang H, Ma W, Johnson DS, Myers RM, Wong WH. An integrated software system for analyzing ChIP-chip and ChIP-seq data. *Nat Biotechnol* 2008, 26(11):1293–1300.
22. Jothi R, Cuddapah S, Barski A, Cui K, Zhao K. Genome-wide identification of in vivo protein-DNA binding sites from ChIP-Seq data. *Nucleic Acids Res* 2008, 36(16):5221–5231.
23. Feng X, Grossman R, Stein L. PeakRanger: a cloud-enabled peak caller for ChIP-seq data. *BMC Bioinformatics* 2011, 12:139.
24. Rashid NU, Giresi PG, Ibrahim JG, Sun W, Lieb JD. ZINBA integrates local covariates with DNA-seq data to identify broad and narrow regions of enrichment, even within amplified genomic regions. *Genome Biol* 2011, 12(7):R67.
25. Carroll TS, Liang Z, Salama R, Stark R, de Santiago I. Impact of artifact removal on ChIP quality metrics in ChIP-seq and ChIP-exo data. *Front Genet* 2014, 5:75.
26. Amemiya HM, Kundaje A, Boyle AP. The ENCODE Blacklist: Identification of Problematic Regions of the Genome. *Sci Rep* 2019, 9(1):9354.
27. Chen X, Xu H, Yuan P, Fang F, Huss M, Vega VB, Wong E, Orlov YL, Zhang W, Jiang J *et al.* Integration of external signaling pathways with the core transcriptional network in embryonic stem cells. *Cell* 2008, 133(6):1106–1117.
28. Shen L, Shao NY, Liu X, Maze I, Feng J, Nestler EJ. diffReps: detecting differential chromatin modification sites from ChIP-seq data with biological replicates. *PLoS One* 2013, 8(6):e65598.
29. Manser P, Reimers M. A simple scaling normalization for comparing ChIP-Seq samples. *PeerJ PrePrints* 2014, 1.
30. Tu S, Li M, Chen H, Tan F, Xu J, Waxman DJ, Zhang Y, Shao Z. MAnorm2 for quantitatively comparing groups of ChIP-seq samples. *Genome Res* 2021, 31(1):131–145.
31. Nair NU, Sahu AD, Bucher P, Moret BM. ChIPnorm: a statistical method for normalizing and identifying differential regions in histone modification ChIP-seq libraries. *PLoS One* 2012, 7(8):e39573.
32. Taslim C, Wu J, Yan P, Singer G, Parvin J, Huang T, Lin S, Huang K. Comparative study on ChIP-seq data: normalization and binding pattern characterization. *Bioinformatics* 2009, 25(18):2334–2340.
33. Orlando DA, Chen MW, Brown VE, Solanki S, Choi YJ, Olson ER, Fritz CC, Bradner JE, Guenther MG. Quantitative ChIP-Seq normalization reveals global modulation of the epigenome. *Cell Rep* 2014, 9(3):1163–1170.
34. Lun AT, Smyth GK. csaw: a Bioconductor package for differential binding analysis of ChIP-seq data using sliding windows. *Nucleic Acids Res* 2016, 44(5):e45.
35. Zhang Y, Lin YH, Johnson TD, Rozek LS, Sartor MA. PePr: a peak-calling prioritization pipeline to identify consistent or differential peaks from replicated ChIP-Seq data. *Bioinformatics* 2014, 30(18):2568–2575.

36. Xu H, Wei CL, Lin F, Sung WK. An HMM approach to genome-wide identification of differential histone modification sites from ChIP-seq data. *Bioinformatics* 2008, 24(20):2344–2349.
37. Allhoff M, Sere K, J FP, Zenke M, I GC. Differential peak calling of ChIP-seq signals with replicates with THOR. *Nucleic Acids Res* 2016, 44(20):e153.
38. Stark R, Brown G. DiffBind: differential binding analysis of ChIP-Seq peak data. *In R package version* 2011, 100.
39. Liang K, Keles S. Detecting differential binding of transcription factors with ChIP-seq. *Bioinformatics* 2012, 28(1):121–122.
40. Steinhauser S, Kurzawa N, Eils R, Herrmann C. A comprehensive comparison of tools for differential ChIP-seq analysis. *Brief Bioinform* 2016, 17(6):953–966.
41. Eder T, Grebien F. Comprehensive assessment of differential ChIP-seq tools guides optimal algorithm selection. *Genome Biol* 2022, 23(1):119.
42. Chen L, Wang C, Qin ZS, Wu H. A novel statistical method for quantitative comparison of multiple ChIP-seq datasets. *Bioinformatics* 2015.
43. Taslim C, Huang T, Lin S. DIME: R-package for identifying differential ChIP-seq based on an ensemble of mixture models. *Bioinformatics* 2011, 27(11):1569–1570.
44. Schweikert G, Kuo D. MMDiff2: statistical testing for ChIP-Seq data sets. In., vol. R package version 1.24.0; 2022.
45. Song Q, Smith AD. Identifying dispersed epigenomic domains from ChIP-Seq data. *Bioinformatics* 2011, 27(6):870–871.
46. Yu G, Wang LG, He QY. ChIPseeker: an R/Bioconductor package for ChIP peak annotation, comparison and visualization. *Bioinformatics* 2015, 31(14):2382–2383.
47. McLean CY, Bristor D, Hiller M, Clarke SL, Schaar BT, Lowe CB, Wenger AM, Bejerano G. GREAT improves functional interpretation of cis-regulatory regions. *Nat Biotechnol* 2010, 28(5):495–501.
48. Zhu LJ, Gazin C, Lawson ND, Pages H, Lin SM, Lapointe DS, Green MR. ChIPpeakAnno: a Bioconductor package to annotate ChIP-seq and ChIP-chip data. *BMC Bioinformatics* 2010, 11:237.
49. Welch RP, Lee C, Imbriano PM, Patil S, Weymouth TE, Smith RA, Scott LJ, Sartor MA. ChIP-Enrich: gene set enrichment testing for ChIP-seq data. *Nucleic Acids Res* 2014.
50. Liu T, Ortiz JA, Taing L, Meyer CA, Lee B, Zhang Y, Shin H, Wong SS, Ma J, Lei Y *et al.* Cistrome: an integrative platform for transcriptional regulation studies. *Genome Biol* 2011, 12(8):R83.
51. Machanick P, Bailey TL. MEME-ChIP: motif analysis of large DNA datasets. *Bioinformatics* 2011, 27(12):1696–1697.
52. Bailey TL, Johnson J, Grant CE, Noble WS. The MEME Suite. *Nucleic Acids Res* 2015, 43(W1):W39–49.
53. Droit A, Gottardo R, Robertson G, Li L. rGADEM: de novo motif discovery. R package version 2.44.0. . In.; 2022.
54. Thomas-Chollier M, Herrmann C, Defrance M, Sand O, Thieffry D, van Helden J. RSAT peak-motifs: motif analysis in full-size ChIP-seq datasets. *Nucleic Acids Res* 2012, 40(4):e31.
55. Castro-Mondragon JA, Riudavets-Puig R, Rauluseviciute I, Lemma RB, Turchi L, Blanc-Mathieu R, Lucas J, Boddie P, Khan A, Manosalva Perez N

et al. JASPAR 2022: the 9th release of the open-access database of transcription factor binding profiles. *Nucleic Acids Res* 2022, 50(D1):D165–D173.
56. Mahony S, Benos PV. STAMP: a web tool for exploring DNA-binding motif similarities. *Nucleic Acids Res* 2007, 35(Web Server issue):W253–258.
57. Gupta S, Stamatoyannopoulos JA, Bailey TL, Noble WS. Quantifying similarity between motifs. *Genome Biol* 2007, 8(2):R24.
58. McLeay RC, Bailey TL. Motif Enrichment Analysis: a unified framework and an evaluation on ChIP data. *BMC Bioinformatics* 2010, 11:165.
59. Bailey TL, Machanick P. Inferring direct DNA binding from ChIP-seq. *Nucleic Acids Res* 2012, 40(17):e128.
60. Bailey TL, Grant CE. SEA: Simple Enrichment Analysis of motifs. *bioRxiv* 2021, doi: https://doi.org/10.1101/2021.08.23.457422.
61. Grant CE, Bailey TL, Noble WS. FIMO: scanning for occurrences of a given motif. *Bioinformatics* 2011, 27(7):1017–1018.
62. The MEME Suite (http://meme.nbcr.net/meme/)
63. Ernst J, Kellis M. Discovery and characterization of chromatin states for systematic annotation of the human genome. *Nat Biotechnol* 2010, 28(8):817–825.
64. Klein HU, Schafer M, Porse BT, Hasemann MS, Ickstadt K, Dugas M. Integrative analysis of histone ChIP-seq and transcription data using Bayesian mixture models. *Bioinformatics* 2014, 30(8):1154–1162.
65. Schafer M, Klein HU, Schwender H. Integrative analysis of multiple genomic variables using a hierarchical Bayesian model. *Bioinformatics* 2017, 33(20):3220–3227.
66. Wang S, Sun H, Ma J, Zang C, Wang C, Wang J, Tang Q, Meyer CA, Zhang Y, Liu XS. Target analysis by integration of transcriptome and ChIP-seq data with BETA. *Nature protocols* 2013, 8(12):2502–2515.
67. Shin H, Liu T, Manrai AK, Liu XS. CEAS: cis-regulatory element annotation system. *Bioinformatics* 2009, 25(19):2605–2606.

14

Epigenomics by DNA Methylation Sequencing

The genomic information embedded in the primary nucleotide sequence of DNA is modulated by epigenomic code generated from chemical modifications of DNA bases and key DNA-interacting proteins such as the histones. The methylation of cytosines leading to the formation of 5-methylcytosines (5mCs), for example, provides a major means for the modification of the primary DNA code. As detailed in Chapter 2, DNA methylation plays an important role in many biological functions such as embryonic development, cell differentiation, and stem cell pluripotency, through regulating gene expression and chromatin remodeling. Abnormal patterns of DNA methylation, on the other hand, lead to diseases such as cancer. DNA methylome analysis, as a key component of epigenomics, has for many years been conducted with the use of microarrays (such as the Illumina Infinium MethylationEPIC BeadChips). While microarrays are low-cost and easy to use, their inherent constraints, such as limited genomic coverage from the use of pre-selected probes, and being available for only a few model organisms, have limited their use. In comparison, NGS offers a more unbiased, comprehensive, and quantitative approach for the study of DNA methylation status in a wide array of species. This chapter focuses on DNA methylation sequencing data generation and analysis. For epigenomic studies that involve interrogation of histone modifications, ChIP-seq (covered in Chapter 13) can be used.

14.1 DNA Methylation Sequencing Strategies

Because the DNA polymerases used in the regular NGS sequencing library construction process cannot distinguish methylated from unmethylated cytosines, DNA methylation pattern is usually not retained in the process. In order to study DNA methylation status with NGS, different strategies need to be used, including bisulfite conversion, methylated DNA enrichment, and more recently enzymatic conversion-based methyl-seq. Bisulfite conversion employs a chemical conversion process, which uses sodium bisulfite to deaminate unmethylated cytosines. After the conversion, unmethylated

DOI: 10.1201/9780429329180-17

cytosines in a DNA molecule are converted to uracils, while 5mCs in the same molecule are retained since they are non-reactive. The subsequent sequencing of the converted DNA, therefore, reads unmethylated cytosines as thymines, and methylated cytosines as cytosines. The efficiency and specificity of this process can be monitored and optimized through the use of certain methylated and unmethylated DNAs as controls. Based on genomic coverage, this bisulfite sequencing approach can be further divided into two subcategories as covered next. The more recently developed enzymatic methyl-seq is based on a similar approach, but the conversion of unmethylated cytosines to uracils is based on enzymatic reactions, which is not DNA damaging compared to the chemical method. The methylated DNA enrichment approach relies on capturing of methylated DNA for targeted sequencing, with the use of 5mC antibodies or proteins that bind to methylated cytosines. This section details these major DNA methyl-seq approaches.

14.1.1 Bisulfite Conversion Methyl-Seq

14.1.1.1 Whole-Genome Bisulfite Sequencing (WGBS)

As the name suggests, WGBS analyzes cytosine methylation in the entire genome, i.e., the methylome. In preparing WGBS libraries from total genomic DNA, regular DNA library construction protocols need to be modified. For example, if adaptors are added prior to the bisulfite conversion step, they must not contain unmethylated cytosines, i.e., all cytosines in the adaptor sequence must be methylated. In the PCR step, a polymerase that can tolerate uracil residues needs to be used. As a result of the conversion and subsequent PCR amplification, the two DNA strands that were originally complementary are no longer complementary. Instead, four strands that are distinct from the original complementary strands are generated (Figure 14.1). Furthermore, the conversion leads to reduced sequence complexity due to underrepresentation of cytosines in the generated reads. Without the use of an external sequencing library to create a calibration table for basecaller, the reduced sequence complexity will lead to high basecall error rate. Therefore, use of a calibration library, such as the phiX174 control library for Illumina sequencing, is needed for bisulfite sequencing data generation.

The power to detect DNA methylation levels, and differentially methylated sites or regions between different experimental groups (e.g., disease vs. normal), is dependent on sequencing depth and the number of biological replicates in each group. In addition, key statistical factors, including mean DNA methylation level at a given site, within-group biological variation, and between-condition difference, also heavily affect detection power [1]. For WGBS studies, a general guideline from the U.S. NIH Roadmap Epigenomics Project recommends at least two replicates per condition with a combined depth of at least 30× [2]. Consistent with this recommendation, Ziller and colleagues suggested to use a sequencing depth of 5–15× per WGBS sample,

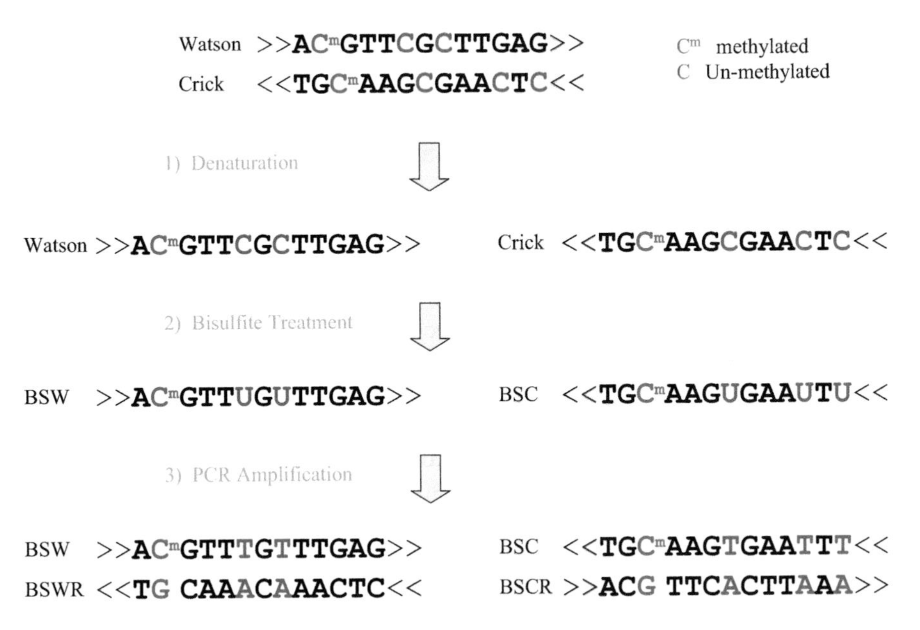

FIGURE 14.1
Major steps of bisulfite sequencing. Prior to bisulfite treatment, the two strands of DNA are first separated by denaturation. The bisulfite treatment then converts unmethylated, but not methylated, cytosines to uracils. The two strands from the treatment, BSW and BSC, are then subjected to PCR amplification. This leads to the generation of four strands (BSW, BSWR, BSC, and BSCR), all of which are distinct from the original Watson and Crick strands. (From Xi Y. and Li W. (2009) BSMAP: whole genome bisulfite sequence MAPping program. *BMC Bioinformatics*, 10, 232. Used under the terms of the Creative Commons Attribution License (http://creativecommons.org/licenses/by/2.0). © 2009 Xi and Li.)

and that sequencing at higher depth may not be as cost effective as adding more biological replicates in reaching higher detection power [3].

14.1.1.2 Reduced Representation Bisulfite Sequencing (RRBS)

While WGBS enables detection of methylation in the entire genome, the cost associated with such analyses is relatively high. To reduce the cost, strategies such as RRBS [4] were devised. To perform RRBS, genomic DNA is first digested with a methylation-insensitive restriction enzyme (such as MspI) that recognizes CpG-containing restriction site. The digested DNA products are then separated and size selected to pick fragments in a certain size range for bisulfite conversion and then sequencing. While it has limited coverage of CpG-poor regions of the genome, RRBS provides a rough survey of DNA methylation by examining 4–17% of the 28 million CpG dinucleotides in the human genome [5]. If particular region(s) of the genome are found to be of special interest, they can be captured for subsequent sequencing using approaches such as ligation capture [6, 7], bisulfite padlock probes [8], or liquid hybridization capture [9, 10].

14.1.2 Enzymatic Conversion Methyl-Seq

Besides the bisulfite-based chemical method, the conversion of cytosine to uracil through deamination can also be achieved using enzymes such as APOBEC (apolipoprotein B mRNA editing enzyme, catalytic polypeptide-like). Besides unmethylated cytosine, however, APOBEC can also deaminate 5mC and 5hmC (Chapter 2). To protect them from deamination in order to quantify their presence, two other enzymes TET2 (Tet methylcytosine dioxygenase 2) and BGT (β-glucosyltransferase) need to be used first to convert 5mC to 5caC, and 5hmC to 5-ghmC (5-glucosylhydroxymethylcytosine), respectively. After such enzymatic modifications, 5caC and 5-ghmC are no longer substrates of APOBEC. In the subsequent APOBEC-based deamination step, only unmethylated cytosines are converted to uracils, allowing determination of 5mCs and 5hmCs in the original sample. Compared to bisulfite treatment, enzyme-based conversion does not cause DNA damage and sample loss. This leads to increased detection accuracy, sensitivity, and coverage of CpGs and high GC content regions [11–13]. The increased sensitivity makes it possible to study single cells, cell-free DNA, or FFPE DNA. With enzymatic conversion, long-range methylation information can also be retained and studied using long-read NGS technologies [14].

14.1.3 Enrichment-Based Methyl-Seq

Different from the above bisulfite conversion-based methods, the methylated DNA enrichment strategy captures methylated DNA for targeted sequencing. One of the methods based on this strategy is MeDIP-seq, or methylated DNA immunoprecipitation coupled with NGS. In this method, antibodies against 5mC are used to precipitate methylated single-stranded DNA fragments for sequencing. Another commonly used method is MBD-seq, or methyl-CpG-binding domain capture (MBDCap) followed by NGS. MBD-seq utilizes proteins such as MBD2 or MECP2 that contain the methyl-CpG-binding domain to enrich for methylated double-stranded DNA fragments. In one type of MBDCap method called MIRA (Methylated-CpG Island Recovery Assay), a protein complex of MBD2 and MBD3L1 (methyl-CpG-binding domain protein 3-like-1) is used to achieve enhanced affinity to methylated CpG regions. While MeDIP-seq and MBD-aeq usually generate highly concordant results, there are some differences between these two approaches. MeDIP-seq can detect both CpG and non-CpG methylation, while MBD-seq is focused on methylated CpG sites because of the binding affinity of MBD. At methylated CpG sites, MeDIP tends to enrich at regions that have low CpG density, while MBD-seq favors regions of relatively higher CpG content [15, 16].

In principle, these enrichment-based methods are very similar to ChIP-seq (Chapter 13), based on the same process of specific protein-based DNA capture, protein-DNA complex affinity binding, and target DNA elution. Likewise, their sequencing data generation and subsequent analysis are also similar to those in ChIP-seq. Therefore, the data analysis methods covered in Chapter 13 equally apply to the analysis of sequencing data generated by MeDIP-seq, MBD-seq, or other methylated DNA enrichment-based NGS methods. This chapter is, therefore, mostly focused on the analysis of chemical or enzymatic conversion-based methyl-seq data.

14.1.4 Differentiation of Cytosine Methylation from Demethylation Products

Besides 5mC, cytosine demethylation products, such as 5hmC (Figure 14.2), have also received growing interest because of their potential role as a new epigenetic marker [17]. Although the regular enzymatic conversion-based methyl-seq procedure detects both 5mC and 5hmC indiscriminately, a modified version can be used to detect 5hmC alone, in which the step of TET2 protection is skipped leaving 5mC unprotected while the glucosylation of 5hmC by BGT still protects 5-hmC from deamination by APOBEC. By combining results from the regular and the modified versions of the enzymatic methyl-seq, separate detection of both 5mC and 5hmC is achieved.

Other bisulfite conversion-based approaches to distinguish 5hmC from 5mC include TAB-seq (or TET-assisted bisulfite sequencing) [18] and oxBS-seq (or oxidative bisulfite sequencing) [19]. Among the three 5mC demethylation intermediate products (5hmC, 5fC, and 5caC, see Chapter 2), 5fC and 5caC are converted to uracils by sodium bisulfite but 5hmC (like 5mC) is not. During subsequent sequencing, as a consequence, 5hmC cannot be differentiated from 5mC, while 5fC/5caC cannot be differentiated from unmethylated cytosine. In the instance of oxBS-seq, 5hmC is first oxidized to 5fC before the bisulfite conversion step. Since the 5fC is then converted into uracil, subsequent sequencing and analysis provides information on 5mC

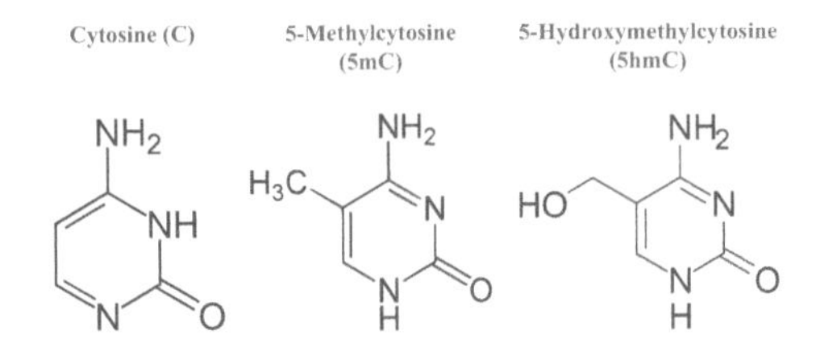

FIGURE 14.2
The chemical structures of cytosine, 5-methylcytosine (5mC), and 5-hydroxymethylcytosine (5hmC).

alone. After subtracting oxBS-seq result from regular bisulfite sequencing result, information on 5hmC is revealed.

Differentiation of 5fC and 5caC has also been made possible with recent method development [20, 21], but their levels are found to be typically low. Some third-generation single-molecule sequencing technologies, such as the Pacific Biosciences' SMRT sequencing and nanopore sequencing, have been shown to be capable of differentially detecting these different modifications without relying on bisulfite or enzymatic conversion [22–26]. The SMRT platform detects various base modifications based on polymerase dynamics as reflected in the pulses of light emitted from incorporation of nucleotides, while the nanopore platform is based on recognition of specific patterns in the ionic current signal.

14.2 DNA Methylation Sequencing Data Analysis

14.2.1 Quality Control and Preprocessing

After raw data generation, the quality control (QC) step (Figure 14.3) removes low-quality reads or basecalls as they directly affect subsequent alignment to the reference genome and DNA methylation site identification. The general data QC steps detailed in Chapter 5 should be performed for their removal. Other QC steps include adapter trimming as some sequencing reactions may run through DNA inserts into adapters. In addition, for MspI-digested RRBS libraries, the DNA fragment end repair step during the library construction artificially introduces two bases (an unmethylated cytosine and a guanine) to both ends, both of which should be trimmed off as well. Tools such as Trim Galore (a wrapper tool using Cutadapt and FastQC) [27] can be used for these trimming steps, especially removing the two artificially introduced bases in RRBS reads derived from MspI digestion. Besides these general-purpose QC tools, some packages designed for bisulfite sequencing reads processing, including BSmooth [28] and WBSA [29], also contain QC modules.

14.2.2 Read Mapping

In order to identify methylated DNA sites, sequencing reads derived from bisulfite or enzymatic conversion, or methylated DNA enrichment, need to be first mapped to the reference genome. Mapping of reads generated from the enrichment-based methods is rather straightforward, and like mapping ChIP-seq reads, is usually conducted with general aligners, such as Bowtie, BWA, or SOAP. Mapping of bisulfite or enzymatic conversion-based methyl-seq reads, however, is less straightforward. This is because through the

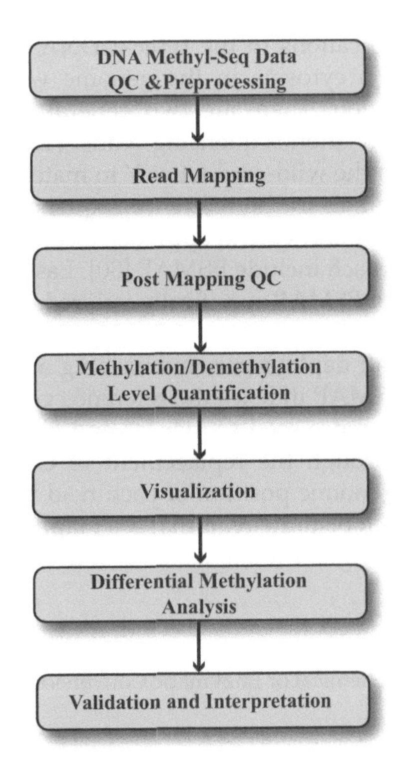

FIGURE 14.3
Major steps of chemical or enzymatic conversion-based DNA methyl-seq data analysis.

bisulfite or enzymatic conversion and the subsequent sequencing process, a converted unmethylated cytosine is read as a thymine (T), or an adenine (A) on the complementary strand, while a methylated cytosine remains as a cytosine (C) or a guanine (G) on the complementary strand (see Figure 14.1). Such base changes have several implications for the read mapping process:

- *Fuzziness in mapping*: A T in the reads could be mapped to a C or T in the reference sequence, thus complicating the searching process.
- *Increase in search space*: This is partly caused by the non one-to-one mapping, and more seriously, by the generation of the four bisulfite-converted strands that are distinct from the reference strands (also illustrated in Figure 14.1), leading to significant increase in search space.
- *Reduction in sequence complexity*: The amount of Cs in the bisulfite reads is significantly reduced, and this reduction in sequence complexity leads to higher levels of mapping ambiguity. Consequently, aligning bisulfite sequencing reads to the reference genome is not as straightforward as that for ChIP-seq or other DNA deep sequencing data.

Without making modifications to the typical DNA alignment process, the methylation status of a cytosine in the genome will affect alignment of reads covering the region. There are two general strategies for mapping conversion-based methylation sequencing reads: (1) replacing all Cs in the reference genome with the wild-card letter Y to match both Cs and Ts in the reads; (2) converting all Cs in the reference sequence and reads to Ts, and then aligning with a seed-and-extend approach (see Table 14.1). Aligners that use the wild-card approach include BSMAP [30], Last [31], GSNAP [32], and RRBSMAP (a version of BSMAP specifically tailored for RRBS reads, merged into BSMAP-2.0) [33]. In the example of BSMAP, it uses SOAP for carrying out read alignment, and deploys genome hashing and bitwise masking for speed and accuracy. BSMAP indexes the reference genome using hash table containing original reference seed sequences and all their possible bisulfite conversion variants through the replacement of Cs with Ts. After determining the potential genomic position of each read by looking up the hash table, for the T(s) in each bisulfite read that are mapped to reference genome position(s) where the original reference base(s) are C(s), BSMAP masks as

TABLE 14.1

Read Mapping Tools for Chemical or Enzymatic Conversion-Based DNA Methylation Sequencing

Name	Description	Reference
Three-Letter Aligners		
Bismark	Deploys Bowtie 2 (or HISAT2) for alignment, and performs cytosine methylation calling	[34]
bwa-meth	Wraps BWA-MEM, provides local alignment for speed and accurary even without trimming	[35]
BS-Seeker2/ BS-Seeker3	Incorporates major aligners, such as Bowtie 2, to achieve gapped local alignment. BS-Seeker3 further improves speed and accuracy, and offers post-alignment analysis including QC and visualization	[36, 37]
BSmooth	Uses Bowtie 2 (or Merman) for alignment. Also provides QC, smoothing-based methylation quantification, and differential methylation detection	[28]
Wild-Card Aligners		
BSMAP/ RRBSMAP	Combines hash table seeding incorporated in SOAP, and bitwise masking to achieve speed and accuracy	[30]
GSNAP	Employs hash tables built for plus and minus strands using C-T/G-A substitutions	[32]
Last	Builds on the traditional alignment strategy of seed-and-extend (like Blast), but with use of adaptive seeds	[31]
BatMeth2	Performs indel-sensitive mapping, DNA methylation quantification, differential methylation detection, annotation, and visualization	[41]

C(s). The masked bisulfite reads are then mapped again to the reference genome.

Aligners such as Bismark [34], bwa-meth [35], and BS-Seeker2/BS-Seeker3 [36, 37] use the other (three-letter) approach. One advantage of this approach is that fast mapping algorithms such as BWA-MEM or bowtie2 can be used. For example, Bismark carries out alignment by first converting Cs in the reads into Ts, and Gs into As (equivalent of the C-to-T conversion on the complementary strand) (Figure 14.4). This conversion process is also performed on the reference genome. The converted reads are then aligned, using Bowtie or Bowtie2, to the converted reference genome in four parallel processes (also refer to Figure 14.1), out of which a unique best alignment is determined [alignment (1) in Figure 14.4]. Among the above wild-card and three-letter methods, benchmark studies [38–40] found that bwa-meth, Bismark, and BSMAP offer a good combination of accuracy, speed, and genomic region coverage.

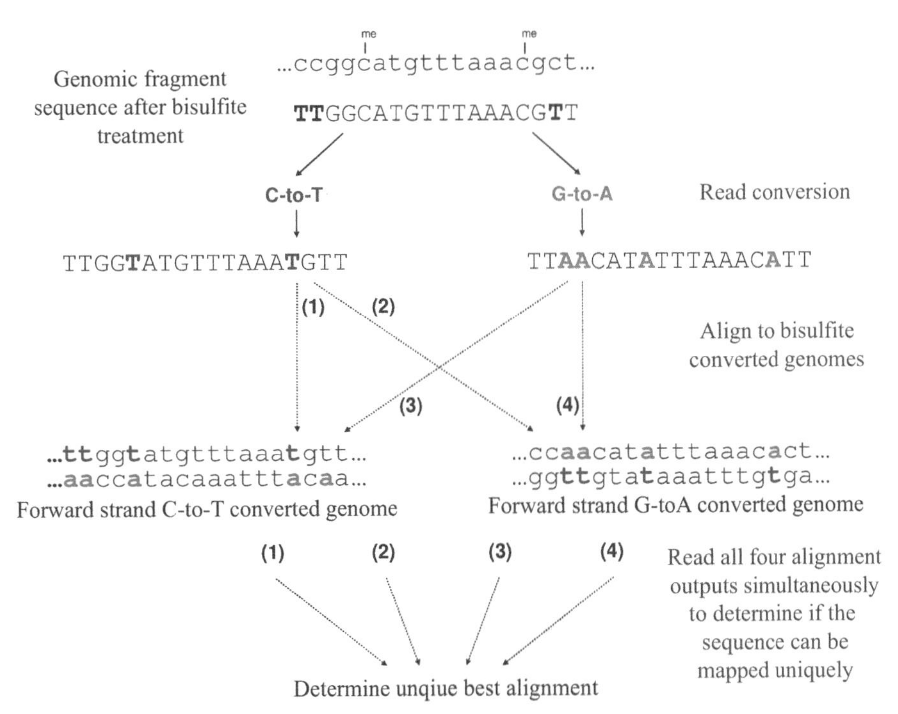

FIGURE 14.4
The "three-letter" bisulfite sequencing read alignment approach used by Bismark. (Adapted from Krueger F. and Andrews S.R. (2011) Bismark: a flexible aligner and methylation caller for Bisulfite-Seq applications. *Bioinformatics*, 27, 1571–2. Used under the terms of the Creative Commons Attribution Non-Commercial License (http://creativecommons.org/licenses/by-nc/2.5). © 2011 Krueger and Andrews.)

After mapping, additional QC is needed prior to extraction of methylation information. For example, reads that have low MAPQ (e.g., < 10) should be removed. Duplicate reads that map exactly to the same position are most likely PCR artifacts and should be removed from further analysis (except in the case of RRBS). Methylation bias (or mbias) should also be checked and corrected. This bias is caused by the step of end repair during library prep post fragmentation, which incorporates unmethylated Cs at both ends of repaired DNA fragments leading to artificially low methylation levels at the affected sites [42]. Mbias plot, showing mean methylation levels at all base positions across the entire length of sequence reads, can be used to reveal the magnitude of the bias, based on which affected base positions should be excluded from further analysis. In addition, polymorphic sites which involve C→T variation (G→A on the complementary strand) will also affect analysis and should also be excluded. As part of post-mapping QC, distribution of the mapped reads in the genome should be visually examined. This provides an initial survey of the results, and at the same time may reveal other abnormalities such as significantly unbalanced numbers of reads mapped to the two DNA strands in a genomic region (this should be inspected with caution and the reads may need to be filtered out). Methyl-seq QC tools such as BSeQC [43], and post-mapping analysis tools such as MethylDackel [44], can be used to carry out these post-mapping QC steps.

14.2.3 Quantification of DNA Methylation/Demethylation Products

After read mapping, uniquely mapped reads need to be aggregated to quantify the methylation level (also called β-value) at individual cytosine sites in the reference genome, based on the frequency of Cs (i.e. methylated cytosines) and Ts (unmethylated cytosines) in reads mapped to each of these sites. This quantitative step can be performed by dividing the total number of Cs by the total combined number of Cs and Ts that are mapped to each site. All of the sequence mappers introduced in the previous section generate this information. MethylDackel and other post-mapping tools such as methylKit [45] can also be used for methylation quantification. If separate quantification of 5mC and other demethylation products (such as 5hmC) is needed after applying approaches such as oxBS-seq, TAB-seq, or enzymatic methyl-seq, simultaneous estimation of methylation/demethylation product levels can be achieved using methods such as MLML [46], which is part of the MethPipe package [47]. For this quantification step, it should be noted that the involved calculations usually require a minimum depth, e.g., at least three reads, at the individual sites to avoid deriving unreliable methylation levels from too few reads.

Besides quantifying methylation levels at individual cytosine sites, DNA methylation quantification is also calculated on a region-by-region basis, usually performed to facilitate comparisons between multiple samples.

Different approaches can be used for regional DNA methylation quantification. These include approaches to segment the genome into bins, sliding windows of user-defined size (e.g. 100 bp) [48], or predefined regions (such as promoters, CpG islands, gene bodies, introns, etc.). The mean of methylation levels of individual cytosine sites, or alternatively the overall proportion of methylated cytosines among all cytosines, within each of such regions can be used to represent each region.

These calculations, however, do not take into consideration the possible existence of SNPs that involve the change from C to T. Some algorithms, such as Bis-SNP [49], remove this potential confounding factor through distinguishing conversion-caused base changes from genetic variants. The use of sequence reads from the complementary strand makes this possible, because a T produced from bisulfite conversion will have a G on the opposite strand while a C→T SNP will have an A on the other strand.

Different from the conversion-based sequencing methods, the methylated DNA enrichment sequencing approaches such as MeDIP-seq and MBD-seq cannot quantify methylation at the single-nucleotide resolution. In addition, the absolute levels of DNA methylation cannot be obtained from the enrichment-based methods, as the sequence read counts from these methods are a function of both absolute DNA methylation levels and regional CpG content. Since these approaches are based on affinity immunoprecipitation and more similar to ChIP-seq, analytical methods developed for ChIP-seq data analysis, including background determination, normalization, and peak detection, can be applied for quantification of DNA methylation by these approaches. As an output, the degree of DNA methylation can be summarized as coverage over a predefined region, such as per gene, promoter, or certain sized bin.

14.2.4 Visualization

Visualizing DNA methylation data serves at least two purposes. Firstly, distribution pattern of DNA methylation may be discerned through visualization. Secondly, visual examination of known DNA methylation regions and other randomly selected regions also offers data validation and a quick estimate of data quality. One method to visualize DNA methyl-seq data and associated information, such as depth of coverage, is through the use of bedGraph files. This standard format (Figure 14.5), compatible with most genome browsers and tools including the Washington University EpiGenome Browser [50], can be directly generated from many of the methylation quantification tools such as Bismark and methylKit. Figure 14.6 shows an example of displaying methylation level along with read depth in the genome.

Alternatively, DNA methylation quantification results can be saved in tab-delimited files and then converted to bigBed or bigWig formats [51]. Both formats are compatible with and enable visualization of the methylation

```
track type=bedGraph
chr19 45408804 45408805 1.0
chr19 45408806 45408807 0.75
chr19 45408854 45408855 0.3
chr19 45408855 45408856 0.5
```

FIGURE 14.5
An example of the bedGraph file format. It includes a track definition line (the first line), followed by track data lines in four column format, i.e. chromosome, chromosome start position, chromosome end position, and data value.

results in web-based genome browsers such as the UCSC Genome Browser, or desktop-based ones such as IGV or IGB. An additional option is to export DNA methylation data to the VCF format using tools such as GobyWeb [52], and then visualize with genome browsers such as IGV or Savant.

14.3 Detection of Differentially Methylated Cytosines and Regions

One frequent goal of DNA methylation analysis is to compare and identify specific cytosines and genomic regions that show differential methylation between conditions. To identify differentially methylated cytosines and regions (DMCs/DMRs), different tools employing different statistical methods have been developed (see Table 14.2). The statistical methods employed by these tools include parametric tests such as *t*-test or ANOVA, and nonparametric tests such as Fisher's exact test, Wilcoxon test, Chi-square test, or Kruskal–Wallis test. The parametric tests assume normal distribution, which is likely to be violated for DNA methylation data as it tends to follow bimodal distribution. As a result, most currently available tools use nonparametric tests. For example, methylKit and RnBeads [53, 54] use Fisher's exact test for comparison of groups without replicates. This test can also be applied directly often with good performance. For comparisons involving multiple samples per group, methylKit and RnBeads use logistic and linear regression, respectively. WBSA employs Wilcoxon test, and BSmooth applies a modified *t*-test with local data smoothing to increase detection power. Another package called Methy-Pipe [55] detects DMRs using the Mann-Whitney U test with a sliding window approach. More sophisticated approaches include the use of a beta-binomial hierarchical model in MOABS [56], and Shannon entropy in QDMR [57]. Besides these different statistical tests or models, another notable difference among these tools is on how biological replicates are handled. Earlier methods tend to pool replicate data for DMC/DMR detection, leading to the loss of information on sample-to-sample variation. Newer methods, such as BSmooth and MOABS, are more

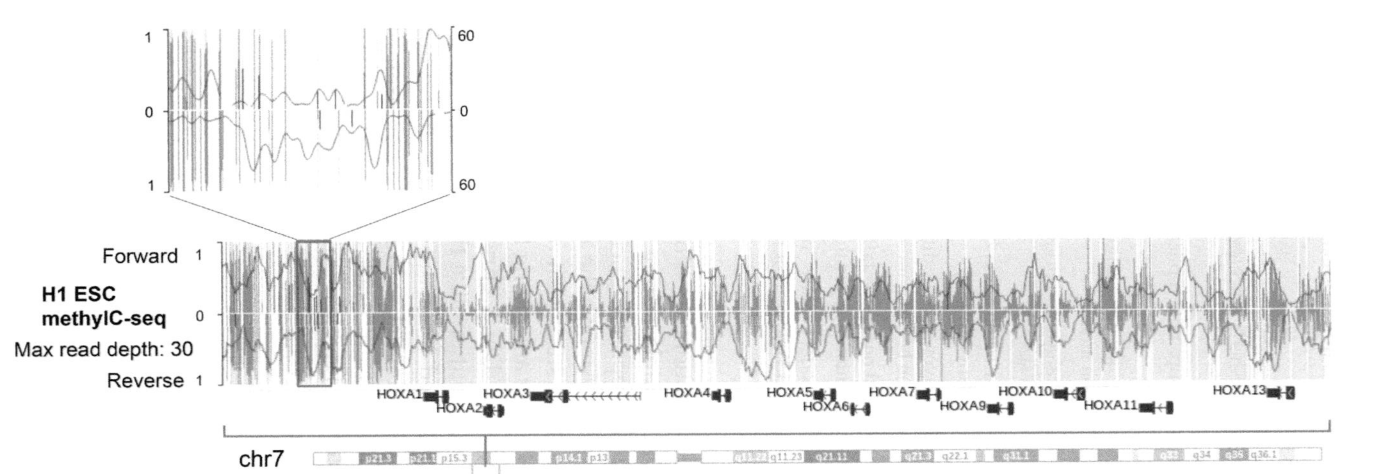

FIGURE 14.6

Visualization of DNA methylation data in a genome browser. Shown here is the methylC track in the Washington University EpiGenome Browser, for a region of the human chromosome 7 where the *HOXA* gene cluster is located. The original WGBS data was collected from H1 human embryonic stem cells. Both DNA methylation levels (represented by vertical bars) and read depth (the smoothed black curve) are displayed in a strand-specific fashion. The different foreground/background colors of the vertical bars represent different cytosine contexts, with CG represented by blue/grey, CHG by orange/light orange, and CHH by magenta/light magenta. A zoomed-in view of the red boxed region is shown on the top. The left axis marks DNA methylation level, while the right marks read depth. (From Zhou X. et al. (2014) methylC Track: visual integration of single-base resolution DNA methylation data on the WashU EpiGenome Browser. *Bioinformatics*, 30, 2206–7. Modified and used with permission from Oxford University Press. © 2014 Zhou et al.)

TABLE 14.2

Tools for Detection of Differentially Methylated Cytosines/Regions

Name	Description	Reference
Fisher's	Applies directly the classical Fisher's exact test for DMC/DMR recognition	N/A
methylKit	Uses logistic regression for groups with replicates, and Fisher's exact test if without replicates	[45]
methylSig	Identifies DMCs/DMRs using likelihood ratio estimation based on a beta-binomial distribution model	[61]
RnBeads	Combines statistical testing *p*-values, and priority ranking based on absolute and relative effect size	[53, 54]
RADMeth	Uses log-likelihood ratio test based on a beta-binomial regression model for differential methylation testing	[62]
DMRFinder	Applies Wald and empirical Bayes tests for differential methylation detection based on beta-binomial modelling	[63]
Metilene	A nonparametric method based on segmentation of the genome using a circular binary segmentation algorithm	[64]
DSS	Uses a Bayesian hierarchical model to allow information sharing across different CpG sites, and Wald test for DMC detection	[65]

replicate-aware and provide estimation on biological variation, thereby increasing detection power. On multiple testing correction, FDR is mostly used, while other methods are also reported, such as a sliding linear model (SLIM) method used by methylKit. Among the currently available DMC/DMR detection tools, benchmarking studies show that methylKit, Fisher's exact test, methylSig, and DMRFinder are among the top performers [58, 59].

Data obtained from methylated DNA enrichment-based approaches follows the negative binomial distribution, like the ChIP-seq and RNA-seq data. Therefore, they can be analyzed to identify DMRs using algorithms developed for RNA-seq-based differential expression. For example, tools such as EdgeR and DESeq can be directly used. In some DNA methylation analysis tools, such as Repitools [60], EdgeR is directly called.

14.4 Data Verification, Validation, and Interpretation

The DMCs/DMRs identified in the previous step need to be verified and further validated. Verification is usually conducted on the same set of samples as those used for DNA methylation sequencing data generation. Further validation, on the other hand, is carried out on a new set of samples. For DNA methylation sequencing data verification and validation, the following techniques are often used: methylation-specific PCR (such as MethyLight), or methylation-independent PCR coupled with pyrosequencing, mass spectrometry, or combined bisulfite restriction analysis (COBRA).

Data interpretation is a key step to translate a list of DMCs/DMRs into mechanistic understanding of the biological process under study. Most potential effects of the DMCs/DMRs can only be revealed through examining them in their genomic context. Tools such as EpiExplorer [66], methylKit, or WBSA can be very helpful in this regard via placing them in the context of other genomic features such as CpG islands, transcription start sites, histone modification marks, or repetitive regions. DMCs/DMRs can also be mapped to nearby genes, which can then be subjected to gene set enrichment, biological pathway, and gene networking analyses. In this regard, the web-based Genomic Regions Enrichment of Annotations Tool (or GREAT) [67] can be used to map DMCs/DMRs to nearby genes, while controlling for gene size difference and distance, for functional annotation and interpretation.

References

1. Seiler Vellame D, Castanho I, Dahir A, Mill J, Hannon E. Characterizing the properties of bisulfite sequencing data: maximizing power and sensitivity to identify between-group differences in DNA methylation. *BMC Genomics* 2021, 22(1):446.
2. Standards and Guidelines for Whole Genome Shotgun Bisulfite Sequencing (www.roadmapepigenomics.org/protocols)
3. Ziller MJ, Hansen KD, Meissner A, Aryee MJ. Coverage recommendations for methylation analysis by whole-genome bisulfite sequencing. *Nat Methods* 2015, 12(3):230–232.
4. Meissner A, Gnirke A, Bell GW, Ramsahoye B, Lander ES, Jaenisch R. Reduced representation bisulfite sequencing for comparative high-resolution DNA methylation analysis. *Nucleic Acids Res* 2005, 33(18):5868–5877.
5. Sun Z, Cunningham J, Slager S, Kocher JP. Base resolution methylome profiling: considerations in platform selection, data preprocessing and analysis. *Epigenomics* 2015, 7(5):813–828.
6. Nautiyal S, Carlton VE, Lu Y, Ireland JS, Flaucher D, Moorhead M, Gray JW, Spellman P, Mindrinos M, Berg P *et al.* High-throughput method for analyzing methylation of CpGs in targeted genomic regions. *Proc Natl Acad Sci U S A* 2010, 107(28):12587–12592.
7. Varley KE, Mitra RD. Bisulfite Patch PCR enables multiplexed sequencing of promoter methylation across cancer samples. *Genome Res* 2010, 20(9):1279–1287.
8. Deng J, Shoemaker R, Xie B, Gore A, LeProust EM, Antosiewicz-Bourget J, Egli D, Maherali N, Park IH, Yu J *et al.* Targeted bisulfite sequencing reveals changes in DNA methylation associated with nuclear reprogramming. *Nat Biotechnol* 2009, 27(4):353–360.

9. Ivanov M, Kals M, Kacevska M, Metspalu A, Ingelman-Sundberg M, Milani L. In-solution hybrid capture of bisulfite-converted DNA for targeted bisulfite sequencing of 174 ADME genes. *Nucleic Acids Res* 2013, 41(6):e72.
10. Liu MC, Oxnard GR, Klein EA, Swanton C, Seiden MV, Consortium C. Sensitive and specific multi-cancer detection and localization using methylation signatures in cell-free DNA. *Ann Oncol* 2020, 31(6):745–759.
11. Han Y, Zheleznyakova GY, Marincevic-Zuniga Y, Kakhki MP, Raine A, Needhamsen M, Jagodic M. Comparison of EM-seq and PBAT methylome library methods for low-input DNA. *Epigenetics* 2021, 17(10):1195–1204.
12. Vaisvila R, Ponnaluri VKC, Sun Z, Langhorst BW, Saleh L, Guan S, Dai N, Campbell MA, Sexton BS, Marks K *et al.* Enzymatic methyl sequencing detects DNA methylation at single-base resolution from picograms of DNA. *Genome Res* 2021, 31(7):1280–1289.
13. Feng S, Zhong Z, Wang M, Jacobsen SE. Efficient and accurate determination of genome-wide DNA methylation patterns in Arabidopsis thaliana with enzymatic methyl sequencing. *Epigenetics Chromatin* 2020, 13(1):42.
14. Sun Z, Vaisvila R, Hussong LM, Yan B, Baum C, Saleh L, Samaranayake M, Guan S, Dai N, Correa IR, Jr. *et al.* Nondestructive enzymatic deamination enables single-molecule long-read amplicon sequencing for the determination of 5-methylcytosine and 5-hydroxymethylcytosine at single-base resolution. *Genome Res* 2021, 31(2):291–300.
15. Harris RA, Wang T, Coarfa C, Nagarajan RP, Hong C, Downey SL, Johnson BE, Fouse SD, Delaney A, Zhao Y *et al.* Comparison of sequencing-based methods to profile DNA methylation and identification of monoallelic epigenetic modifications. *Nat Biotechnol* 2010, 28(10):1097–1105.
16. Nair SS, Coolen MW, Stirzaker C, Song JZ, Statham AL, Strbenac D, Robinson MD, Clark SJ. Comparison of methyl-DNA immunoprecipitation (MeDIP) and methyl-CpG binding domain (MBD) protein capture for genome-wide DNA methylation analysis reveal CpG sequence coverage bias. *Epigenetics* 2011, 6(1):34–44.
17. Rodriguez-Aguilera JR, Ecsedi S, Goldsmith C, Cros MP, Dominguez-Lopez M, Guerrero-Celis N, Perez-Cabeza de Vaca R, Chemin I, Recillas-Targa F, Chagoya de Sanchez V *et al.* Genome-wide 5-hydroxymethylcytosine (5hmC) emerges at early stage of in vitro differentiation of a putative hepatocyte progenitor. *Sci Rep* 2020, 10(1):7822.
18. Yu M, Han D, Hon GC, He C. Tet-Assisted Bisulfite Sequencing (TAB-seq). *Methods Mol Biol* 2018, 1708 :645–663.
19. Booth MJ, Branco MR, Ficz G, Oxley D, Krueger F, Reik W, Balasubramanian S. Quantitative sequencing of 5-methylcytosine and 5-hydroxymethylcytosine at single-base resolution. *Science* 2012, 336(6083):934–937.
20. Booth MJ, Marsico G, Bachman M, Beraldi D, Balasubramanian S. Quantitative sequencing of 5-formylcytosine in DNA at single-base resolution. *Nat Chem* 2014, 6(5):435–440.
21. Liu Y, Hu Z, Cheng J, Siejka-Zielinska P, Chen J, Inoue M, Ahmed AA, Song CX. Subtraction-free and bisulfite-free specific sequencing of 5-methylcytosine and its oxidized derivatives at base resolution. *Nat Commun* 2021, 12(1):618.

22. Flusberg BA, Webster DR, Lee JH, Travers KJ, Olivares EC, Clark TA, Korlach J, Turner SW. Direct detection of DNA methylation during single-molecule, real-time sequencing. *Nat Methods* 2010, 7(6):461–465.
23. Laszlo AH, Derrington IM, Brinkerhoff H, Langford KW, Nova IC, Samson JM, Bartlett JJ, Pavlenok M, Gundlach JH. Detection and mapping of 5-methylcytosine and 5-hydroxymethylcytosine with nanopore MspA. *Proc Natl Acad Sci U S A* 2013, 110(47):18904–18909.
24. Schreiber J, Wescoe ZL, Abu-Shumays R, Vivian JT, Baatar B, Karplus K, Akeson M. Error rates for nanopore discrimination among cytosine, methylcytosine, and hydroxymethylcytosine along individual DNA strands. *Proc Natl Acad Sci U S A* 2013, 110(47):18910–18915.
25. Tse OYO, Jiang P, Cheng SH, Peng W, Shang H, Wong J, Chan SL, Poon LCY, Leung TY, Chan KCA *et al.* Genome-wide detection of cytosine methylation by single molecule real-time sequencing. *Proc Natl Acad Sci U S A* 2021, 118(5):e2019768118.
26. Rand AC, Jain M, Eizenga JM, Musselman-Brown A, Olsen HE, Akeson M, Paten B. Mapping DNA methylation with high-throughput nanopore sequencing. *Nat Methods* 2017, 14(4):411–413.
27. Trim Galore! (www.bioinformatics.babraham.ac.uk/projects/trim_galore/)
28. Hansen KD, Langmead B, Irizarry RA. BSmooth: from whole genome bisulfite sequencing reads to differentially methylated regions. *Genome Biol* 2012, 13(10):R83.
29. Liang F, Tang B, Wang Y, Wang J, Yu C, Chen X, Zhu J, Yan J, Zhao W, Li R. WBSA: web service for bisulfite sequencing data analysis. *PLoS One* 2014, 9(1):e86707.
30. Xi Y, Li W. BSMAP: whole genome bisulfite sequence MAPping program. *BMC Bioinformatics* 2009, 10:232.
31. Frith MC, Mori R, Asai K. A mostly traditional approach improves alignment of bisulfite-converted DNA. *Nucleic Acids Res* 2012, 40(13):e100.
32. Wu TD, Nacu S. Fast and SNP-tolerant detection of complex variants and splicing in short reads. *Bioinformatics* 2010, 26(7):873–881.
33. Xi Y, Bock C, Muller F, Sun D, Meissner A, Li W. RRBSMAP: a fast, accurate and user-friendly alignment tool for reduced representation bisulfite sequencing. *Bioinformatics* 2012, 28(3):430–432.
34. Krueger F, Andrews SR. Bismark: a flexible aligner and methylation caller for Bisulfite-Seq applications. *Bioinformatics* 2011, 27(11):1571–1572.
35. Pedersen BS, Eyring K, De S, Yang IV, Schwartz DA. Fast and accurate alignment of long bisulfite-seq reads. *arXiv preprint arXiv:14011129* 2014.
36. Huang KYY, Huang YJ, Chen PY. BS-Seeker3: ultrafast pipeline for bisulfite sequencing. *BMC Bioinformatics* 2018, 19(1):111.
37. Guo W, Fiziev P, Yan W, Cokus S, Sun X, Zhang MQ, Chen PY, Pellegrini M. BS-Seeker2: a versatile aligning pipeline for bisulfite sequencing data. *BMC Genomics* 2013, 14 :774.
38. Kunde-Ramamoorthy G, Coarfa C, Laritsky E, Kessler NJ, Harris RA, Xu M, Chen R, Shen L, Milosavljevic A, Waterland RA. Comparison and quantitative verification of mapping algorithms for whole-genome bisulfite sequencing. *Nucleic Acids Res* 2014, 42(6):e43.

39. Nunn A, Otto C, Stadler PF, Langenberger D. Comprehensive benchmarking of software for mapping whole genome bisulfite data: from read alignment to DNA methylation analysis. *Brief Bioinform* 2021, 22(5):bbab021.
40. Sun X, Han Y, Zhou L, Chen E, Lu B, Liu Y, Pan X, Cowley AW, Jr., Liang M, Wu Q *et al.* A comprehensive evaluation of alignment software for reduced representation bisulfite sequencing data. *Bioinformatics* 2018, 34(16):2715–2723.
41. Zhou Q, Lim JQ, Sung WK, Li G. An integrated package for bisulfite DNA methylation data analysis with indel-sensitive mapping. *BMC Bioinformatics* 2019, 20(1):47.
42. Bock C. Analysing and interpreting DNA methylation data. *Nat Rev Genet* 2012, 13(10):705–719.
43. Lin X, Sun D, Rodriguez B, Zhao Q, Sun H, Zhang Y, Li W. BSeQC: quality control of bisulfite sequencing experiments. *Bioinformatics* 2013, 29(24):3227–3229.
44. MethylDackel (https://github.com/dpryan79/MethylDackel)
45. Akalin A, Kormaksson M, Li S, Garrett-Bakelman FE, Figueroa ME, Melnick A, Mason CE. methylKit: a comprehensive R package for the analysis of genome-wide DNA methylation profiles. *Genome Biol* 2012, 13(10):R87.
46. Qu J, Zhou M, Song Q, Hong EE, Smith AD. MLML: consistent simultaneous estimates of DNA methylation and hydroxymethylation. *Bioinformatics* 2013, 29(20):2645–2646.
47. MethPipe (https://github.com/smithlabcode/methpipe/)
48. Smith ZD, Chan MM, Mikkelsen TS, Gu H, Gnirke A, Regev A, Meissner A. A unique regulatory phase of DNA methylation in the early mammalian embryo. *Nature* 2012, 484(7394):339–344.
49. Liu Y, Siegmund KD, Laird PW, Berman BP. Bis-SNP: Combined DNA methylation and SNP calling for Bisulfite-seq data. *Genome Biol* 2012, 13(7):R61.
50. Washington University EpiGenome Browser (http://epigenomegateway.wustl.edu/browser/)
51. Kent WJ, Zweig AS, Barber G, Hinrichs AS, Karolchik D. BigWig and BigBed: enabling browsing of large distributed datasets. *Bioinformatics* 2010, 26(17):2204–2207.
52. Dorff KC, Chambwe N, Zeno Z, Simi M, Shaknovich R, Campagne F. GobyWeb: simplified management and analysis of gene expression and DNA methylation sequencing data. *PLoS One* 2013, 8(7):e69666.
53. Muller F, Scherer M, Assenov Y, Lutsik P, Walter J, Lengauer T, Bock C. RnBeads 2.0: comprehensive analysis of DNA methylation data. *Genome Biol* 2019, 20(1):55.
54. Assenov Y, Muller F, Lutsik P, Walter J, Lengauer T, Bock C. Comprehensive analysis of DNA methylation data with RnBeads. *Nat Methods* 2014, 11(11):1138–1140.
55. Jiang P, Sun K, Lun FM, Guo AM, Wang H, Chan KC, Chiu RW, Lo YM, Sun H. Methy-Pipe: an integrated bioinformatics pipeline for whole genome bisulfite sequencing data analysis. *PLoS One* 2014, 9(6):e100360.
56. Sun D, Xi Y, Rodriguez B, Park HJ, Tong P, Meong M, Goodell MA, Li W. MOABS: model based analysis of bisulfite sequencing data. *Genome Biol* 2014, 15(2):R38.
57. Zhang Y, Liu H, Lv J, Xiao X, Zhu J, Liu X, Su J, Li X, Wu Q, Wang F *et al.* QDMR: a quantitative method for identification of differentially methylated regions by entropy. *Nucleic Acids Res* 2011, 39(9):e58.

58. Piao Y, Xu W, Park KH, Ryu KH, Xiang R. Comprehensive evaluation of differential methylation analysis methods for bisulfite sequencing data. *Int J Environ Res Public Health* 2021, 18(15):7975.
59. Liu Y, Han Y, Zhou L, Pan X, Sun X, Liu Y, Liang M, Qin J, Lu Y, Liu P. A comprehensive evaluation of computational tools to identify differential methylation regions using RRBS data. *Genomics* 2020, 112(6):4567–4576.
60. Statham AL, Strbenac D, Coolen MW, Stirzaker C, Clark SJ, Robinson MD. Repitools: an R package for the analysis of enrichment-based epigenomic data. *Bioinformatics* 2010, 26(13):1662–1663.
61. Park Y, Figueroa ME, Rozek LS, Sartor MA. MethylSig: a whole genome DNA methylation analysis pipeline. *Bioinformatics* 2014, 30(17):2414–2422.
62. Dolzhenko E, Smith AD. Using beta-binomial regression for high-precision differential methylation analysis in multifactor whole-genome bisulfite sequencing experiments. *BMC Bioinformatics* 2014, 15:215.
63. Gaspar JM, Hart RP. DMRfinder: efficiently identifying differentially methylated regions from MethylC-seq data. *BMC Bioinformatics* 2017, 18(1):528.
64. Juhling F, Kretzmer H, Bernhart SH, Otto C, Stadler PF, Hoffmann S. metilene: fast and sensitive calling of differentially methylated regions from bisulfite sequencing data. *Genome Res* 2016, 26(2):256–262.
65. Feng H, Conneely KN, Wu H. A Bayesian hierarchical model to detect differentially methylated loci from single nucleotide resolution sequencing data. *Nucleic Acids Res* 2014, 42(8):e69.
66. Halachev K, Bast H, Albrecht F, Lengauer T, Bock C. EpiExplorer: live exploration and global analysis of large epigenomic datasets. *Genome Biol* 2012, 13(10):R96.
67. McLean CY, Bristor D, Hiller M, Clarke SL, Schaar BT, Lowe CB, Wenger AM, Bejerano G. GREAT improves functional interpretation of cis-regulatory regions. *Nat Biotechnol* 2010, 28(5):495–501.

15

Whole Metagenome Sequencing for Microbial Community Analysis

A small amount of environmental sample, such as a handful of soil, is rich in microbial life, but the number of microbial species in such a sample is unknown. The microbiome on or within our body contains tens of thousands, if not more, species of bacteria, fungi, and archaea. Besides their tremendous species diversity, the composition, as well as function, of such microbial communities is not static but constantly changing according to the status of their environment. Our current understanding of these diverse and dynamic microbial communities is still significantly lacking, as most of our knowledge comes from culturable species. For those that still cannot be cultured in the lab, which comprise the majority of microorganisms on earth, we know very little. Metagenomics offers an important approach to study microbial diversity in these environmental communities without relying on artificial culturing. Also referred to as environmental or community genomics, metagenomics examines all genomes existing in a microbial community as a whole without the need to capture or amplify individual genomes. Through simultaneous analysis of all DNA molecules present in a microbial community, metagenomics provides a profile of taxonomic composition and functional status of the community and its environment.

Before the advent of NGS, metagenomics studies were usually conducted with DNA cloning combined with Sanger sequencing. In this approach, DNA extracted from a microbial community is first fragmented, and then the DNA fragments are cloned into plasmid vectors for amplification in order to produce enough materials for Sanger sequencing. With the continuous development and significant cost drop in NGS technologies, massively parallel metagenomic sequencing has quickly replaced this traditional low-throughput approach and become a major approach for studying various microbial communities. The high sensitivity offered by the NGS approach provides direct access to the unculturable microbial majority that were previously "invisible" to analysis [1]. The Human Microbiome Project exemplifies the use of NGS in interrogating complex metagenomes, such as those at different sites of the human body including the gastrointestinal tract. The application of NGS in metagenomic analysis of a large variety of other microbial communities, such as those in soil, the phyllosphere, the ocean, and those

DOI: 10.1201/9780429329180-18

associated with bioremediation and biofuel generation, has led to exponential increase in the number of metagenomes studied.

Compared to the NGS data generated from a single species (most chapters in this book deal with individual species), the metagenomics data from microbial community sequencing is much more complicated. Each metagenome contains DNA sequences from a large but unknown number of species, including viruses, bacteria, archaea, fungi, and microscopic eukaryotes. To further complicate the situation, the relative abundance of these species varies widely. In comparison to sequencing reads collected from a single species, metagenomic sequencing reads contain much higher heterogeneity because of the tremendous genome diversity in each microbial community. Also because of the tremendous DNA sequence complexity contained in the metagenome, most metagenomic sequencing effort can only sample part of the DNA pool. As a result of this limited sampling in a highly diverse DNA space, metagenomic NGS data is highly fragmented and has low redundancy. Due to the lack of redundant (i.e., partially overlapping, not duplicate) reads, metagenomic NGS data has an inherently higher error rate when compared to single-genome sequencing. All these differences between metagenomic and monogenomic NGS data require an entirely different set of tools for NGS-based metagenome data analysis for microbial community compositional and functional profiling.

15.1 Experimental Design and Sample Preparation

Metagenomics studies aim to determine identities and relative abundance of different members, or taxa, in a microbial community, and how environmental factors affect the composition and function of these communities. To achieve this by sequencing, there are two general approaches: whole genome shotgun metagenomic sequencing and targeted metagenomic sequencing. The shotgun approach provides random sampling of all genomes contained in an environmental or host-associated microbial sample. To carry out shotgun sequencing, total DNA extracted from such a sample is first broken into small fragments for short-read sequencing, or high-molecular-weight (HMW) DNA directly sequenced using long-read sequencing.

In the targeted approach, genomic component(s) that are shared among different species are PCR amplified and the amplicons are sequenced. The most commonly used target in this approach is the 16S rRNA gene, while other genes that code for specific protein functions (such as resistance to specific antibiotics) or non-coding genes are also used. The 16S rRNA gene, being considered as the universal clock of life [2], is usually used as a surrogate marker for measuring the relative abundance of different operational taxonomic units (OTUs, a metagenomics term to describe a species or a group of species when only DNA sequence information is available). By focusing on the 16S rRNA gene or other specific genomic target(s), this approach greatly reduces complexity in the generated data, thereby achieving deeper coverage and accommodating more samples. It should be noted that the 16S rRNA-based approach only produces approximate estimation of relative taxonomic abundance, due to 16S rRNA copy number variation in some species and the fact that the standard 16S rRNA PCR primers may not bind to their supposed target sites in all cases because of random mutation. In comparison, the shotgun approach, while requiring significantly more sequencing, takes an unbiased path to offer a comprehensive assessment of genome content in the community, and thereby provides in-depth information on community composition and function. This chapter focuses on shotgun metagenome sequencing data analysis.

15.1.1 Metagenome Sample Collection

The success of a metagenomics project is to a degree dependent on factors that are not related to genomics. One such factor is how much is known about the habitat where study samples will be collected. The more physically, chemically, and ecologically characterized the habitat is, the more knowledge will be gathered from the metagenomic NGS data. In-depth characterization and detailed description of the sampling environment is one foundation of a successful metagenomics experiment. Keeping detailed metadata on the habitat and the sampling process, such as characteristics of the general environment, geographical location and specific features of the sampling locales, and the sampling method, is of great importance to downstream data interpretation.

As the composition and complexity of a metagenome sample are determined by the habitat and the sampling site, the unique characteristics of a sampling environment, along with the question to be answered or specific hypothesis to be tested, eventually determine how many reads are required. It should also be emphasized that since where the samples are collected directly shapes the outcome, the sampling sites must be representative of the habitat under study. In order to collect representative samples, information on spatial and temporal variation in the habitat must be known prior to sample collection. If this information is not available, a small-scale trial shotgun sequencing run might prove helpful with a small number of samples sequenced. Alternatively, a targeted 16S rRNA amplicon sequencing can also be used to survey the diversity of the microbial community.

15.1.2 Metagenome Sample Processing

DNA extraction is the first and also a key step in metagenome sequencing sample preparation. The DNA extracted from this step should represent all, or at least most, members of the sampling community and their relative abundance, be of high purity and free of contaminants that might interfere with the subsequent sequencing library construction. While this step might be routine in conventional genome sequencing for a single organism, extracting high-quality DNA from microbial community samples collected from various habitats poses challenges. For example, humic acids, polysaccharides, tannins, and other compounds are major contaminants in environmental samples such as those from the soil, which if not removed can lead to inhibition of enzymes used in library construction. In host-associated habitats such as the human gut, host DNA is the major potential contaminant.

Besides purity, extracting DNA in equal efficiency from different community members is another challenge, as optimal condition of cell lysis for DNA release from one group of microbes may not be ideal for another. For example, mechanical disruption is often used for breaking up cells in metagenomics studies, but by using this method DNA released from easily lysed cells may be sheared to fragments when tougher cells are eventually disrupted. While these challenging issues should be acknowledged and addressed, they are not insurmountable and robust extraction protocols are available for various habitats (e.g., [3]).

Advancements in sequencing library preparation protocols have reduced the amounts of DNA required considerably to lower nanograms level (e.g., the Nextera XT protocol needs only 1 ng DNA to start). This should accommodate DNA extracted from most habitats. In situations where only very limited amount of DNA is available, amplification of the DNA might be needed to generate enough material for creating sequencing libraries. To maintain the relative abundance level between community members, strategies such as multiple displacement amplification can be used. Such amplification can generate more than enough DNA for library construction from femtograms of starting DNA.

15.2 Sequencing Approaches

There are several key factors that need to be considered before the sequencing process starts. These include sequencing depth, read length, and sequencing platforms. The depth of sequencing is dependent on the species richness and abundances in the samples, and the goal to be pursued. For example, a study that attempts to locate rare members in a highly diverse microbial community requires deeper sequencing than one that is only focused on more abundant members in a less diverse environment. With regard to read length, longer reads are always better than shorter reads in metagenomics for sorting out the inherent sequence complexity. The read length from most current short read sequencers can reach 150 bp from each end for paired-end sequencing.

As overviewed in Chapter 4, long-read sequencing technologies, such as PacBio's SMRT and ONT's nanopore sequencing, generate much longer (but fewer and less accurate) reads. With read throughput and sequencing accuracy steadily increasing, there are increasing numbers of samples being sequenced on these systems. A hybrid approach is also often used to take advantage of the different strength of these technologies, with the use of short reads to generate an in-depth survey of the community and long reads to provide scaffolding for assembling contigs (see next). Future advancements in sequencing technologies will undoubtedly lead to continuous increase in read length and drop in cost making the goals of metagenomics more achievable.

15.3 Overview of Shotgun Metagenome Sequencing Data Analysis

For microbial community profiling, whole-metagenome shotgun sequencing provides rich information on a community's taxonomic composition and functional status, without requiring pre-existing knowledge of all genomes contained in the community. Figure 15.1 shows an overview of a general shotgun metagenomic sequencing data analysis workflow. Most of this workflow involves assembly of reads to reconstruct the so-called metagenome-assembled genomes (or MAGs), while an assembly-free approach uses sequencing reads directly. For both approaches, to perform taxonomic profiling and functional analysis, sequence homology and other feature search against genes of microbes in the currently known taxonomy, as catalogued in various public databases, are key steps. While results from these key steps are limited to the currently known genomes and catalogued sequences, the rapid increase in the number of sequenced microbial genomes will gradually alleviate this limitation. The rapid increase in the employment of long-read sequencing to the metagenomics field also helps with reconstruction of more MAGs for subsequent analyses. Besides taxonomic profiling and functional analysis in one condition or habitat, comparative metagenomics analysis between conditions or habitats is usually performed to achieve the final goal of studying the effects of environmental factors on a microbial community.

The following sections cover the various steps of metagenomics data analysis. Because of the great diversity in sampling habitats/conditions and the specific questions asked in each study, there is no fixed workflow for metagenomics data analysis. The steps outlined in Figure 15.1 and covered next are not necessarily arranged in the most appropriate order for a particular project, and they can be used in different combinations and/or with

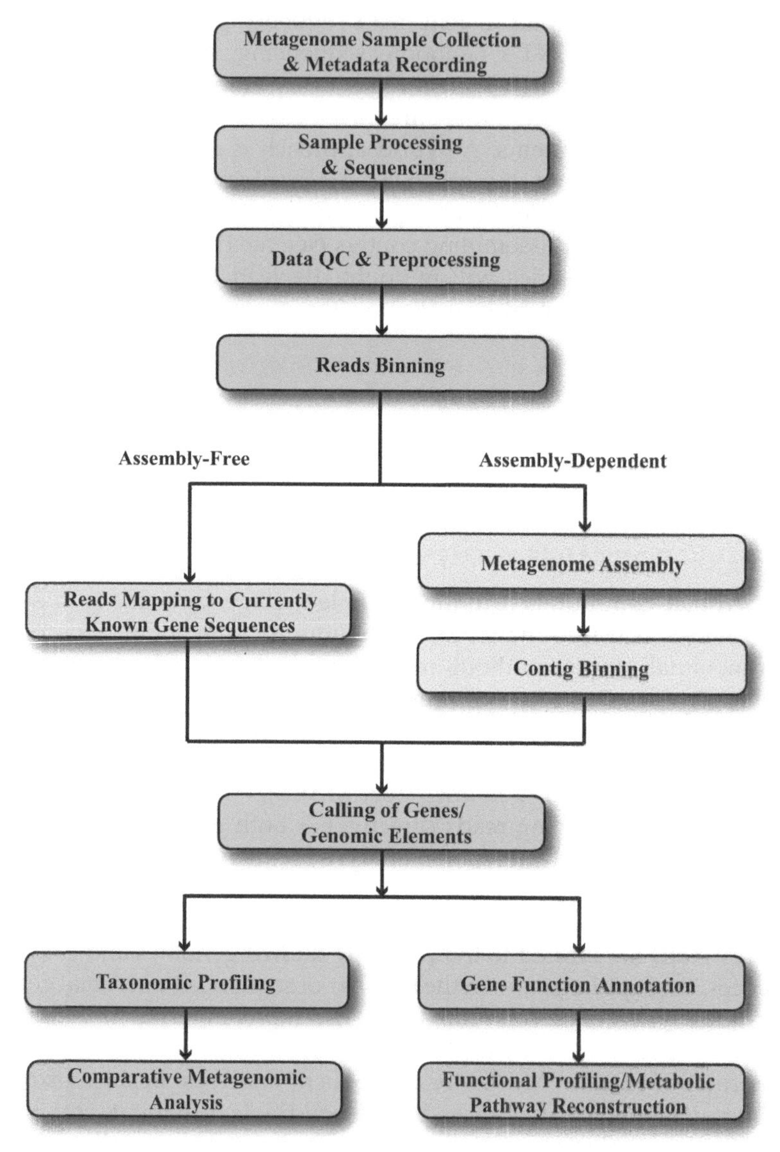

FIGURE 15.1
Major steps of metagenome analysis.

some step(s) omitted. Since the publication of the first edition of this book, there has been rapid increase in the number of tools available for metagenome analysis. Some of the currently available tools, such as those required for taxonomic profiling based on search against multiple databases, require considerable computing resources and power.

15.4 Sequencing Data Quality Control and Preprocessing

To ensure data quality and avoid erroneous results, metagenomic shotgun sequencing reads should first be examined and preprocessed prior to conducting downstream analysis. Using the tools introduced in Chapter 5, reads of low quality should be filtered out, and low-quality bases and adapter sequences trimmed off. In addition, for samples from host-associated habitats, contaminating host sequences need to be marked and excluded from further analysis. This can be realized by simply using mappers such as bowtie2 (the --un-conc output option), or using bowtie2 to output all reads and then samtools to extract reads unmapped to the host genome. Specially developed tools for marking and removing DNA contamination sequences include KneadData [4], BMTagger [5], and DeconSeq [6]. Additional data preprocessing also includes removal of duplicated reads. This can be conducted with tools such as the Dedupe tool in the BBTools suite [7], FastUniq [8], and the Picard module called EstimateLibraryComplexity, all of which identify and remove duplicate reads without the need to align reads to a reference genome.

15.5 Taxonomic Characterization of a Microbial Community

15.5.1 Metagenome Assembly

While the ultimate goal of metagenomics is to assemble each genome in a microbial community, this is currently still far from achievable because of several reasons. The number of organisms in a metagenome is unknown, and there are wide variations in their relative abundance and therefore sequencing depth among the organisms. This is especially the case for samples collected from highly complex microbial communities. The large number of species in these samples, and the concomitant low sequencing depth for most species, make metagenome assembly extremely challenging. Sequence similarity between closely related species poses further challenges to assemblers, often leading to chimeric assemblies that contain reads from different OTUs. Despite the challenges, metagenome reads assembly is an important step in metagenomic sequencing data analysis. It has led to rapid increase in the total number of MAGs recovered from a wide range of environments, such as the human gut [9, 10], and a diversity of habitats on Earth encompassing all continents and oceans [11].

For *de novo* metagenome assembly, the assemblers introduced in Chapter 12 for single-genome *de novo* assembly, such as SOAPdenovo and Velvet, were initially applied but with limited success. As a result, assemblers tailored for metagenome reads have been developed. For assembling long reads such as those generated from the PacBio and ONT platforms, assemblers such as metaFlye [12], Raven [13], Canu [14], and Hifiasm-meta [15] can be used. For short reads, more assemblers are available, including SPAdes/metaSPAdes [16, 17], MEGAHIT [18], IDBA-UD [19], MetaVelvet/MetaVelvet-SL [20, 21], and Ray Meta [22]. Similar to single-genome assemblers, many of these short read metagenome assemblers, such as metaSPAdes, MEGAHIT, IDBA-UD, and Ray Meta, are based on the de Bruijn graph approach (see Chapter 12). In addition, these methods use multi-*k*-mer sizes, instead of a fixed *k*-mer size, in order to improve assemblies. The difference from the single-genome assemblers, though, is that they attempt to identify subgraphs within a mixed de Bruijn graph, each of which is expected to represent an individual genome. For example, metaSPAdes first builds a large de Bruijn graph from all metagenomic reads using SPAdes and then transforms it into an assembly graph. Within the assembly graph, subgraphs that contain alternative paths are identified, corresponding to large fragments from individual genomes. Besides these assemblers that are designed for either long or short reads, other assemblers combine long and short reads in an effort to increased assembly quality. These hybrid assemblers include MaSuRCA [23], hybridSPAdes [24], and OPERA-MS [25].

After the assembly process, a metagenome usually comprises mostly of contigs of various sizes. To evaluate the assembly quality, traditional evaluation metrics, such as N50, are not as informative and representative as in evaluating single-genome assemblies. Instead, aggregate statistics such as the total number of contigs, the percentage of reads mapped to them, and the maximum, median, and average lengths of the contigs are often used. Further inspection of the assembly quality includes looking for chimeric or mis-assemblies. There are currently a number of tools available to assess MAG quality, including CheckM [26], MetaQUAST [27], and BUSCO [28], all of which rely on reference genomes. Reference-free tools include DeepMAsED [29] and ALE (Assembly Likelihood Evaluation) [30].

After contig assembly, if paired reads are available, metagenome scaffolds can be built from the contigs. Many of the metagenome assemblers have a module to carry out scaffolding. Besides these modules, dedicated metagenome scaffolding tools like Bambus 2 [31] may be used to determine if additional scaffolding is needed. Bambus 2 accepts contigs constructed with most assemblers using reads from all sequencing platforms. In the process of building scaffolds from contigs, ambiguous and inconsistent contigs may also be identified. Besides scaffolding, another approach for the assembly of MAGs is contig binning, which places contigs derived from the same genome into the same bin. Contigs in the same bin are then reassembled into a MAG.

15.5.2 Sequence Binning

As indicated above, metagenomic sequence binning refers to the process of clustering sequence fragments in a mixture into different "bins" corresponding to their genomic or taxonomic origins. This process can be conducted on contigs (most often), reads, or genes. With longer reads or contigs, high-resolution binning can be achieved at the levels of family, genus, species, or even strains. Short sequences may be binned only to the level of phylum due to the limited information carried in the sequences. Since sequence binning reduces the complexity inherent in the metagenomics data, each set of binned sequences can then be subjected to independent analysis in other steps, such as MAG reconstruction for identification of new archaebacterial and eubacterial specie or viruses.

Two binning approaches are usually used. One approach, called genome binning, bins contigs or reads on the basis of sequence composition and abundance. Sequence composition encompasses characteristics such as oligonucleotide (often tetramer) frequency, G/C content, and codon usage. The genome binning approach is built on the assumption that sequences from the same species or closely related species/organisms are more similar to each other in these characteristics and abundance than to distantly or non-related species/organisms. Methods that use this strategy include MetaBAT 2 [32], MaxBin 2.0 [33], GroopM [34] (now replaced by Rosella) [35], CONCOCT [36], MetaWatt [37], and VAMB [38] (Table 15.1). CONCOCT, as an example for genome binning, places contigs into bins based on sequence composition and coverage across samples. After combining *k*-mer frequency and coverage information, CONCOCT applies PCA to reduce the dimensionality of the matrix and reveal clusters. This is followed by the utilization of a Gaussian mixture model (GMM) for binning the contigs. VAMB is another example of genome binning method that is based on machine learning. It uses deep variational autoencoders to encode sequence co-abundance and *k*-mer distribution before clustering of sequences. Besides the individual methods, there are also ensemble methods that integrate binning results from the different individual genome binning methods. Examples of such ensemble binning methods are MetaBinner [39], MetaWRAP [40], and DAS tool [41]. The Critical Assessment of Metagenome Interpretation (CAMI) challenge round II shows that some of these ensemble methods can achieve better results than individual methods [42]. In general, the genome binning approach is more reliable for contigs and long reads, as short reads carry less information due to their limited length. While it has the advantage of being fast as it does not rely on aligning metagenomic sequences to reference databases, variation in the distribution of sequence composition and coverage can lead to inaccuracies.

The other approach, called taxonomy binning, assigns metagenomic sequences to their taxonomic sources of origin by comparing against microbial sequence database(s) that are taxonomically annotated. Methods that use this approach first place sequences into different bins, followed by labeling of the bins with taxonomic identifiers. Examples of taxonomy binning methods include MEGAN/MEGAN-LR (for long reads) [43, 44], Kraken 2 [45], and PhyloPythiaS+ [46] (Table 15.1). MEGAN (for Metagenome Analyzer), as an example, first aligns reads against a reference database such as BLAST-nr using an alignment tool called DIAMOND [47] that achieves fast alignment speed. The alignment result is then processed by a program called MEGANIZER for taxonomic binning that aims to assign a taxon ID to each sequence. This binning process uses an algorithm called naïve LCA, which assigns each read to a node representing the lowest common ancestor. Based on this algorithm, a read that aligns to a widely conserved gene is assigned

TABLE 15.1

Commonly Used Binning Algorithms

Name	Description	Reference
Genome Binning		
MaxBin 2.0	Classifies contigs into different bins using an Expectation-Maximization algorithm on the basis of their tetranucleotide frequencies and coverages	[33]
CONCOCT	Combines coverage and tetramer frequency for contig binning using GMM	[36]
MetaBAT 2	Uses adaptive binning to group the most reliable contigs first (such as those of high length), and then gradually add remaining contigs	[32]
MetaWatt	Uses multivariate statistics of tetramer frequencies and differential coverage information for binning. Also assesses binning quality using taxonomic annotation of contigs in each bin	[37]
VAMB	A machine-learning based binner that encodes *k*-mer distribution and sequence co-abundance information using variational autoencoders for subsequent binning	[38]
GroopM	Bins sequences by primarily leveraging differential coverage information	[34]
MetaBinner	An ensemble binner that integrates component binning results generated with multiple features and initiations	[39]
MetaWRAP	An ensemble binner to generate hybrid bin sets from other binners and select final bins based on CheckM results	[40]
Taxonomy Binning		
MEGAN	Aligns sequences against the NCBI-nr reference database and then performs taxonomic binning using the naïve LCA algorithm	[43]
Kraken 2	Assigns taxonomic labels to sequences based on search of *k*-mers within the sequences against a database of indexed and sorted *k*-mers (or their minimizers) extracted from all genomes	[45, 48]
PhyloPythiaS+	Achieves taxonomic binning through building sample-specific support vector machine taxonomic classifier using most relevant taxa and training sequences determined automatically	[46]

to a taxon at a higher level (e.g., phylum or class), while another read that aligns to a less conserved gene that is limited to a select group of organisms is assigned to a lower-level taxon (such as genus or species). As it is based on the current annotation of catalogued sequences, this approach is not suitable to find currently unknown species or taxa.

15.5.3 Calling of Genes and Other Genomic Elements from Metagenomic Sequences

To answer the questions of what taxonomic groups are in a microbial community and what they are doing, identification of genes from assembled contigs or MAGs is an essential step. For gene coding region identification, since ORF-containing metagenomic sequences may not carry full-length ORFs, metagenome ORF-calling algorithms do not penalize for their incompleteness. Many metagenome ORF callers employ machine learning strategies. For example, Prodigal [49] and MetaGeneAnnotator [50] employ dynamic programming. GeneMarkS-2 [51], FragGeneScan (FGS) [52], and Glimmer-MG [53] are based on Markov models. CNN-MGP [54] and Balrog [55] use convolutional neural networks. Among these methods, Prodigal is an unsupervised algorithm that does not require training data. It predicts protein-coding genes using inherent properties in the sequences themselves, such as start codon usage, ribosomal-binding site motif usage, genetic code usage, G/C content, and hexamer coding statistics, among other information. Calling of other genomic elements, such as ncRNAs and CRISPRs, may require long reads or contigs as well as more computational resources. Currently a limited number of tools are available to identify these elements, such as tRNAscan-SE [56], ARAGORN [57], MinCED (a modified version of the CRISPR Recognition Tool or CRT) [58], CRISPRFinder [59], and CRISPRDetect [60]. Besides providing answers to the composition and function of a microbial community, calling of genes and other genomic elements also helps identify mis-assembled reads or locate adjoining contigs that are not yet placed into the same scaffold.

15.5.4 Taxonomic Profiling

One goal of metagenome analysis is profile taxonomic composition and relative abundance of each taxon in a microbial community. This is related to but different from taxonomic binning, which aims to group metagenomic sequences into different bins. The CAMI II challenge finds that taxonomic profilers MetaPhlAn [61] and mOTUs [62] had the best overall performance, both of which are based on the use of phylogenetic gene markers not taxonomic binning of reads. Phylogenetic gene markers are composed of ubiquitous but phylogenetically diverse genes, with good examples being the rRNA genes (e.g., 16S), *recA* (DNA recombinase A), *rpoB* (RNA polymerase beta subunit), *fusA* (protein chain elongation factor), and *gyrB* (DNA gyrase subunit B). MetaPhlAn, as an example, conducts homology search of metagenomic reads against an extensive list of clade-specific gene markers. From the mapping of query reads to the marker genes, the number of mapped reads is normalized prior to calculation of relative abundance levels of different taxa identified in the sample [61]. Other commonly used taxonomic profilers include Kaiju [63], Centrifuge [64], Bracken [65], GTDB-tk [66], and IGGsearch [9]. Among these tools, GTDB-tk is a toolkit that assigns taxonomic classification of MAGs using the Genome Taxonomy Database (GTDB). It first calls genes from MAGs using Prodigal, then identifies marker genes, followed by subsequent marker-gene-based phylogenetic inference [67]. Bracken, which is built upon taxonomic binning results from Kraken, uses a Bayesian approach to estimate the abundance levels of different taxa. Taxonomic profilers developed specifically for long reads include MetaMaps [68] and BugSeq [69]. Currently available comparisons on the use of such long-read taxonomic profilers vs. the aforementioned short-read methods suggest that the long-range sequence information provided by long reads and the methods designed to take advantage of such information lead to improved taxonomic profiling [70].

15.6 Functional Characterization of a Microbial Community

15.6.1 Gene Function Annotation

Gene calling from metagenomic sequences provides substrate for functional analysis of the underlying community, i.e., answering the question of "what they are doing." Functional annotation of called genes can reveal the full repertoire of protein functions in a habitat, including metabolism, signal transduction, stress tolerance, virulence, etc. Uncommon functions may suggest unusual lifestyle and activity in a community. The relative abundance of different types of genes also reveals specificity about a community and how organisms in the community deal with environmental factors in the habitat.

To conduct functional annotation, predicted protein sequences from called ORFs are searched against a database of reference protein sequences, or HMMs that describe protein families. Protein sequence databases such as UniProt [71], InterPro [72], COG [73], and eggNOG [74], are among the most commonly used databases. This task of database searching to identify all possible peptides coded by the metagenome is a computationally intensive process. If local computing resources permit, locally installed stand-alone tools such as Prokka [75], DRAM [76], DFAST [77], or NCBI's Prokaryotic Genome Annotation Pipeline (PGAP) [78] can be used. Alternatively, the task can also be submitted to a web-based system such as the metagenomics RAST server (MG-RAST) [79], the Integrated Microbial Genomes & Microbiomes system (IMG/M) [80], GhostKOALA [81], or MGnify [82]. Some tools, such as eggNOG-mapper [83] provide both online and stand-alone versions.

15.6.2 Gene Function Profiling and Metabolic Pathway Reconstruction

To further answer the question of "what they are doing," functional profiling of a microbial community can be performed at the levels of annotated genes (or gene families) and metabolic pathways. Analysis at the gene/gene family level relies on function annotations with the use of the databases detailed above, while at the metabolic pathway level this requires databases that capture currently known metabolic pathways, including KEGG [84] and MetaCyc [85]. Based on reference information from these databases, functional profiling tools such as HUMAnN [61] and Carnelian [86] provide the abundance of each gene/gene family, or metabolic pathway detected in a community. This abundance profile can be further stratified to reveal the contribution of each identified OTU in the community. As an example, HUMAnN employs MBLASTX to search metagenomic reads against the KEGG Orthology database to determine the abundances of individual orthologous protein families. At the pathway level, HUMAnN reconstructs pathways using MinPath [87], which is a maximum parsimony approach to explain the observed families and their abundances with a minimal set of pathways. After further noise reduction and smoothing, the output from HUMAnN displays pathway coverage (i.e., whether each pathway is present or absent), and the relative abundance of each pathway in the metagenomic samples.

Besides quantifying present metabolic pathways, continuous methods development also makes it possible to semi-automatically reconstruct genome-scale metabolic models, or GSMMs, to reveal the metabolic potential of a genome. This reconstruction process requires integration of gene coding information in an assembled individual microbial genome and metabolic pathway information catalogued in reference databases. Examples of such methods are ModelSEED [88], CarveMe [89], Pathway Tools [90], Merlin [91], and RAVEN [92]. From the reconstructed single-species GSMMs and the relative species abundance measured from a metagenomic sample, the many GSMMs can be integrated to predict the metabolic status of the underlying community. MAMBO, for Metabolomic Analysis of Metagenomes using fBa and Optimization, is a good exemplary tool for this task [93].

In addition, the increasing availability of metabolomics data also makes it possible to correlate metagenomic sequence and metabolite data. For example, MIMOSA is a reference-based tool to correlate metagenomic and metabolomic data [94]. With this tool, community metabolic potential scores can be calculated from metagenomic sequencing data to describe the potential ability of a microbial community to metabolize small molecules. Abundances of metabolites, as measured by metabolomics, can then be correlated with the community's metabolic potential scores. Furthermore, the specific taxa, genes, or reactions that may contribute to the formation of the metabolites can be identified. MelonnPan is another tool designed to predict metabolomic profiles from microbial sequencing features. This method does not rely on reference database search, instead it uses a machine learning approach [95].

15.7 Comparative Metagenomic Analysis

Comparative metagenomic analysis between habitats or conditions can lead to insights about the underlying microbial communities and their dynamics. However, statistical comparison between metagenomes is not as straightforward as other NGS-based comparative analyses (such as RNA-seq). This is mostly due to the tremendous amount of variability involved in comparative metagenomic analysis. One source of this variability is biological, as microbial composition can vary greatly between different samples. Another source is technical, due to insufficient sequencing depth and therefore undersampling of low-abundance species. These species generate fewer reads and are more affected by stochastic factors in the sequence sampling process, as in general the number of reads from a species is dependent on a number of factors, including relative abundance of the species, genome size, genome copy number, within-species heterogeneity, and DNA extraction efficiency. Due to these biological and technical factors, many species or OTUs detected in one sample or condition are often absent in another sample or condition. If rare species need to be studied in a metagenome study, it is more cost effective to artificially increase their abundance using cell enrichment technologies such as flow cell sorting rather than increasing sequencing depth. In a typical metagenomics project that does not artificially increase the abundance of rare species, their undersampling can lead to significant biases in subsequent data normalization and detection of significant differences between samples. Compared to other steps in the metagenomic data analysis pipeline, there has been relatively less method development in comparative metagenomic analysis.

15.7.1 Metagenome Sequencing Data Normalization

Similar to RNA-seq data, metagenomic abundance data needs to be normalized prior to comparative analysis. Currently there is still no consensus as to how metagenomics data should be normalized. Among the normalization approaches that have been reported, total-sum scaling (TSS), equivalent to the Total Count approach in RNA-seq (Chapter 7), is performed by dividing the raw count of reads assigned to a certain species or OTU by the total number of reads in the same sample. Another approach is cumulative-sum scaling (CSS), which, similar to the Upper Quartile approach in RNA-seq, is calculated by dividing the raw count of reads assigned to a species or OTU by the cumulative sum of counts up to a certain percentile. Other normalization methods introduced in Chapter 7, such as TMM and RLE, can also be used on metagenomic abundance data.

15.7.2 Identification of Differentially Abundant Species or OTUs

To identify species or OTUs that are differentially abundant between habitats or conditions, currently available tools include metagenomeSeq [96], phyloseq [97], LEfSe [98], STAMP [99], ANCOM and ANCOM-BC [100, 101], corncob [102], and MaAsLin 2 [103]. These tools use different methods and statistics to detect differential abundance between metagenomes. For example, metagenomeSeq implements the CSS normalization, and a distribution mixture statistical model to deal with the biases caused by the undersampling issue that confound comparative metagenomic analyses. LEfSe uses the Kruskal–Wallis rank-sum test to detect features that display significant differential abundance between conditions. Among the newer methods, MaAsLin 2 uses general linear models to detect associations between abundances of microbial features, such as taxa (or genes), and environmental or other phenotypic metadata. It offers a number of normalization methods including TSS, CSS, TMM, etc.

15.8 Integrated Metagenomics Data Analysis Pipelines

Besides the tools developed for each of the individual steps above, pipelines designed for integrated comprehensive analysis of metagenomics data are also available. These pipelines, including Quantitative Insight Into Microbial Ecology (QIIME2) [104], IMG/M, MEGAN, MG-RAST, metaWRAP, and bioBakery [61], contain large collections of tools that encompass the many aspects of metagenomics data mining from preprocessing, binning, feature identification, functional annotation, to cross-condition comparison. For example, MG-RAST directly takes sequencing and metadata files as input, conducts reads QC and preprocessing, gene calling, protein identification, annotation mapping, abundance profiling, comparative analysis, and metabolic reconstruction.

15.9 Metagenomics Data Repositories

In the United States, like for other NGS data, the NCBI SRA database provides the official repository for all metagenomic data collected by NGS technologies.

In Europe, MGnify (previously EBI Metagenomics) offers archiving and analysis of metagenomics data. The data archived by MGnify is also accessible through ENA (or SRA). In China, a database called gcMeta, or Global Catalogue of Metagenomics [105], has been established more recently. Besides these official metagenomics data repositories, MG-RAST and IMG/M are two *de facto* metagenomic data repositories that also enable data sharing in a collaborative environment and with the entire research community. The value of these repositories will become more apparent when more and more metagenomics data becomes available. For example, they can accelerate the discovery of new genes and species through providing opportunities to compare currently unknown sequences that exist in multiple metagenomes. In a typical shotgun metagenomics study, many sequences are previously unknown and may represent novel genes or sequences from currently uncatalogued species. To discover novel genes and new species, meta-analysis of data (including metadata) is needed, which is only enabled by these repositories.

References

1. Lloyd KG, Steen AD, Ladau J, Yin J, Crosby L. Phylogenetically novel uncultured microbial cells dominate earth microbiomes. *mSystems* 2018, 3(5):e00055-18.
2. Yarza P, Ludwig W, Euzeby J, Amann R, Schleifer KH, Glockner FO, Rossello-Mora R. Update of the All-Species Living Tree Project based on 16S and 23S rRNA sequence analyses. *Syst Appl Microbiol* 2010, 33(6):291–299.
3. Delmont TO, Robe P, Clark I, Simonet P, Vogel TM. Metagenomic comparison of direct and indirect soil DNA extraction approaches. *J Microbiol Methods* 2011, 86(3):397–400.
4. McIver LJ, Abu-Ali G, Franzosa EA, Schwager R, Morgan XC, Waldron L, Segata N, Huttenhower C. bioBakery: a meta'omic analysis environment. *Bioinformatics* 2018, 34(7):1235–1237.
5. BMTagger (http://biowulf.nih.gov/apps/bmtagger.html)
6. Schmieder R, Edwards R. Fast identification and removal of sequence contamination from genomic and metagenomic datasets. *PLoS One* 2011, 6(3):e17288.
7. Bushnell B. BBMap: a fast, accurate, splice-aware aligner. In.: Lawrence Berkeley National Lab.(LBNL), Berkeley, CA (United States); 2014.
8. Xu H, Luo X, Qian J, Pang X, Song J, Qian G, Chen J, Chen S. FastUniq: a fast de novo duplicates removal tool for paired short reads. *PLoS One* 2012, 7(12):e52249.
9. Nayfach S, Shi ZJ, Seshadri R, Pollard KS, Kyrpides NC. New insights from uncultivated genomes of the global human gut microbiome. *Nature* 2019, 568(7753):505–510.
10. Almeida A, Mitchell AL, Boland M, Forster SC, Gloor GB, Tarkowska A, Lawley TD, Finn RD. A new genomic blueprint of the human gut microbiota. *Nature* 2019, 568(7753):499–504.

11. Nayfach S, Roux S, Seshadri R, Udwary D, Varghese N, Schulz F, Wu D, Paez-Espino D, Chen IM, Huntemann M *et al.* A genomic catalog of Earth's microbiomes. *Nat Biotechnol* 2021, 39(4):499–509.
12. Kolmogorov M, Bickhart DM, Behsaz B, Gurevich A, Rayko M, Shin SB, Kuhn K, Yuan J, Polevikov E, Smith TPL *et al.* metaFlye: scalable long-read metagenome assembly using repeat graphs. *Nat Methods* 2020, 17(11): 1103–1110.
13. Vaser R, Šikić M. Time- and memory-efficient genome assembly with Raven. *Nat Comput Sci* 2021, 1(5):332–336.
14. Koren S, Walenz BP, Berlin K, Miller JR, Bergman NH, Phillippy AM. Canu: scalable and accurate long-read assembly via adaptive k-mer weighting and repeat separation. *Genome Res* 2017, 27(5):722–736.
15. Feng X, Cheng H, Portik D, Li H. Metagenome assembly of high-fidelity long reads with Hifiasm-meta. *Nat Methods* 2022, 19(6):671–674.
16. Nurk S, Meleshko D, Korobeynikov A, Pevzner PA. metaSPAdes: a new versatile metagenomic assembler. *Genome Res* 2017, 27(5):824–834.
17. Bankevich A, Nurk S, Antipov D, Gurevich AA, Dvorkin M, Kulikov AS, Lesin VM, Nikolenko SI, Pham S, Prjibelski AD *et al.* SPAdes: a new genome assembly algorithm and its applications to single-cell sequencing. *J Comput Biol* 2012, 19(5):455–477.
18. Li D, Liu CM, Luo R, Sadakane K, Lam TW. MEGAHIT: an ultra-fast single-node solution for large and complex metagenomics assembly via succinct de Bruijn graph. *Bioinformatics* 2015, 31(10):1674–1676.
19. Peng Y, Leung HC, Yiu SM, Chin FY. IDBA-UD: a de novo assembler for single-cell and metagenomic sequencing data with highly uneven depth. *Bioinformatics* 2012, 28(11):1420–1428.
20. Namiki T, Hachiya T, Tanaka H, Sakakibara Y. MetaVelvet: an extension of Velvet assembler to de novo metagenome assembly from short sequence reads. *Nucleic Acids Res* 2012, 40(20):e155.
21. Afiahayati, Sato K, Sakakibara Y. MetaVelvet-SL: an extension of the Velvet assembler to a de novo metagenomic assembler utilizing supervised learning. *DNA Res* 2015, 22(1):69–77.
22. Boisvert S, Raymond F, Godzaridis E, Laviolette F, Corbeil J. Ray Meta: scalable de novo metagenome assembly and profiling. *Genome Biol* 2012, 13(12):R122.
23. Zimin AV, Marcais G, Puiu D, Roberts M, Salzberg SL, Yorke JA. The MaSuRCA genome assembler. *Bioinformatics* 2013, 29(21):2669–2677.
24. Antipov D, Korobeynikov A, McLean JS, Pevzner PA. hybridSPAdes: an algorithm for hybrid assembly of short and long reads. *Bioinformatics* 2016, 32(7):1009–1015.
25. Bertrand D, Shaw J, Kalathiyappan M, Ng AHQ, Kumar MS, Li C, Dvornicic M, Soldo JP, Koh JY, Tong C *et al.* Hybrid metagenomic assembly enables high-resolution analysis of resistance determinants and mobile elements in human microbiomes. *Nat Biotechnol* 2019, 37(8):937–944.
26. Parks DH, Imelfort M, Skennerton CT, Hugenholtz P, Tyson GW. CheckM: assessing the quality of microbial genomes recovered from isolates, single cells, and metagenomes. *Genome Res* 2015, 25(7):1043–1055.
27. Mikheenko A, Saveliev V, Gurevich A. MetaQUAST: evaluation of metagenome assemblies. *Bioinformatics* 2016, 32(7):1088–1090.

28. Simao FA, Waterhouse RM, Ioannidis P, Kriventseva EV, Zdobnov EM. BUSCO: assessing genome assembly and annotation completeness with single-copy orthologs. *Bioinformatics* 2015, 31(19):3210–3212.
29. Mineeva O, Rojas-Carulla M, Ley RE, Scholkopf B, Youngblut ND. DeepMAsED: evaluating the quality of metagenomic assemblies. *Bioinformatics* 2020, 36(10):3011–3017.
30. Clark SC, Egan R, Frazier PI, Wang Z. ALE: a generic assembly likelihood evaluation framework for assessing the accuracy of genome and metagenome assemblies. *Bioinformatics* 2013, 29(4):435–443.
31. Koren S, Treangen TJ, Pop M. Bambus 2: scaffolding metagenomes. *Bioinformatics* 2011, 27(21):2964–2971.
32. Kang DD, Li F, Kirton E, Thomas A, Egan R, An H, Wang Z. MetaBAT 2: an adaptive binning algorithm for robust and efficient genome reconstruction from metagenome assemblies. *PeerJ* 2019, 7:e7359.
33. Wu YW, Simmons BA, Singer SW. MaxBin 2.0: an automated binning algorithm to recover genomes from multiple metagenomic datasets. *Bioinformatics* 2016, 32(4):605–607.
34. Imelfort M, Parks D, Woodcroft BJ, Dennis P, Hugenholtz P, Tyson GW. GroopM: an automated tool for the recovery of population genomes from related metagenomes. *PeerJ* 2014, 2:e603.
35. Rosella (https://github.com/rhysnewell/rosella)
36. Alneberg J, Bjarnason BS, de Bruijn I, Schirmer M, Quick J, Ijaz UZ, Lahti L, Loman NJ, Andersson AF, Quince C. Binning metagenomic contigs by coverage and composition. *Nat Methods* 2014, 11(11):1144–1146.
37. Strous M, Kraft B, Bisdorf R, Tegetmeyer HE. The binning of metagenomic contigs for microbial physiology of mixed cultures. *Front Microbiol* 2012, 3:410.
38. Nissen JN, Johansen J, Allesoe RL, Sonderby CK, Armenteros JJA, Gronbech CH, Jensen LJ, Nielsen HB, Petersen TN, Winther O *et al.* Improved metagenome binning and assembly using deep variational autoencoders. *Nat Biotechnol* 2021, 39(5):555–560.
39. Wang Z, Huang P, You R, Sun F, Zhu S. MetaBinner: a high-performance and stand-alone ensemble binning method to recover individual genomes from complex microbial communities. *Genome Biol* 2023, 24(1):1.
40. Uritskiy GV, DiRuggiero J, Taylor J. MetaWRAP-a flexible pipeline for genome-resolved metagenomic data analysis. *Microbiome* 2018, 6(1):158.
41. Sieber CMK, Probst AJ, Sharrar A, Thomas BC, Hess M, Tringe SG, Banfield JF. Recovery of genomes from metagenomes via a dereplication, aggregation and scoring strategy. *Nat Microbiol* 2018, 3(7):836–843.
42. Meyer F, Fritz A, Deng ZL, Koslicki D, Lesker TR, Gurevich A, Robertson G, Alser M, Antipov D, Beghini F *et al.* Critical Assessment of Metagenome Interpretation: the second round of challenges. *Nat Methods* 2022, 19(4):429–440.
43. Huson DH, Beier S, Flade I, Gorska A, El-Hadidi M, Mitra S, Ruscheweyh HJ, Tappu R. MEGAN Community Edition – Interactive Exploration and Analysis of Large-Scale Microbiome Sequencing Data. *PLoS Comput Biol* 2016, 12(6):e1004957.
44. Huson DH, Albrecht B, Bagci C, Bessarab I, Gorska A, Jolic D, Williams RBH. MEGAN-LR: new algorithms allow accurate binning and easy interactive exploration of metagenomic long reads and contigs. *Biol Direct* 2018, 13(1):6.

45. Wood DE, Lu J, Langmead B. Improved metagenomic analysis with Kraken 2. *Genome Biol* 2019, 20(1):257.
46. Gregor I, Droge J, Schirmer M, Quince C, McHardy AC. PhyloPythiaS+: a self-training method for the rapid reconstruction of low-ranking taxonomic bins from metagenomes. *PeerJ* 2016, 4:e1603.
47. Buchfink B, Xie C, Huson DH. Fast and sensitive protein alignment using DIAMOND. *Nat Methods* 2015, 12(1):59–60.
48. Wood DE, Salzberg SL. Kraken: ultrafast metagenomic sequence classification using exact alignments. *Genome Biol* 2014, 15(3):R46.
49. Hyatt D, Chen GL, Locascio PF, Land ML, Larimer FW, Hauser LJ. Prodigal: prokaryotic gene recognition and translation initiation site identification. *BMC Bioinformatics* 2010, 11:119.
50. Noguchi H, Taniguchi T, Itoh T. MetaGeneAnnotator: detecting species-specific patterns of ribosomal binding site for precise gene prediction in anonymous prokaryotic and phage genomes. *DNA Res* 2008, 15(6):387–396.
51. Lomsadze A, Gemayel K, Tang S, Borodovsky M. Modeling leaderless transcription and atypical genes results in more accurate gene prediction in prokaryotes. *Genome Res* 2018, 28(7):1079–1089.
52. Rho M, Tang H, Ye Y. FragGeneScan: predicting genes in short and error-prone reads. *Nucleic Acids Res* 2010, 38(20):e191.
53. Kelley DR, Liu B, Delcher AL, Pop M, Salzberg SL. Gene prediction with Glimmer for metagenomic sequences augmented by classification and clustering. *Nucleic Acids Res* 2012, 40(1):e9.
54. Al-Ajlan A, El Allali A. CNN-MGP: Convolutional Neural Networks for Metagenomics Gene Prediction. *Interdiscip Sci* 2019, 11(4):628–635.
55. Sommer MJ, Salzberg SL. Balrog: A universal protein model for prokaryotic gene prediction. *PLoS Comput Biol* 2021, 17(2):e1008727.
56. Lowe TM, Eddy SR. tRNAscan-SE: a program for improved detection of transfer RNA genes in genomic sequence. *Nucleic Acids Res* 1997, 25(5):955–964.
57. Laslett D, Canback B. ARAGORN, a program to detect tRNA genes and tmRNA genes in nucleotide sequences. *Nucleic Acids Res* 2004, 32(1):11–16.
58. Bland C, Ramsey TL, Sabree F, Lowe M, Brown K, Kyrpides NC, Hugenholtz P. CRISPR recognition tool (CRT): a tool for automatic detection of clustered regularly interspaced palindromic repeats. *BMC Bioinformatics* 2007, 8:209.
59. Grissa I, Vergnaud G, Pourcel C. CRISPRFinder: a web tool to identify clustered regularly interspaced short palindromic repeats. *Nucleic Acids Res* 2007, 35(Web Server issue):W52–57.
60. Biswas A, Staals RH, Morales SE, Fineran PC, Brown CM. CRISPRDetect: A flexible algorithm to define CRISPR arrays. *BMC Genomics* 2016, 17:356.
61. Beghini F, McIver LJ, Blanco-Miguez A, Dubois L, Asnicar F, Maharjan S, Mailyan A, Manghi P, Scholz M, Thomas AM *et al.* Integrating taxonomic, functional, and strain-level profiling of diverse microbial communities with bioBakery 3. *Elife* 2021, 10:e65088.
62. Milanese A, Mende DR, Paoli L, Salazar G, Ruscheweyh HJ, Cuenca M, Hingamp P, Alves R, Costea PI, Coelho LP *et al.* Microbial abundance, activity and population genomic profiling with mOTUs2. *Nat Commun* 2019, 10(1):1014.

63. Menzel P, Ng KL, Krogh A. Fast and sensitive taxonomic classification for metagenomics with Kaiju. *Nat Commun* 2016, 7:11257.
64. Kim D, Song L, Breitwieser FP, Salzberg SL. Centrifuge: rapid and sensitive classification of metagenomic sequences. *Genome Res* 2016, 26(12):1721–1729.
65. Lu J, Breitwieser FP, Thielen P, Salzberg SL. Bracken: estimating species abundance in metagenomics data. *PeerJ Comput Sci* 2017, 30:e104.
66. Chaumeil PA, Mussig AJ, Hugenholtz P, Parks DH. GTDB-Tk: a toolkit to classify genomes with the Genome Taxonomy Database. *Bioinformatics* 2019, 36(6):1925–1927.
67. Parks DH, Chuvochina M, Waite DW, Rinke C, Skarshewski A, Chaumeil PA, Hugenholtz P. A standardized bacterial taxonomy based on genome phylogeny substantially revises the tree of life. *Nat Biotechnol* 2018, 36(10):996–1004.
68. Dilthey AT, Jain C, Koren S, Phillippy AM. Strain-level metagenomic assignment and compositional estimation for long reads with MetaMaps. *Nat Commun* 2019, 10(1):3066.
69. Fan J, Huang S, Chorlton SD. BugSeq: a highly accurate cloud platform for long-read metagenomic analyses. *BMC Bioinformatics* 2021, 22(1):160.
70. Portik DM, Brown CT, Pierce-Ward NT. Evaluation of taxonomic classification and profiling methods for long-read shotgun metagenomic sequencing datasets. *BMC Bioinformatics* 2022, 23(1):541.
71. UniProt C. UniProt: the universal protein knowledgebase in 2021. *Nucleic Acids Res* 2021, 49(D1):D480–D489.
72. Blum M, Chang HY, Chuguransky S, Grego T, Kandasaamy S, Mitchell A, Nuka G, Paysan-Lafosse T, Qureshi M, Raj S *et al.* The InterPro protein families and domains database: 20 years on. *Nucleic Acids Res* 2021, 49(D1):D 344–D354.
73. Galperin MY, Wolf YI, Makarova KS, Vera Alvarez R, Landsman D, Koonin EV. COG database update: focus on microbial diversity, model organisms, and widespread pathogens. *Nucleic Acids Res* 2021, 49(D1):D274–D281.
74. Huerta-Cepas J, Szklarczyk D, Heller D, Hernandez-Plaza A, Forslund SK, Cook H, Mende DR, Letunic I, Rattei T, Jensen LJ *et al.* eggNOG 5.0: a hierarchical, functionally and phylogenetically annotated orthology resource based on 5090 organisms and 2502 viruses. *Nucleic Acids Res* 2019, 47(D1):D309–D314.
75. Seemann T. Prokka: rapid prokaryotic genome annotation. *Bioinformatics* 2014, 30(14):2068–2069.
76. Shaffer M, Borton MA, McGivern BB, Zayed AA, La Rosa SL, Solden LM, Liu P, Narrowe AB, Rodriguez-Ramos J, Bolduc B *et al.* DRAM for distilling microbial metabolism to automate the curation of microbiome function. *Nucleic Acids Res* 2020, 48(16):8883–8900.
77. Tanizawa Y, Fujisawa T, Nakamura Y. DFAST: a flexible prokaryotic genome annotation pipeline for faster genome publication. *Bioinformatics* 2018, 34(6):1037–1039.
78. Tatusova T, DiCuccio M, Badretdin A, Chetvernin V, Nawrocki EP, Zaslavsky L, Lomsadze A, Pruitt KD, Borodovsky M, Ostell J. NCBI prokaryotic genome annotation pipeline. *Nucleic Acids Res* 2016, 44(14):6614–6624.
79. Keegan KP, Glass EM, Meyer F. MG-RAST, a Metagenomics Service for Analysis of Microbial Community Structure and Function. *Methods Mol Biol* 2016, 1399:207–233.

80. Chen IA, Chu K, Palaniappan K, Ratner A, Huang J, Huntemann M, Hajek P, Ritter S, Varghese N, Seshadri R *et al.* The IMG/M data management and analysis system v.6.0: new tools and advanced capabilities. *Nucleic Acids Res* 2021, 49(D1):D751–D763.
81. Kanehisa M, Sato Y, Morishima K. BlastKOALA and GhostKOALA: KEGG tools for functional characterization of genome and metagenome sequences. *J Mol Biol* 2016, 428(4):726–731.
82. Mitchell AL, Almeida A, Beracochea M, Boland M, Burgin J, Cochrane G, Crusoe MR, Kale V, Potter SC, Richardson LJ *et al.* MGnify: the microbiome analysis resource in 2020. *Nucleic Acids Res* 2020, 48(D1):D570–D578.
83. Cantalapiedra CP, Hernandez-Plaza A, Letunic I, Bork P, Huerta-Cepas J. eggNOG-mapper v2: functional annotation, orthology assignments, and domain prediction at the metagenomic scale. *Mol Biol Evol* 2021, 38(12):5825–5829.
84. Kanehisa M, Furumichi M, Tanabe M, Sato Y, Morishima K. KEGG: new perspectives on genomes, pathways, diseases and drugs. *Nucleic Acids Res* 2017, 45(D1):D353–D361.
85. Caspi R, Billington R, Keseler IM, Kothari A, Krummenacker M, Midford PE, Ong WK, Paley S, Subhraveti P, Karp PD. The MetaCyc database of metabolic pathways and enzymes – a 2019 update. *Nucleic Acids Res* 2020, 48(D1):D445–D453.
86. Nazeen S, Yu YW, Berger B. Carnelian uncovers hidden functional patterns across diverse study populations from whole metagenome sequencing reads. *Genome Biol* 2020, 21(1):47.
87. Ye Y, Doak TG. A parsimony approach to biological pathway reconstruction/inference for genomes and metagenomes. *PLoS Comput Biol* 2009, 5(8):e1000465.
88. Seaver SMD, Liu F, Zhang Q, Jeffryes J, Faria JP, Edirisinghe JN, Mundy M, Chia N, Noor E, Beber ME *et al.* The ModelSEED Biochemistry Database for the integration of metabolic annotations and the reconstruction, comparison and analysis of metabolic models for plants, fungi and microbes. *Nucleic Acids Res* 2021, 49(D1):D1555.
89. Machado D, Andrejev S, Tramontano M, Patil KR. Fast automated reconstruction of genome-scale metabolic models for microbial species and communities. *Nucleic Acids Res* 2018, 46(15):7542–7553.
90. Paley S, Billington R, Herson J, Krummenacker M, Karp PD. Pathway tools visualization of organism-scale metabolic networks. *Metabolites* 2021, 11(2):64.
91. Capela J, Lagoa D, Rodrigues R, Cunha E, Cruz F, Barbosa A, Bastos J, Lima D, Ferreira EC, Rocha M *et al.* merlin, an improved framework for the reconstruction of high-quality genome-scale metabolic models. *Nucleic Acids Res* 2022, 50(11):6052–6066.
92. Wang H, Marcisauskas S, Sanchez BJ, Domenzain I, Hermansson D, Agren R, Nielsen J, Kerkhoven EJ. RAVEN 2.0: a versatile toolbox for metabolic network reconstruction and a case study on Streptomyces coelicolor. *PLoS Comput Biol* 2018, 14(10):e1006541.
93. Garza DR, van Verk MC, Huynen MA, Dutilh BE. Towards predicting the environmental metabolome from metagenomics with a mechanistic model. *Nat Microbiol* 2018, 3(4):456–460.

94. Noecker C, Eng A, Srinivasan S, Theriot CM, Young VB, Jansson JK, Fredricks DN, Borenstein E. Metabolic model-based integration of microbiome taxonomic and metabolomic profiles elucidates mechanistic links between ecological and metabolic variation. *mSystems* 2016, 1(1):e00013– e00015.
95. Mallick H, Franzosa EA, McLver LJ, Banerjee S, Sirota-Madi A, Kostic AD, Clish CB, Vlamakis H, Xavier RJ, Huttenhower C. Predictive metabolomic profiling of microbial communities using amplicon or metagenomic sequences. *Nat Commun* 2019, 10(1):3136.
96. Paulson JN, Stine OC, Bravo HC, Pop M. Differential abundance analysis for microbial marker-gene surveys. *Nat Methods* 2013, 10(12):1200–1202.
97. McMurdie PJ, Holmes S. Waste not, want not: why rarefying microbiome data is inadmissible. *PLoS Comput Biol* 2014, 10(4):e1003531.
98. Segata N, Izard J, Waldron L, Gevers D, Miropolsky L, Garrett WS, Huttenhower C. Metagenomic biomarker discovery and explanation. *Genome Biol* 2011, 12(6):R60.
99. Parks DH, Tyson GW, Hugenholtz P, Beiko RG. STAMP: statistical analysis of taxonomic and functional profiles. *Bioinformatics* 2014, 30(21):3123–3124.
100. Mandal S, Van Treuren W, White RA, Eggesbo M, Knight R, Peddada SD. Analysis of composition of microbiomes: a novel method for studying microbial composition. *Microb Ecol Health Dis* 2015, 26:27663.
101. Lin H, Peddada SD. Analysis of compositions of microbiomes with bias correction. *Nat Commun* 2020, 11(1):3514.
102. Martin BD, Witten D, Willis AD. Modeling microbial abundances and dysbiosis with beta-binomial regression. *Ann Appl Stat* 2020, 14(1):94–115.
103. Mallick H, Rahnavard A, McIver LJ, Ma S, Zhang Y, Nguyen LH, Tickle TL, Weingart G, Ren B, Schwager EH *et al.* Multivariable association discovery in population-scale meta-omics studies. *PLoS Comput Biol* 2021, 17(11):e1009442.
104. Bolyen E, Rideout JR, Dillon MR, Bokulich NA, Abnet CC, Al-Ghalith GA, Alexander H, Alm EJ, Arumugam M, Asnicar F *et al.* Reproducible, interactive, scalable and extensible microbiome data science using QIIME 2. *Nat Biotechnol* 2019, 37(8):852–857.
105. Shi W, Qi H, Sun Q, Fan G, Liu S, Wang J, Zhu B, Liu H, Zhao F, Wang X *et al.* gcMeta: a Global Catalogue of Metagenomics platform to support the archiving, standardization and analysis of microbiome data. *Nucleic Acids Res* 2019, 47(D1):D637–D648.

Part IV

The Changing Landscape of NGS Technologies and Data Analysis

16

What's Next for Next-Generation Sequencing (NGS)?

16.1 The Changing Landscape of Next-Generation Sequencing (NGS)

Since its invention, massively parallel sequencing has been the major driving force in moving life science and medicine forward. After more than a decade, NGS continues to be the most exciting and dynamic area of genomics, and the technology as a whole continues to evolve. As NGS is used ever more widely in research, clinical, agricultural, and industrial applications, the drive for technological advancements will only become greater and the competition among existing platforms more intense. Furthermore, new sequencing technologies continue to emerge and join the fray. As a result, the landscape of NGS is constantly changing.

Since the publication of the first edition of this book in early 2016, the general trends in the NGS arena can be briefly summarized as follows:

1) Continued drop in sequencing cost: As of this writing (late 2022) the sequencing cost per human genome has dropped to $300 compared to $300,000 in 2007 when NGS started to emerge. Emerging new systems will sequence the human genome at $100 or less.
2) Continued improvement on accuracy: Short reads technologies, such as Illumina's SBS, continue to reduce sequencing errors with the use of new detection dyes, polymerases, and reaction blockers. The advancements achieved by long-read technologies are even more substantial (see Chapter 4 for details on the PacBio and ONT platforms), which has led to their increased adoption by the community.
3) Increased use of single DNA (or RNA) molecule sequencing: Single molecule sequencing offers the capability to directly read individual target DNA molecules without relying on template amplification, or conversion to cDNA in the case of RNA.
4) Increased representation of different read lengths on different platforms: As short and long read lengths have their strengths and also limitations, short reads platforms such as the Illumina SBS continue to

DOI: 10.1201/9780429329180-20

make inroads into building long reads from short reads, while long-read platforms have also started to provide short reads.

5) Reduction in sequencing and sample preparation time: Different platforms use different strategies from chemistry updates, hardware upgrades, to algorithmic improvements, to achieve quicker sequencing turnaround time. To cut library prep time, constant improvements are made to chemistries and protocols with increased amenability to automation.

6) Decreased requirement on the amount of starting material: Historically NGS requires large amounts of DNA or RNA to start. With the drive to accommodate more sample types, such as those that do not generate much DNA or RNA (e.g., liquid biopsy and single cells), the sensitivity of library making reagents and procedures has been significantly improved.

16.2 Newer Sequencing Technologies

In Chapter 4, major NGS technologies that are widely used as the time of writing this book are detailed. This chapter presents some of the newer technologies that are in increasing usage or in active development. Among those in the short-read category, MGI, the instruments manufacturing arm of BGI, has been providing a technology called DNBSEQ. Originally developed by Complete Genomics [1], this technology is based on the use of DNA nanoball or DNB, which is a large mass of amplified template DNA formed through rolling circle amplification (Figure 16.1). The sequencing process is based on the use of a procedure originally called combinatorial probe-anchor ligation (or cPAL), and later refined by BGI/MGI to become combinatorial probe-anchor synthesis (or cPAS) [2]. This process consists of iterative cycles of hybridization of an anchor sequence to the template, then ligation of fluorescence-labeled probes, imaging for on-board basecalling, and removal of the anchor-probe complex to prepare for the next cycle with a new set of anchors and probes to interrogate the next base position. The current lineup of MGI sequencers include DNBSEQ-T7, DNBSEQ-G400, and DNBSEQ-G50, which can produce up to 6,000 Gb, 1,440 Gb, and 150 Gb data, respectively. The error profile of DNB sequencing is similar to that of the Illumina platform, with the error rate slightly higher but still less than 1% [3]. Its cost per Gb is among the lowest among currently available platforms [4]. In 2020, MGI released a new sequencing chemistry called CoolMPS, which is based on the use of nucleotide-specific antibodies to detect nucleotide incorporation. This chemistry uses unlabeled reversibly terminated nucleotides for incorporation during synthesis, and four natural nucleotide-specific antibodies with

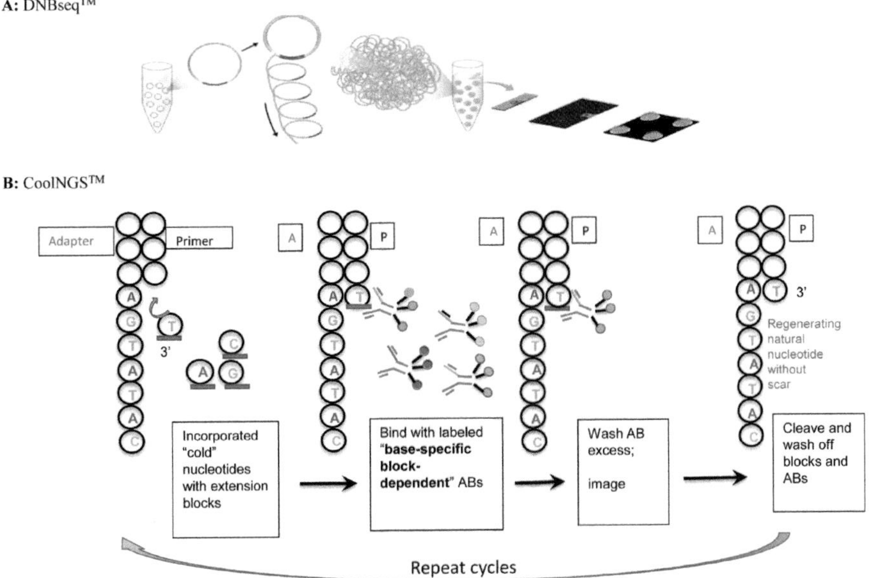

FIGURE 16.1
MGI/BGI nanoball sequencing and CoolMPS chemistry. A. Nanoball sequencing starts from circularization of DNA target molecules. After rolling circle amplification, nanoballs are formed from circularized targets, and subsequently deposited onto a silicon-based sequencing chip for sequencing using the cPAS process. B. In the CoolMPS chemistry (initially called CoolNGS), nucleotide-specific antibodies are used for detection of incorporated nucleotides. This detection mechanism avoids DNA "scars" derived from labeling of nucleotides with fluorescent tags, potentially leading to increased read length. (From Gao, G., Smith, D.I. Clinical Massively Parallel Sequencing. *Clinical Chemistry*, 2020, 66(1): 77–88, by permission from Oxford University Press.)

each labeled with a specific type of fluorescent dye for the detection step in each cycle (Figure 16.1). To increase sequencing signal to background noise ratio, multiple fluorescence dyes are attached to each antibody molecule. It has been shown to have the potential to produce longer and more accurate reads at lower cost [5].

Among emerging technologies, Element Biosciences offers a system based on Avidity chemistry. Although still based on the same basic SBS process, on the Element system the signal detection and incorporation of nucleotides are separate, as the system does not collect sequencing signal from the nucleotide incorporation process. Instead the signal is collected from binding of nucleotides to the sequencing template. With Avidity chemistry [6], each nucleotide attaches to a fluorescence-emitting core, and each type of nucleotide (A, C, G, or T) attaches to their own cores that emit specific fluorescence signal for detection. The most innovative aspect of this chemistry is that each core contains multiple fluorophores and connects to multiple copies of a certain type of nucleotide, which enables signal amplification to boost signal-to-noise ratio. After multivalent binding of the nucleotide complexes (also called avidites, after which the sequencing chemistry is named) to the template DNA and signal detection, the bound avidites are removed, which is followed by a "dark" (i.e., no imaging) nucleotide incorporation step to produce scarless DNA for a new round of avidite binding, detection, and nucleotide incorporation. Besides the sequencing chemistry, the flow cell surface where sequencing reactions take place also has ultra-low non-specific binding to reduce background fluorescence and thereby further increases signal-to-noise ratio. Besides reducing sequencing errors, the increased detection sensitivity also leads to lower reagent consumption and thereby decreased sequencing cost. Based on currently available data (as of November 2022) [6], the Element system is capable of producing 300 Gb sequencing data with 90% of reads having quality over Q30.

Other emerging short-read technologies include those from Ultima, Singular Genomics, and Omniome (part of PacBio). The Ultima system is based on the use of (1) a new chemistry called mostly natural sequencing-by-synthesis (or mnSBS); (2) an open circular wafer with a large surface for accelerated fluidics and imaging as well as reduced reagent usage; and (3) an AI-based basecalling process. With the mnSBS chemistry, in each sequencing cycle only one nucleobase is used from a reagent mix that contains mostly natural (i.e., unlabeled and non-terminated) nucleotides and a minority (<20%) of fluorescently labeled, non-terminated nucleotides. This allows production of mostly scarless DNA and at the same time a minority of fluorescently labeled DNA molecules for imaging and basecalling. This platform has the capability to generate billions of 300 bp reads, at an accuracy of Q30 > 85%, in 20 hours [7]. Singular Genomics provides an SBS platform based on the use of fluorescently labeled nucleotides and 4-color optical imaging to improve speed and flexibility. Its first release is a G4 benchtop sequencer that produces up to 100 Gb data with read length of up to 150 bases from each end and error rate of <1% (mostly substitutions), in 6–19 hours. Technical details for some of the emerging platforms are still not available at the time of writing. Among the currently accessible platforms, some benchmark data has become available to provide side-by-side comparisons on genome coverage, error rate, alignment, and detection of various variants including SNVs, indels, and SVs [8].

Among those in the long-read category, there are technologies that generate native long reads similar to those from PacBio and ONT, and also those that produce synthetic long reads. The sequencing platforms currently being developed by Base4 and Quantapore are two examples of native long-reads technologies. Base4 sequencing is based on the use of a fundamentally different process called pyrophosphorolysis. In the SBS scheme, a polymerase incorporates a nucleotide into an elongating DNA (or RNA) strand and releases a pyrophosphate as a side product. With pyrophosphorolysis sequencing, the reaction direction is reversed, i.e., the dNMP at the 3′ end of a DNA strand reacts with pyrophosphate to be released from the strand as dNTP. This sequencing-by-desynthesis process used by Base4 works through detection of the released nucleotide [9]. The Quantapore platform uses a nanopore-based approach. On this platform, DNA templates need to be labeled first with fluorophores, as instead of detecting changes in electrical current as in the case of ONT, this technology detects optical signals emitted from the fluorophores when they are activated upon passing through the pore [10]. Other emerging native long-read technologies are also based on the use of nanopores, including those developed by Genia and Stratos Genomics (now parts of Roche).

Synthetic long-read technologies are based on innovative utilization of short-read sequencing that links short reads together to create artificial long reads. Technologies in this category include 10× Genomics' linked reads [11], MGI's single tube Long Fragment Read (stLFR) [12], Transposase Enzyme Linked Long-read Sequencing (TELL-seq) from Universal Sequencing Technology [13], LoopSeq from Loop Genomics (now part of Element Biosciences) [14], and Illumina's Complete Long Reads. Many of these technologies use a clonal barcoding approach, through which a large DNA molecule is first barcoded, then shorter fragments are generated from the large molecule to be sequenced using short-read sequencing. In this process each short fragment generated from the same large molecule carries the original molecule's unique barcode. This barcode is subsequently used to group and order short reads to reconstruct the original large DNA molecule. Among the aforementioned synthetic long-read technologies, 10× Genomics discontinued its linked reads offering in 2020, while Illumina has announced Complete Long Reads in late 2022 with no technical specifics available at the time of writing.

16.3 Continued Evolution and Growth of Bioinformatics Tools for NGS Data Analysis

Bioinformatics tool development will continue to evolve with new sequencing technologies as well as applications becoming available and getting adopted. The emergence of new technologies, in turn, will lead to adaptation and evolution of existing technologies. The increased change in read length and other aspects of sequencing output, such as diversity of sequencing error models form different platforms, will lead to development of new tools and revision of existing tools. As higher read length increases sequence information content and uniqueness, which in turn leads to increased "assemblability" or "mappability," new alignment algorithms or updated versions of existing ones will continue to be developed to harness the power afforded by this increase in read length. For example, SPAdes, originally developed for *de novo* assembly of Illumina short reads [15], has been revised to enable use of both short and long but less accurate PacBio and ONT reads to achieve hybrid assembly [16]. New long-read *de novo* genome assemblers, such as miniasm [17], Flye [18], NECAT [19], and Raven [20], have been developed more recently. To accommodate mapping of long reads to a reference genome, BWA-MEM was a first attempt to align long reads with the widely used BWA algorithm. With ultra-long but noisier reads becoming more common, additional aligners, such as minimap [17] and subsequently minimap2 [21], are developed and updated constantly [22] with rapid advancements in sequencing technologies.

Significantly longer reads associated with third- or future-generation sequencing technologies, as well as synthetic long reads, will not only improve *de novo* genome assembly and mapping to reference sequences but also all other NGS applications. For example, increased read length in RNA-seq can lead to recognition of different transcripts that are produced from the same gene, and therefore facilitate studies of alternative splicing. Algorithms and tools for other applications or steps, from basecalling, variant calling, ChIP-seq peaking calling, to DNA methylation sequencing, and metagenome characterization, are under constant development. While new ones are continuously being introduced, many existing algorithms and tools are also under constant revision. As basecalling is highly platform-dependent, basecallers are usually developed as part of the sequencing platform development process. While there are also third-party basecallers being developed in an attempt to further improve performance, efforts on algorithmic and software tool development are mostly focused on downstream analyses.

To help illustrate how data analytic tool development closely follows the development of new technologies and applications, Figure 16.2 shows the total number of publications on scRNA-seq data analysis algorithms every year from 2013 to 2021. As a new application, single-cell RNA-seq has seen continuous growth in algorithmic development and utilization since 2013. While the numbers do not directly measure the total number of new or updated scRNA-seq algorithms, they do to a large degree reflect the amount of algorithmic development efforts as well as the demand in this direction. Algorithmic development and utilization in other new applications and technologies show a similar trend.

16.4 Efficient Management of NGS Analytic Workflows

With the active development of algorithms and the wide array of bioinformatic tools becoming available, efficient workflow management

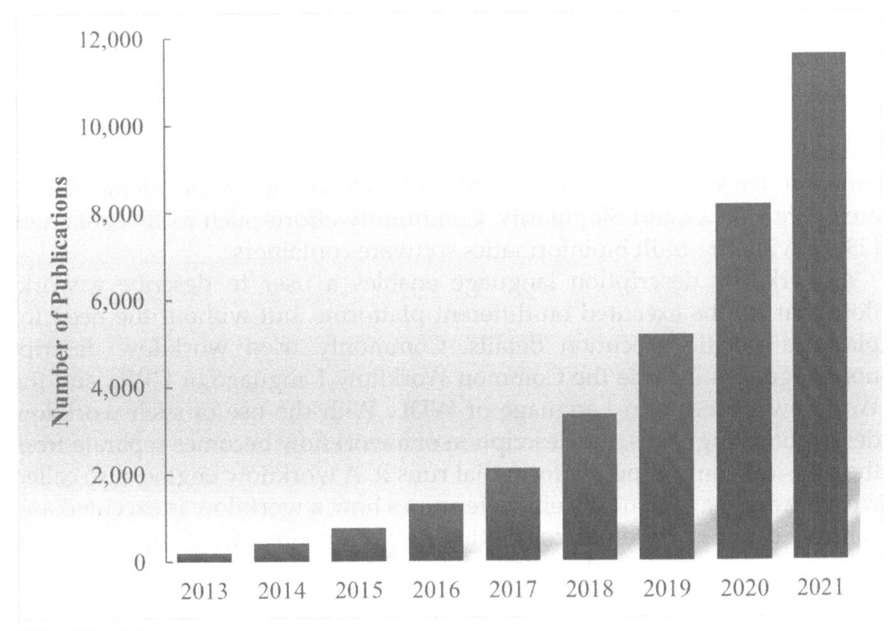

FIGURE 16.2
The increase in the number of single-cell RNA-seq algorithm-related publications from 2013 till 2021. (Data source: Google Scholar, using *"single cell RNA-seq" AND (algorithm OR method OR tool)* as query term.)

has become increasingly important. The clinical use of NGS on diagnosis, genetic risk assessment, and patient management places further demands on NGS data workflow management. The quest for increased reproducibility is another impetus for effective workflow management, as the tool(s) used, their version, and their configuration all have effects on final results. Commercial software systems, such as CLC Genomics Workbench, DNAnexus, DNAstar, PierianDx, QIAGEN Clinical Insight Interpret, Seven Bridges, and SciDAP, tend to streamline their analytical workflows in one package. Because bioinformatic tools developed out of academic settings tend to be more specialized and as a result a bit fragmented, effective workflow management strategies and systems are becoming increasingly needed. As a result, usage of workflow management systems has been on the rise to overcome the limitations of using the traditional way of downloading and deploying tools one by one for a workflow. Such limitations involve the difficulty to deploy associated system libraries (including dependencies), compilers or interpreters of different versions, for different computing environments. Several key technologies in computer science, including the use of containers, workflow description languages, and workflow engines, have been adopted for managing NGS analytic workflows.

In computer science terms, a container is an executable unit of software package that contains everything an application needs to run, including all code and requisite libraries and dependencies. This technology allows the software in a container to run identically in different computing environments, overcoming the issues associated with running the same tool on different platforms. Two widely used container engines in the bioinformatics arena are Docker and Singularity. Community efforts such as Biocontainers [23] provide pre-built bioinformatics software containers.

A workflow description language enables a user to describe a workflow that can be executed on different platforms, but without the need for platform-specific execution details. Commonly used workflow description languages include the Common Workflow Language or CWL, and the Workflow Description Language or WDL. With the use of such workflow description languages, the description of a workflow becomes separate from the physical computing platform that runs it. A workflow engine, also called workflow management system, determines how a workflow is executed and controlled under different computing environments. Workflow engines in the NGS field include Galaxy [24], Cromwell as used in Terra [25], Nextflow [26], and CWL-Airflow [27]. These workflow management systems enable seamless execution of tools assembled into an integrated pipeline (e.g., RNA-seq quantification followed by differential expression analysis), as well as provide infrastructure and guidelines for management and distribution of NGS and other bioinformatics packages. With increased use of containers and workflow management systems, NGS data analysis will become more streamlined and standardized, thereby improving reproducibility, as well as portability (for porting workflows across different computing environments) and scalability (for efficient use of large-scale computational resources).

16.5 Deepening Applications of NGS to Single-Cell and Spatial Sequencing

Among the various new directions that NGS has empowered and grown into is single-cell sequencing. In this revision, a new chapter is dedicated to single-cell transcriptomics. Beyond transcriptomic information, information on other -omic aspects of single cells is also required to reveal their inner working mechanisms and the tremendous heterogeneity among them. NGS has been making inroads into these aspects, with examples from single-cell ATAC-seq, single-cell whole genome/exome sequencing, and single-cell epigenomics. Single-cell ATAC-seq, or Assay for Transposase-Accessible Chromatin using sequencing, is a technique to reveal chromatin accessibility landscape for transcription control. Available exemplary studies have demonstrated the rich information it can provide for analysis of gene regulatory programs [28]. Compared to tools developed for scRNA-seq, scATAC-seq data analysis tools are still low in numbers and not as well developed. Currently available tools include SnapATAC [29], cisTopic [30], SCALE [31], and Signac [32].

Single-cell genome/exome sequencing, with the goal of tracking somatic evolution and revealing genetic heterogeneity at the single-cell level, has been made possible with whole genome amplification methods. These methods include multiple displacement amplification (MDA) [33], multiple annealing and loop-based amplification cycles (MALBAC) [34], degenerate oligonucleotide-primed PCR (DOP-PCR) [35], or commercially available kits such as PicoPLEX and RepliG [36]. As each diploid cell only has two sets of chromosomes and therefore a very low amount of DNA (~6.5 pg in a typical human cell), the coverage of whole genome amplification is typically uneven across the genome due to stochastic effects, amplification errors, and locus-specific amplification bias. Such issues have prevented scaling-up of single-cell genomics to a level that can be comparable to single-cell transcriptomics. Progress has been made to overcome these issues with the development of emulsion MDA (or eMDA) [37], single droplet MDA (or sd-MDA) [38], and direct library preparation (DLP) [39]. To call SNVs from single-cell genomics data, currently a relatively short list of tools is available, including SCcaller [40], SCAN-SNV [41], and Single Cell Genotyper [42]. SNV Calling can also be made from scRNA-seq data, or coupled scDNA-seq and scRNA-seq data, using tools such as SSrGE [43]. Tools for calling CNVs or structural variants include AneuFinder [44] and Ginkgo [45].

Single-cell epigenomics offers another dimension for single-cell sequencing. Strategies such as single-cell whole genome bisulfite sequencing (scWGBS) [46], single-cell reduced representation bisulfite sequencing (scRRBS) [47], and single-nucleus methylome sequencing ver2 (snmC-seq2) [48] have been used to detect DNA methylation as a epigenetic marker for cell typing. Tools such as scBS-map [49] can be used for reads alignment, and Methylpy [50] for calling of unmethylated and methylated cytosines. Simultaneous interrogation of both the epigenome and genome of single cells has also been made possible with methods such as epi-gSCAR [51]. Compared to scRNA-seq and scATAC-seq conducted on high-throughput platforms such as 10× Chromium, the throughput of single-cell genome and epigenome sequencing is still lower, although this may well change over time.

While single-cell sequencing offers unprecedented resolution, isolation of single cells (see Chapter 8) typically leads to the loss of contextual information about their original location in their native tissue microenvironment. Investigation into such spatial information provides insights on regional specificity and cross-region heterogeneity, e.g., when comparing a pathogenic region with the surrounding normal region on the same slide. Spatial transcriptomics, enabled by rapid technology development from both academia and industry [52], is increasingly used to provide this additional layer of information. While this new field is still rapidly evolving, a number of platforms have been gradually adopted by the community, included among which are academically developed platforms such as Slide-seqV2 [53], and commercial ones such as Visium from 10× Genomics, GeoMx Digital Spatial Profiler from Nanostring, and MERFISH from Vizgen. These systems, based on a diversity of detection mechanisms, have different technical capabilities in terms of spatial resolution, physical dimensions of the area to be assayed, and transcriptomic coverage (whole transcriptome vs. targeted genes). To analyze data from these systems, there are currently some tools available, such as Space Ranger from 10× Genomics, Giotto [54], and BayesSpace [55], to help identify cell types, characterize gene expression profiles and infer cell-cell interactions in a spatial context. To meet the challenges in spatial transcriptomics towards the goal of integrating gene expression profile, spatial context, and cell morphology for deeper cell typing, more tools are needed. More comprehensive tools are also needed to integrate spatial transcriptomics data, with other omics data including spatial proteomics and metabolomics.

16.6 Increasing Use of Machine Learning in NGS Data Analytics

As a branch of artificial intelligence, machine learning (ML) trains algorithms to uncover patterns in large data sets and thereby makes classifications or predictions without being explicitly programmed. Within ML, deep learning (DL) with the use of neural networks is well suited for tasks that involve large volumes of highly complex and information-rich data. As NGS has become a major producer of big data, besides astronomy, YouTube, and social media, application of what ML and DL can offer is warranted [56, 57]. In earlier chapters, such as Chapters 8 and 11, the applications of ML and DL to single-cell sequencing data analysis and detection of clinically relevant genetic variants have been covered. Besides these areas, nearly all aspects of NGS data analysis have seen increased use of ML and DL. Read mapping and variant calling, the two most basic steps, have been greatly improved by the application of ML and DL approaches. In the precisionFDA Truth Challenge V2, the majority of read mapping and variant calling pipelines employ ML and DL approaches [58]. The Illumina DRAGEN reads mapper + variant caller system, for example, applies graph-based mapping and supervised ML-based variant calling to improve accuracy as well as speed. For the mapping step, DRAGEN uses phased graph reference genome built through stitching alternate haplotypes into the standard reference genome, and thereby creating alternative graph paths for seed mapping and alignment of reads. Such a reference genome based on the use of graph representation, a subfield of ML, represents more accurately the polymorphic nature of a species' or population's genome. In practice, this alt-aware approach leads to improved mapping performance by reducing ambiguity when reads containing alternate haplotypes are mapped more accurately than using the standard reference genome alone. For the variant calling step, DRAGEN uses a supervised ML model trained to incorporate read-based and contextual features that is then employed to recalibrate the QUAL and GQ fields from a standard VCF output to increase variant calling accuracy. Among other examples, the Seven Bridges GRAF pipeline also uses a pan-genome graph reference, created by incorporating known alternate haplotypes and integrating genomic polymorphic information from multiple variant databases, to improve reads alignment. To improve variant calling, the pipeline uses an adaptive ML model to remove false-positive variants. Similarly Sentieon's DNAscope and TNscope pipelines also use ML-enhanced variant filtering to improve the accuracy of calling germline and somatic variants, respectively.

DeepVariant, an open-source variant caller, is the first variant caller among all variant callers that uses a DL-based framework. Originally developed for classifying images [59], this Tensorflow-based framework uses a convolutional neural network (CNN) for calling small germline variants from short reads. Since its introduction, DL-based methods have grown significantly overtaking statistical model-based methods. Open-source tools for calling variants from long reads, such as those generated from ONT and PacBio, have been dominated by ML/DL-based methods since the beginning, largely out of the need to deal with the higher sequencing error rates on these platforms. The PEPPER-Margin-DeepVariant pipeline, for example, uses a recurrent neural network (RNN) for variant filtering and HMM for phasing and haplotyping, prior to variant calling by DeepVariant [60]. Examples of other such tools include Clair/Clair3 based on RNNs [61, 62], NanoCaller on deep CNN [63], and Longshot on pair-HMM [64]. For structural variant calling, the use of DL has not been as wide at the time of writing, but existing tools such as DeepSV [65] and dudeML [66] have shown great promise.

Besides read mapping and variant calling, many of the NGS applications covered in this book have also seen increasing use of ML and DL. For example, in performing bulk RNA-seq, statistical models are usually used to identify differentially expressed genes. Emerging ML-based methods have been shown to be able to identify DEGs, some of which may be missed by statistical methods. For example, an ML model based on the use of InfoGain feature selection and Logistic Regression classification was found to be robust and have improved sensitivity [67]. One benefit of ML-based gene feature selections is that they do not make assumptions on the distribution of gene transcript counts. In ChIP-seq, peak detection is a major step to reveal regions of the genome where mapped reads are enriched that indicate epigenetic interactions. As covered in Chapter 13, peaks are typically called using statistical methods. To overcome the challenges these methods face, such as uneven background noise and high false-positive rates, DL-based approaches have been employed. For example, a CNN-based pipeline called CNN-Peaks uses partially labeled peaks inspected by researchers and genome annotations as input to first train a model for subsequent peak detection. This supervised learning process takes into consideration variations in local context and background noise and sets specific peak detection cut-off values accordingly for each genomic region to reduce false-positive rates. In methyl-seq, currently available ML/DL-based methods are mostly designed to detect epigenetic base modifications from data generated on long-read platforms, such as DeepMP [68] and DeepSignal-plant [69] for ONT data, and a holistic kinetic (HK) model for PacBio data [70].

Beyond the various NGS steps and applications, ML and DL can also greatly facilitate integration of data collected from multiple -omics technologies, including genomics, transcriptomics, epigenomics, proteomics, and metabolomics. Data collected using these -omics technologies informs on the different layers of a biological system, and integration of such complementary information leads to a more holistic view of the system. Towards this direction, beyond basic research ML, especially DL, has been accelerating clinical application of multi-omics data to disease subtyping, biomarker discovery, disease prognosis, treatment outcome prediction, and drug repurposing. For example, to predict severity of pediatric irritable bowel syndrome (IBS), a Random Forest-based classifier is developed from integration of shotgun metagenomics and metabolomics data [71]. This classifier is optimized based on the selection and use of key features derived from the metagenomics and metabolomics data, including microbial diversity, functional pathways, and fecal metabolites. Through the use of such gut microbiome and metabolites features as integrated diagnostic biomarkers, the classifier can reliably predict IBS status and thereby help to stratify patients for improved disease management. As another example, an oncology molecular classifier integrates genome and transcriptome sequencing data to identify tissue-of-origin (TOO) for cancers of unknown primary (or CUP), a syndrome of metastatic cancers with no clinically detectable primary sites of origin. Without knowing their TOO, CUP has very poor prognosis. By combining DNA-level mutation and RNA-level gene expression information, the DL model makes predictions on tumor TOO across a broad spectrum of tumor types at high accuracy, thereby providing diagnostic utility and guiding therapeutic intervention [72].

References

1. Drmanac R, Sparks AB, Callow MJ, Halpern AL, Burns NL, Kermani BG, Carnevali P, Nazarenko I, Nilsen GB, Yeung G *et al.* Human genome sequencing using unchained base reads on self-assembling DNA nanoarrays. *Science* 2010, 327(5961):78–81.

2. Fehlmann T, Reinheimer S, Geng C, Su X, Drmanac S, Alexeev A, Zhang C, Backes C, Ludwig N, Hart M *et al.* cPAS-based sequencing on the BGISEQ-500 to explore small non-coding RNAs. *Clin Epigenetics* 2016, 8:123.
3. Zhu K, Du P, Xiong J, Ren X, Sun C, Tao Y, Ding Y, Xu Y, Meng H, Wang CC *et al.* Comparative Performance of the MGISEQ-2000 and Illumina X-Ten Sequencing Platforms for Paleogenomics. *Front Genet* 2021, 12:745508.
4. Tedersoo L, Albertsen M, Anslan S, Callahan B. Perspectives and Benefits of High-Throughput Long-Read Sequencing in Microbial Ecology. *Appl Environ Microbiol* 2021, 87(17):e0062621.
5. Drmanac S, Callow M, Chen L, Zhou P, Eckhardt L, Xu C, Gong M, Gablenz S, Rajagopal J, Yang Q *et al.* CoolMPS™: Advanced massively parallel sequencing using antibodies specific to each natural nucleobase. *bioRxiv* 2020, doi: https://doi.org/10.1101/2020.02.19.953307
6. Arslan S, Garcia FJ, Guo M, Kellinger MW, Kruglyak S, LeVieux JA, Mah AH, Wang H, Zhao J, Zhou C. Sequencing by avidity enables high accuracy with low reagent consumption. *bioRxiv* 2022, doi: https://doi.org/10.1101/2022.11.03.514117
7. Almogy G, Pratt M, Oberstrass F, Lee L, Mazur D, Beckett N, Barad O, Soifer I, Perelman E, Etzioni Y *et al.* Cost-efficient whole genome-sequencing using novel mostly natural sequencing-by-synthesis chemistry and open fluidics platform. *bioRxiv* 2022, doi: https://doi.org/10.1101/2022.05.29.493900
8. Foox J, Tighe SW, Nicolet CM, Zook JM, Byrska-Bishop M, Clarke WE, Khayat MM, Mahmoud M, Laaguiby PK, Herbert ZT *et al.* Performance assessment of DNA sequencing platforms in the ABRF Next-Generation Sequencing Study. *Nat Biotechnol* 2021, 39(9):1129–1140.
9. Puchtler TJ, Johnson K, Palmer RN, Talbot EL, Ibbotson LA, Powalowska PK, Knox R, Shibahara A, P MSC, Newell OJ *et al.* Single-molecule DNA sequencing of widely varying GC-content using nucleotide release, capture and detection in microdroplets. *Nucleic Acids Res* 2020, 48(22):e132.
10. Marx V. Nanopores: a sequencer in your backpack. *Nat Methods* 2015, 12(11):1015–1018.
11. Zheng GX, Lau BT, Schnall-Levin M, Jarosz M, Bell JM, Hindson CM, Kyriazopoulou-Panagiotopoulou S, Masquelier DA, Merrill L, Terry JM *et al.* Haplotyping germline and cancer genomes with high-throughput linked-read sequencing. *Nat Biotechnol* 2016, 34(3):303–311.
12. Wang O, Chin R, Cheng X, Wu MKY, Mao Q, Tang J, Sun Y, Anderson E, Lam HK, Chen D *et al.* Efficient and unique cobarcoding of second-generation sequencing reads from long DNA molecules enabling cost-effective and accurate sequencing, haplotyping, and de novo assembly. *Genome Res* 2019, 29(5):798–808.
13. Chen Z, Pham L, Wu TC, Mo G, Xia Y, Chang PL, Porter D, Phan T, Che H, Tran H *et al.* Ultralow-input single-tube linked-read library method enables short-read second-generation sequencing systems to routinely generate highly accurate and economical long-range sequencing information. *Genome Res* 2020, 30(6):898–909.
14. Liu S, Wu I, Yu YP, Balamotis M, Ren B, Ben Yehezkel T, Luo JH. Targeted transcriptome analysis using synthetic long read sequencing uncovers isoform reprograming in the progression of colon cancer. *Commun Biol* 2021, 4(1):506.

15. Nurk S, Bankevich A, Antipov D, Gurevich AA, Korobeynikov A, Lapidus A, Prjibelski AD, Pyshkin A, Sirotkin A, Sirotkin Y *et al.* Assembling single-cell genomes and mini-metagenomes from chimeric MDA products. *J Comput Biol* 2013, 20(10):714–737.
16. Antipov D, Korobeynikov A, McLean JS, Pevzner PA. hybridSPAdes: an algorithm for hybrid assembly of short and long reads. *Bioinformatics* 2016, 32(7):1009–1015.
17. Li H. Minimap and miniasm: fast mapping and de novo assembly for noisy long sequences. *Bioinformatics* 2016, 32(14):2103–2110.
18. Kolmogorov M, Yuan J, Lin Y, Pevzner PA. Assembly of long, error-prone reads using repeat graphs. *Nat Biotechnol* 2019, 37(5):540–546.
19. Chen Y, Nie F, Xie SQ, Zheng YF, Dai Q, Bray T, Wang YX, Xing JF, Huang ZJ, Wang DP *et al.* Efficient assembly of nanopore reads via highly accurate and intact error correction. *Nat Commun* 2021, 12(1):60.
20. Vaser R, Šikić M. Time-and memory-efficient genome assembly with Raven. *Nat Comput Sci* 2021, 1(5):332–336.
21. Li H. Minimap2: pairwise alignment for nucleotide sequences. *Bioinformatics* 2018, 34(18):3094–3100.
22. Li H. New strategies to improve minimap2 alignment accuracy. *Bioinformatics* 2021, 37(23):4572–4574.
23. Bai J, Bandla C, Guo J, Vera Alvarez R, Bai M, Vizcaino JA, Moreno P, Gruning B, Sallou O, Perez-Riverol Y. BioContainers Registry: Searching Bioinformatics and Proteomics Tools, Packages, and Containers. *J Proteome Res* 2021, 20(4):2056–2061.
24. Jalili V, Afgan E, Gu Q, Clements D, Blankenberg D, Goecks J, Taylor J, Nekrutenko A. The Galaxy platform for accessible, reproducible and collaborative biomedical analyses: 2020 update. *Nucleic Acids Res* 2020, 48(W1):W395–W402.
25. Terra (https://app.terra.bio/)
26. Di Tommaso P, Chatzou M, Floden EW, Barja PP, Palumbo E, Notredame C. Nextflow enables reproducible computational workflows. *Nat Biotechnol* 2017, 35(4):316–319.
27. Kotliar M, Kartashov AV, Barski A. CWL-Airflow: a lightweight pipeline manager supporting Common Workflow Language. *GigaScience* 2019, 8(7):giz084.
28. Zhang K, Hocker JD, Miller M, Hou X, Chiou J, Poirion OB, Qiu Y, Li YE, Gaulton KJ, Wang A *et al.* A single-cell atlas of chromatin accessibility in the human genome. *Cell* 2021, 184(24):5985–6001 e5919.
29. Fang R, Preissl S, Li Y, Hou X, Lucero J, Wang X, Motamedi A, Shiau AK, Zhou X, Xie F *et al.* Comprehensive analysis of single cell ATAC-seq data with SnapATAC. *Nat Commun* 2021, 12(1):1337.
30. Bravo Gonzalez-Blas C, Minnoye L, Papasokrati D, Aibar S, Hulselmans G, Christiaens V, Davie K, Wouters J, Aerts S. cisTopic: cis-regulatory topic modeling on single-cell ATAC-seq data. *Nat Methods* 2019, 16(5):397–400.
31. Xiong L, Xu K, Tian K, Shao Y, Tang L, Gao G, Zhang M, Jiang T, Zhang QC. SCALE method for single-cell ATAC-seq analysis via latent feature extraction. *Nat Commun* 2019, 10(1):4576.
32. Stuart T, Srivastava A, Madad S, Lareau CA, Satija R. Single-cell chromatin state analysis with Signac. *Nat Methods* 2021, 18(11):1333–1341.

33. Dean FB, Nelson JR, Giesler TL, Lasken RS. Rapid amplification of plasmid and phage DNA using Phi 29 DNA polymerase and multiply-primed rolling circle amplification. *Genome Res* 2001, 11(6):1095–1099.
34. Zong C, Lu S, Chapman AR, Xie XS. Genome-wide detection of single-nucleotide and copy-number variations of a single human cell. *Science* 2012, 338(6114):1622–1626.
35. Telenius H, Carter NP, Bebb CE, Nordenskjold M, Ponder BA, Tunnacliffe A. Degenerate oligonucleotide-primed PCR: general amplification of target DNA by a single degenerate primer. *Genomics* 1992, 13(3):718–725.
36. Imamura H, Monsieurs P, Jara M, Sanders M, Maes I, Vanaerschot M, Berriman M, Cotton JA, Dujardin JC, Domagalska MA. Evaluation of whole genome amplification and bioinformatic methods for the characterization of Leishmania genomes at a single cell level. *Sci Rep* 2020, 10(1):15043.
37. Fu Y, Zhang F, Zhang X, Yin J, Du M, Jiang M, Liu L, Li J, Huang Y, Wang J. High-throughput single-cell whole-genome amplification through centrifugal emulsification and eMDA. *Commun Biol* 2019, 2:147.
38. Hosokawa M, Nishikawa Y, Kogawa M, Takeyama H. Massively parallel whole genome amplification for single-cell sequencing using droplet microfluidics. *Sci Rep* 2017, 7(1):5199.
39. Zahn H, Steif A, Laks E, Eirew P, VanInsberghe M, Shah SP, Aparicio S, Hansen CL. Scalable whole-genome single-cell library preparation without preamplification. *Nat Methods* 2017, 14(2):167–173.
40. Dong X, Zhang L, Milholland B, Lee M, Maslov AY, Wang T, Vijg J. Accurate identification of single-nucleotide variants in whole-genome-amplified single cells. *Nat Methods* 2017, 14(5):491–493.
41. Luquette LJ, Bohrson CL, Sherman MA, Park PJ. Identification of somatic mutations in single cell DNA-seq using a spatial model of allelic imbalance. *Nat Commun* 2019, 10(1):3908.
42. Roth A, McPherson A, Laks E, Biele J, Yap D, Wan A, Smith MA, Nielsen CB, McAlpine JN, Aparicio S *et al.* Clonal genotype and population structure inference from single-cell tumor sequencing. *Nat Methods* 2016, 13(7):573–576.
43. Poirion O, Zhu X, Ching T, Garmire LX. Using single nucleotide variations in single-cell RNA-seq to identify subpopulations and genotype-phenotype linkage. *Nat Commun* 2018, 9(1):4892.
44. Bakker B, Taudt A, Belderbos ME, Porubsky D, Spierings DC, de Jong TV, Halsema N, Kazemier HG, Hoekstra-Wakker K, Bradley A *et al.* Single-cell sequencing reveals karyotype heterogeneity in murine and human malignancies. *Genome Biol* 2016, 17(1):115.
45. Garvin T, Aboukhalil R, Kendall J, Baslan T, Atwal GS, Hicks J, Wigler M, Schatz MC. Interactive analysis and assessment of single-cell copy-number variations. *Nat Methods* 2015, 12(11):1058–1060.
46. Smallwood SA, Lee HJ, Angermueller C, Krueger F, Saadeh H, Peat J, Andrews SR, Stegle O, Reik W, Kelsey G. Single-cell genome-wide bisulfite sequencing for assessing epigenetic heterogeneity. *Nat Methods* 2014, 11(8):817–820.
47. Guo H, Zhu P, Guo F, Li X, Wu X, Fan X, Wen L, Tang F. Profiling DNA methylome landscapes of mammalian cells with single-cell reduced-representation bisulfite sequencing. *Nat Protoc* 2015, 10(5):645–659.

48. Luo C, Rivkin A, Zhou J, Sandoval JP, Kurihara L, Lucero J, Castanon R, Nery JR, Pinto-Duarte A, Bui B *et al.* Robust single-cell DNA methylome profiling with snmC-seq2. *Nat Commun* 2018, 9(1):3824.
49. Wu P, Gao Y, Guo W, Zhu P. Using local alignment to enhance single-cell bisulfite sequencing data efficiency. *Bioinformatics* 2019, 35(18):3273–3278.
50. Schultz MD, He Y, Whitaker JW, Hariharan M, Mukamel EA, Leung D, Rajagopal N, Nery JR, Urich MA, Chen H *et al.* Human body epigenome maps reveal noncanonical DNA methylation variation. *Nature* 2015, 523(7559):212–216.
51. Niemoller C, Wehrle J, Riba J, Claus R, Renz N, Rhein J, Bleul S, Stosch JM, Duyster J, Plass C *et al.* Bisulfite-free epigenomics and genomics of single cells through methylation-sensitive restriction. *Commun Biol* 2021, 4(1):153.
52. Liao J, Lu X, Shao X, Zhu L, Fan X. Uncovering an organ's molecular architecture at single-cell resolution by spatially resolved transcriptomics. *Trends Biotechnol* 2021, 39(1):43–58.
53. Stickels RR, Murray E, Kumar P, Li J, Marshall JL, Di Bella DJ, Arlotta P, Macosko EZ, Chen F. Highly sensitive spatial transcriptomics at near-cellular resolution with Slide-seqV2. *Nat Biotechnol* 2021, 39(3):313–319.
54. Dries R, Zhu Q, Dong R, Linus Eng C-H, Li H, Liu K, Fu Y, Zhao T, Sarkar A, Bao F *et al.* Giotto: a toolbox for integrative analysis and visualization of spatial expression data. *Genome Biol* 2021, 22(1):78.
55. Zhao E, Stone MR, Ren X, Guenthoer J, Smythe KS, Pulliam T, Williams SR, Uytingco CR, Taylor SEB, Nghiem P *et al.* Spatial transcriptomics at subspot resolution with BayesSpace. *Nat Biotechnol* 2021, 39(11):1375–1384.
56. Stephens ZD, Lee SY, Faghri F, Campbell RH, Zhai CX, Efron MJ, Iyer R, Schatz MC, Sinha S, Robinson GE. Big Data: Astronomical or Genomical? *PLoS Biol* 2015, 13(7).
57. Schmidt B, Hildebrandt A. Next-generation sequencing: big data meets high performance computing. *Drug Discov Today* 2017, 22(4):712–717.
58. Olson ND, Wagner J, McDaniel J, Stephens SH, Westreich ST, Prasanna AG, Johanson E, Boja E, Maier EJ, Serang O *et al.* PrecisionFDA Truth Challenge V2: Calling variants from short- and long-reads in difficult-to-map regions. *Cell Genom* 2022, 2(5):100129.
59. Poplin R, Chang PC, Alexander D, Schwartz S, Colthurst T, Ku A, Newburger D, Dijamco J, Nguyen N, Afshar PT *et al.* A universal SNP and small-indel variant caller using deep neural networks. *Nat Biotechnol* 2018, 36(10):983–987.
60. Shafin K, Pesout T, Chang PC, Nattestad M, Kolesnikov A, Goel S, Baid G, Kolmogorov M, Eizenga JM, Miga KH *et al.* Haplotype-aware variant calling with PEPPER-Margin-DeepVariant enables high accuracy in nanopore long-reads. *Nat Methods* 2021, 18(11):1322–1332.
61. Luo R, Wong C-L, Wong Y-S, Tang C-I, Liu C-M, Leung C-M, Lam T-W. Exploring the limit of using a deep neural network on pileup data for germline variant calling. *Nat Mach Intell* 2020, 2(4):220–227.
62. Zheng Z, Li S, Su J, Leung AW-S, Lam T-W, Luo R. Symphonizing pileup and full-alignment for deep learning-based long-read variant calling. *Nat Comput Sci* 2021, 2(12):797–803.
63. Ahsan MU, Liu Q, Fang L, Wang K. NanoCaller for accurate detection of SNPs and indels in difficult-to-map regions from long-read sequencing by haplotype-aware deep neural networks. *Genome Biol* 2021, 22(1):261.

64. Edge P, Bansal V. Longshot enables accurate variant calling in diploid genomes from single-molecule long read sequencing. *Nat Commun* 2019, 10(1):4660.
65. Cai L, Wu Y, Gao J. DeepSV: accurate calling of genomic deletions from high-throughput sequencing data using deep convolutional neural network. *BMC Bioinformatics* 2019, 20(1):665.
66. Hill T, Unckless RL. A Deep Learning Approach for Detecting Copy Number Variation in Next-Generation Sequencing Data. *G3* 2019, 9(11):3575–3582.
67. Wang L, Xi Y, Sung S, Qiao H. RNA-seq assistant: machine learning based methods to identify more transcriptional regulated genes. *BMC Genomics* 2018, 19(1):546.
68. Bonet J, Chen M, Dabad M, Heath S, Gonzalez-Perez A, Lopez-Bigas N, Lagergren J. DeepMP: a deep learning tool to detect DNA base modifications on Nanopore sequencing data. *Bioinformatics* 2022, 38(5):1235–1243.
69. Ni P, Huang N, Nie F, Zhang J, Zhang Z, Wu B, Bai L, Liu W, Xiao CL, Luo F *et al.* Genome-wide detection of cytosine methylations in plant from Nanopore data using deep learning. *Nat Commun* 2021, 12(1):5976.
70. Tse OYO, Jiang P, Cheng SH, Peng W, Shang H, Wong J, Chan SL, Poon LCY, Leung TY, Chan KCA *et al.* Genome-wide detection of cytosine methylation by single molecule real-time sequencing. *Proc Natl Acad Sci U S A* 2021, 118(5):e2019768118.
71. Hollister EB, Oezguen N, Chumpitazi BP, Luna RA, Weidler EM, Rubio-Gonzales M, Dahdouli M, Cope JL, Mistretta TA, Raza S *et al.* Leveraging human microbiome features to diagnose and stratify children with irritable bowel syndrome. *J Mol Diagn* 2019, 21(3):449–461.
72. Abraham J, Heimberger AB, Marshall J, Heath E, Drabick J, Helmstetter A, Xiu J, Magee D, Stafford P, Nabhan C *et al.* Machine learning analysis using 77,044 genomic and transcriptomic profiles to accurately predict tumor type. *Transl Oncol* 2021, 14(3):101016.

Appendix I

Common File Types Used in NGS Data Analysis

BAM: A file format for storing reads alignment data. It is the binary version of the SAM format (see below). Compared to its equivalent SAM file, a BAM file is considerably smaller in size and much faster to load. Unlike SAM files, however, the BAM format is not human-readable. BAM files have a file extension of .bam. Some tools require BAM files to be indexed. Besides the .bam file, an indexed BAM file also has a companion index file of the same name but with a different file extension (.bai).

BCF: Binary VCF (see *VCF*). While it is equivalent to VCF, BCF is much smaller in file size due to compression, and therefore achieves high efficiency in file transfer and parsing.

BCL: Binary basecall files generated from Illumina's proprietary basecalling process.

BED: Browser Extensible Display format used to describe genes or other genomic features in a genome browser. It is a tab-delimited text format that defines how genes or genomic features are displayed as an annotation track in a genome browser such as the UCSC Genome Browser. Each entry line contains three mandatory fields (chrom, chromStart, and chromEnd, specifying for each genomic feature the particular chromosome it is located on and the start and end coordinates) and nine optional fields. Binary PED files (see below) are also referred to as BED files, but this is a totally different file format.

bedGraph: Similar to the BED format, bedGraph provides descriptions of genomic features for their display in a genome browser. Distinctively the bedGraph format allows display of continuous values, such as probability scores and coverage depth, in a genome.

bigBed: A format similar to BED, but bigBed files are binary, compressed, and indexed. Display of bigBed files in a genome browser is significantly faster due to the compression and indexing, which allow transmittal of only the part of the file that is needed for the current view instead of the entire file.

bigWig: A format for visualization of dense, continuous data, such as GC content, in a genome browser. A newer format from the WIG format (see below), bigWig is a compressed and indexed binary file format and loads significantly faster.

CRAM: Standing for Compressed Reference-oriented Alignment Map, CRAM is an alternative reads alignment file format to SAM/BAM. Designed by the European Bioinformatics Institute, CRAM uses a reference-based compression scheme, that is, only bases that are different from the reference sequence are stored. As a result, CRAM files are smaller than equivalent BAM files.

FAST5: Stores raw electrical signals recorded from an Oxford Nanopore sequencer (such as MinION) for sequencing information extraction. Based on the HDF5 file format (see *HDF5*) with an ONT-specific schema. Besides raw detection signals, FAST5 files can also contain basecalls in FASTQ format after analysis and other information such as signal correction. Unlike FASTA (see *FASTA*) or FASTQ (see *FASTQ*) files, FAST5 files are binary and cannot be directly opened with a text editor. FAST5 is expected to be replaced by a new file format, POD5, for improved write performance, decreased file size, and more streamlined downstream data analysis.

FASTA: A text-based format for storing sequences. A sequence stored in the FASTA format contains only two elements: a single-line description (or defline) and the sequence text. The defline starts with the ">" symbol, followed by a sequence identifier and then a short description. The sequence text is usually divided into multiple lines with each less than 80 characters in length. This format has its origin in the FASTA program package developed in the late 1980s. Multiple sequences can be stored in a FASTA file. FASTA files often have file extensions of .fa, .fasta, or .fsa.

FASTQ: The current *de facto* standard for storing sequencing data generated from various NGS systems. It is a compact text-based format containing nucleotide base sequences and their call quality scores. Each read sequence in a FASTQ file is represented by four lines of information. The first line starts with the symbol "@," followed by sequence ID and descriptor. The second line is the read sequence. Line 3 starts with the "+" symbol, which may be followed by the sequence ID and description (optional). Line 4 lists basecall quality scores for each base in the read sequence. This format was originally developed by the Sanger Institute. FASTQ files have file extensions of .fq or .fastq. Compressed FASTQ files also have the suffix .gz or .gzip from the compression utility used to create them.

GFF: General (or Generic) Feature Format. GFF is a tab-delimited text file format that describes how genes or other genomic features are displayed in a genome browser. There are different versions of this format, with GFF3 being the current version and GFF2 now deprecated. The GFF format can be converted to the BED format (see *BED*).

GTF: Gene Transfer Format. A refined GFF format. Identical to GFF2.

GVF: Stands for Genome Variation Format. Used to describe sequence variation information. An alternative to VCF (see *VCF*). GVF is based on the GFF3 format (see *GFF*) with additional pragmas and attributes.

HDF5 (or H5): Standing for Hierarchical Data Format version 5. HDF5 is an open-source file format designed to store and organize large and complex data. The hierarchical structure it uses is similar to a file system, in that its two major objects, groups and datasets, are similar to directories and files, respectively. The FAST5 file format (see *FAST5*) used by Oxford Nanopore sequencers and the single-cell RNA-seq gene-cell matrix data used by the 10× Genomics Cell Ranger software are based on the HDF5 format.

MEX: Market Exchange format. In 10× Genomics single-cell RNA-seq, the MEX file format is used by the Cell Ranger software to output gene-cell data matrix (besides HDF5). This is a sparse matrix format because of the large number of 0's contained in the file. This file format comprises three files, i.e., matrix.mtx that contains the gene-cell barcode matrix, barcodes.tsv for storing cell barcodes, and genes.tsv for genes.

PED: A file format used by PLINK (a toolset for genome-wide association analysis) that contains pedigree/phenotype data.

SAM: Standing for Sequence Alignment/Map, SAM is a standard NGS reads alignment file format, describing how reads are mapped to a reference genome. It is a tab-delimited text format and human-readable. SAM files can be converted into its compressed binary version (BAM) for faster parsing and file size reduction. SAM files have a file extension of .sam. An indexed SAM file also has an accompanying index file that has an file extension of .sai.

SFF (Standard Flowgram Format): A type of binary sequencing file generated by 454 sequencers. Can be converted to the FASTQ format using utilities such as sff2fastq.

VCF: Stands for Variant Call Format. A commonly used file format for storing variant calls. It is a tab-delimited, human-readable text format that contains meta-information lines, a header line, and data lines that describes each variant.

WIG: Wiggle Track Format. It is used for displaying continuous data track, such as GC content, in a genome viewer such as the UCSC Genome Browser. The WIG format is similar to the bedGraph format (see above), but a major difference between the two is that data exported from a WIG track is not as well preserved as that from a bedGraph track. The WIG format can be converted to bigWig (see above) for improved performance.

Appendix II

Glossary

5-Methylcytosine (5-mC): The most frequently observed form of epigenetic DNA modification. Produced by the addition of a methyl group to the fifth carbon of cytosine. Cytosine methylation reduces gene transcription and regulates chromatin remodeling.

Algorithm: A well-defined procedure that comprises a set of instructions for solving a recurrent problem.

Alignment: Similarity-based arrangement of sequences. In NGS data analysis, sequence reads are usually aligned against a reference genome to locate their genomic origins.

Allele: One particular variant form of a gene that has a number of alternative sequence variants.

Annotation: The process of providing biologically relevant information to a piece of DNA or RNA sequence. Also refers to the biological information itself that is attached to a sequence.

ASCII: Standing for American Standard Code for Informational Interchange, ASCII provides a standard for encoding characters. Since a computer only deals with numbers, each human-readable character has to be encoded with a unique number in a computer. An ASCII code is the numerical representation of a character in a computer. For example, in the ASCII table, the character "A" is represented by the number 65.

Assembly: A computational process to reconstruct a longer sequence from short sequences.

Barcode: Unique short artificial sequence(s) attached to DNA molecules in a sequencing sample. The use of barcode sequence(s) enable identification of different samples when they are sequenced together in a mixture (i.e., multiplex sequencing). Also see *Multiplex Sequencing* and *Demultiplexing*.

Basecall Quality Score: A score assigned to each basecall in a sequence read to quantify the confidence level of making the call. In NGS it is defined in the same way as the Phred quality score originally developed for Sanger sequencing. Also see *Phred Quality Score*.

Bisulfite Conversion: A chemical process that leads to the differentiation of methylated cytosines from unmethylated cytosines. The treatment by bisulfite converts unmethylated cytosines in DNA to uracil, while methylated cytosines are not affected by this process. Bisulfite conversion coupled with NGS is a major means to study genome-wide DNA methylation. Also see *Whole-Genome Bisulfite Sequencing*.

Burrows–Wheeler Transform (BWT): A method of permuting the characters of one string into another string. In NGS data analysis, BWT enables fast reference genome searching through providing efficient compression and indexing.

cDNA: Complementary DNA. Refers to DNA that is reversely transcribed from and therefore complementary to an mRNA species.

CDS (Coding DNA Sequence): The region of DNA that is translated into protein.

Cell Ranger: A set of single-cell RNA-seq data analysis pipeline tools developed by 10× Genomics, to carry out steps such as sequence mapping, generation of gene-cell matrix, and downstream analyses such as cell clustering.

ChIP-Seq: Chromatin immunoprecipitation coupled with sequencing. A major application of NGS for studying genome binding of DNA-interacting proteins such as transcription factors.

Codon: A tri-nucleotide sequence of DNA or RNA that codes for a specific amino acid or the signal for protein synthesis termination. There are a total of 64 codons, with 61 specifying amino acids and 3 as termination signals.

Contig: A contiguous segment of RNA or DNA sequence resulted from assembly of a set of overlapping sequence reads.

Copy Number Variation (CNV): One type of genomic variation caused by changes in copy number of a DNA segment, usually as a result of deletion or duplication. CNV is a subcategory of structural variation and involves DNA segments that are usually larger than 50 bp. Also see *Structural Variation*.

Coverage: The average number of times that nucleotides in different genomic positions appear in a sequencing dataset. Also known as *Sequencing Depth* or simply *Depth*.

Demultiplexing: The identification and separation of sequencing reads that are generated from different samples, based on the unique barcode sequence(s) they carry, after a multiplex sequencing run. Also see *Barcode* and *Multiplex Sequencing*.

Depth: Same as *Coverage*.

DNA Polymerase: A class of enzyme that catalyzes the synthesis of new DNA strand from free nucleotides, using an existing DNA strand as template. Many molecular techniques, including PCR and sequencing-by-synthesis, are based on the use of DNA polymerases.

DNase: An enzyme that catalyzes the hydrolysis of DNA into oligonucleotides or nucleotides.

Epigenome: Refers to chemical modifications to DNA and histones, which provides additional regulation to genomic activity.

Exon: A stretch of nucleotide sequence that is part of a gene providing coding information for protein synthesis. Exons are transcribed to and usually retained in mRNA.

Exome: The complete set of exons in an organism's genome.

False Discovery Rate (FDR): A measure of statistical significance after correcting for multiple testing. It estimates the proportion of false discoveries in the final list of findings. Among the various approaches for multiple testing correction, FDR estimation offers a balance between statistical stringency and rate of type II errors and therefore is widely used for high-throughput genomics data analysis. Also see *Multiple Testing Correction*.

GC Content: The percentage of guanines plus cytosines in a DNA/RNA sequence or genome.

Gene Expression: The process by which the information encoded in a gene's nucleotide sequence is used to direct the synthesis of a functional gene product. The level of gene expression in a cell or population of cells is represented by the abundance of its product. The composition of the large number of gene products and their expression levels in a cell or population of cells constitute gene expression profile of the host cell(s).

Genome: The complete set of DNA sequence in a cell or an organism. Contains the complement of information needed to form and maintain the cell or organism. Including both protein-coding and non-coding sequences.

Genotype Posterior Probability: The probability of a genotype given an observed dataset, calculated from NGS reads and often with the use of prior genotype information.

Gene Ontology (GO): An initiative to provide consistent description of gene products using standardized vocabulary. Each gene product is described by three structured ontologies that encompass their associated biological processes, cellular components, and molecular functions.

Hidden Markov Model (HMM): Named after the Russian mathematician Andrei Markov (1856–1922), HMM is a commonly used machine learning and data mining approach for signal processing and pattern recognition. A Markov model is a statistical model that deals with observed sequences and state transitions. In bioinformatics, HMM is often used for basecalling, sequence alignment, and gene prediction.

High-Performance Computing (HPC): A computer system that has the capability to perform over one teraflop (10^{12}) floating-point operations per second by the use of parallel processing.

Indel: A generic term for either insertion or deletion of nucleotide(s) in a DNA sequence. Such insertion/deletion events lead to DNA mutation and sequence length change.

Indexing: The process of creating a data structure for fast search. Techniques of indexing for sequence alignment include hashing (storing information on where a particular subsequence can be found in a reference genome or a large collection of reads), suffix array (that consists of lexicographically sorted genomic DNA sequence suffixes), and BWT (permutation of a genome based on suffix array).

Irreproducible Discovery Rate (IDR): A measure of experimental reproducibility. Developed to evaluate the reproducibility between replicates of a ChIP-seq experiment, it calculates the rate of irreproducible discoveries, i.e., peaks that are called in one replicate but not in another.

***K*-mer**: In genome assembly or sequence alignment, *k*-mer refers to all the possible subsequences of length *k* in a sequence read.

Library: Collection of many different DNA (or RNA) fragments that are systematically modified for target DNA screening or high-throughput analysis (including NGS). Specifically, a sequencing library is a pool of DNA (or RNA) fragments with universal adapters attached to their ends. To construct a sequencing library, DNA (or RNA) molecules extracted from a population of cells are usually randomly fragmented, followed by addition of universal adapters to the two ends of the fragments. Sequences in the adapters enable subsequent enrichment and high-throughput sequencing of the fragments.

Long Non-Coding RNA (lncRNA): Non-protein coding RNA species that are over 200 nucleotides in length. In comparison to small RNAs.

Machine Learning: A branch of Computer Science that focuses on developing software algorithms that provide computers the capability to learn and make predictions on new data. Machine learning is built on computational model construction from existing input data, which is then applied to new data for generating predictions or decisions.

Mapping: The process of searching the sequence of a read against the reference genome sequence to locate its origin in the genome. Also see *Alignment*.

Mapping Quality: An estimation of the probability of mis-aligning a read to a reference genome. It is reported as a Phred-scale quality score. Also see *Phred Quality Score*.

Mate-Pair Reads: Reads generated from two ends of a long DNA fragment. To achieve sequencing of the two ends, the long DNA fragment is first circularized and then fragmented. Paired-end sequencing of the fragment that contains the junction of the two ends generates mate-pair reads.

MeDIP: Methylated DNA immunoprecipitation with anti-5-methylcytosine antibody.

Metagenome: The collection of all the genomes contained in a microbial community that consists of many individual organisms.

Metagenomics: Studies of all the genomes existing in a microbial community as a whole without the need to capture or amplify individual genomes. Also referred to as environmental or community genomics.

Microarray: A high-throughput genomics technology based on the use of predesigned detection probes that are printed or synthesized on a solid surface, such as glass or a silicon chip, in a high-density array format.

Minor Allele Frequency (MAF): Frequency of the least abundant allelic variant in a population.

miRNA: MicroRNA. See *Small RNA*.

mRNA: Messenger RNA, which carries protein-coding information in DNA for protein translation. It acts as the intermediate between DNA and protein. An important component of a transcriptome.

Multiple Testing Correction: Adjustment of statistical confidence based on the number of tests performed. Multiple testing without such an adjustment leads to high levels of false positives. For example, at a *p*-value of 0.05, performing 100 comparisons simultaneously will generate 5 positive outcomes simply by chance if a correction is not applied. Commonly applied multiple testing correction approaches including the Bonferroni adjustment (conservative) and False Discovery Rate estimation. Also see *False Discovery Rate*.

Multiplex Sequencing: Simultaneous sequencing of multiple samples together. The use of artificial barcode sequence(s) enables sample identification. Also see *Barcode* and *Demultiplexing*.

Multireads: Reads that map to multiple genomic locations.

N50: The weighted mean contig size of a genome assembly. To calculate N50, all contigs are first ranked based on their lengths, which is then followed by adding the ranked lengths from the top downward. N50 refers to the length of the contig that makes the total added length equal to or greater than 50% of the assembly size. An often-used metric of *de novo* genome assembly quality.

NAS: Network Attached Storage. Specialized computer data storage server providing data access to a variety of clients through network.

Non-Coding RNA: RNA species that carry out functions other than coding for proteins. Examples include small RNAs and lncRNAs. Also see *Small RNA* and *Long Non-Coding RNA*.

Normalization: A mathematical procedure to correct for unwanted effects of non-intended factors and/or technical bias (such as differences in sequencing depth between samples in RNA-seq). This procedure puts focus on the biological difference of interest, and makes samples in different conditions comparable.

Normalized Strand Correlation (NSC): A measure of signal-to-noise ratio in ChIP-seq. It is calculated as the normalized ratio between the maximum strand cross-correlation (at the fragment-length peak) and the background cross-correlation. Also see *Relative Strand Correlation (RSC)*.

Open Reading Frame (ORF): A continuous segment of DNA containing nucleotide triplet codons that starts with the start codon (ATG) and ends with one of the stop codons (TAA, TAG, or TGA).

Operational Taxonomic Unit (OTU): A common microbial diversity unit used in metagenomics that may represent a species or a group of species. OTUs are clustered together based on DNA sequence information alone.

Paired-End Reads: Reads obtained from the two ends of a DNA fragment. Since the length of the DNA fragment, i.e., the distance between the reads, is known, use of paired-end reads provides additional positional information in mapping or assembly of the reads. In comparison to *Single-End Reads.*

Pathway: A succession of molecular events that leads to a cellular response or product. Each of such events is usually carried out by a gene product. Many biological pathways are involved in metabolism, signal transduction, and gene expression regulation.

PCA: Principal Component Analysis. A dimensionality reduction technique to help summarize and visualize large and complex datasets. PCA is widely used in next-gen sequencing applications such as bulk and single-cell RNA-seq.

PCR Bottleneck Coefficient (PBC): An index of sequencing library complexity. It is calculated after the read mapping step as the ratio between the number of genome locations to which only one unique sequence read maps and the total number of genome locations to which one or more unique reads maps. PBC measures the distribution of read counts towards one read per location.

Phred Quality Score (Q Score): An integer value that is used to estimate the probability of making an error, i.e., calling a base incorrectly. It is calculated as Q = −10xlog(10)P(Err). For example, a Q score of 20 (Q20) means a 1/100 chance of making a wrong call. Q30 represents a 1/1000 chance of making a wrong call, which is considered to be a high-confidence score. Q scores are often represented as ASCII characters for brevity.

Picard: A set of tools written in Java for handling NGS data and file formats.

Pileup: A file format created with SAMtools showing how each genomic coordinate is covered by reference sequence-matching or -unmatching bases from all aligned reads.

piRNA: Piwi-interacting RNA. See *Small RNA.*

Polymerase Chain Reaction (PCR): A molecular biology technique that amplifies the amount of a DNA or RNA fragment, with the use of specific oligonucleotide primers that flank the two ends of the target fragment.

Promoter: DNA sequence upstream of the open reading frame of a gene. The promoter region is recognized by RNA polymerase during initiation of transcription. Contains highly conserved sequence motifs.

Proteome: The complete set of proteins in a cell, tissue, or organ at a certain point of time. Proteomics analyzes a proteome via identifying individual component proteins in the repertoire and their abundance.

Quality Score: See *Basecall Quality Score.*

Read: Sequence readout of a DNA (or RNA) fragment.

RNA-Seq: Stands for RNA sequencing. Also referred to as whole transcriptome shotgun sequencing. RNA-seq is a major technology for transcriptome analysis and a major application of NGS.

RNAi: RNA interference, i.e., inhibition of gene expression. RNAi is usually mediated by small RNAs, which lead to degradation of specific mRNA targets.

RNase: An enzyme that catalyzes the degradation of RNA molecules.

Reduced Representation of Bisulphite Sequencing (RRBS): An NGS-based experimental approach that determines DNA methylation pattern in a reduced genome (usually to save costs). The reduced representation of the genome is usually achieved by the use of restriction enzymes.

rRNA: Ribosomal RNA, i.e., RNA species that are essential components of the ribosome. They play key roles in protein synthesis. By quantity, they are the most abundant RNA species in a cell.

Relative Strand Correlation (RSC): A metric of signal-to-noise ratio in ChIP-seq. RSC is the ratio between background-adjusted cross-correlation coefficient at the fragment-length peak and that at the read-length peak. Also see *Normalized Strand Correlation (NSC)*.

RNA Velocity: A single-cell RNA-seq analysis approach for making predictions on future cellular states in terms of the speed and direction of their movement along a trajectory. This approach is based on detection and comparison of unspliced pre-mature transcripts that still contain introns and spliced mature transcripts.

SAN: Storage Area Network. A type of Local Area Network (LAN) designed to handle large data transfers.

Sanger Sequencing: The first widely adopted DNA sequencing technology. Devised by Dr. Fred Sanger, it is based on the principle of sequencing-by-synthesis with the use of dideoxynucleotides that irreversibly terminate new DNA strand synthesis once incorporated. With the advent of NGS technologies, this sequencing method has become the synonym of first-generation sequencing.

Scaffold: Ordered arrangement of *de novo* assembled contigs. The relative positional relationships between contigs are inferred by mate-pair, paired-end, or long reads. In a scaffold, while the order of contigs is known, sequence gaps still exist between contigs.

Sequencing Depth: See *Coverage*.

Sequencing Library: See *Library*.

Seurat: An R toolkit for comprehensive single-cell RNA-seq data analysis.

Single-End Read: Sequence read generated from one end of a DNA fragment. This is in comparison with paired reads generated from both ends of a DNA fragment. Also see *Paired-End Reads*.

Single Nucleotide Polymorphism (SNP): DNA sequence polymorphism due to variation at a single nucleotide position. Different from the term Single Nucleotide Variation (SNV), SNP only refers to SNV that is relatively common in a population with frequency reaching a certain threshold (usually 1%). Also see *Single Nucleotide Variation*.

Single Nucleotide Variation (SNV): DNA sequence variation that involves change at a single nucleotide position, e.g., the sequence change from ATTGCA to ATCGCA.

siRNA: Small interfering RNA. See *Small RNA*.

Small RNA: Also called small non-coding RNA. The major categories of small RNA are miRNA, siRNA, and piRNA. In comparison to mRNA molecules, these RNA molecules are much smaller in size. Small RNA plays important regulatory roles in cells through mediating RNAi. Also see *RNAi*.

Splicing: The process of removing introns from primary RNA transcripts and joining of exons to form mature mRNAs. Splicing can be conducted in more than one way for many genes, and alternative splicing can lead to the production of different mRNA species from the same gene through retaining different combinations of exons (or even introns sometimes).

SRA: Sequence Read Archive (also called Short Read Archive) maintained by the National Center for Biotechnology Information (NCBI). SRA is one of the major archives of NGS data generated worldwide. Other publicly available NGS data archives include the European Nucleotide Archive (ENA) maintained by the European Bioinformatics Institute (EBI).

Strand Cross-Correlation: In ChIP-seq, there is a shift in base position between reads generated from the forward and reverse strands of DNA. Strand cross-correlation is a measure of this shift, and calculated as the Pearson correlation coefficient between the forward and reverse read counts at each base position when the reads on the two strands are shifted toward and away from each other at different base shift. Also see *Normalized Strand Correlation* and *Relative Strand Correlation*.

Structural Variation (SV): Large-scale genomic change that include large indel, inversion, translocation, or copy number variation. Different from SNPs or small indels, SVs involve DNA segments that are usually larger than 50 bp. Also see *Copy Number Variation*.

Trajectory Inference (TI): Inferring transcriptomic change of an individual cell over time based off of a snapshot of gene expression of a population of cells at a certain point of time that represent a continuum of transitional cellular states. TI is built on the premise that cells in the continuum share many common genes and their gene expression displays gradual change.

Transcript: An RNA molecule transcribed from a segment of DNA.

Transcription Start Site (TSS): The nucleotide site in a segment of DNA from which RNA transcription is initiated.

Transcriptome: The complete set of RNA transcripts in a cell, tissue, or organ at a certain point of time.

Transcriptomics: Studies of the composition of a transcriptome. Encompasses identification of the large number of RNA species in a transcriptome and determination of their abundance levels. Major transcriptomics technologies include microarray and RNA-seq.

Translation: The process of protein synthesis from mRNA. Carried out by ribosomes.

tRNA: Transfer RNA. The function of tRNAs is to transfer amino acids to ribosomes for protein synthesis according to the triplet genetic code.

t-SNE: t-distributed Stochastic Neighbor Embedding. A non-linear dimensionality reduction method. The t-SNE process maps the cells to a dimension-reduced space that best preserves cell-cell similarity in the original data.

UMAP: Uniform Manifold Approximation and Projection. Another non-linear dimensionality reduction method. It has a similar procedure to t-SNE, but it differs in how cell-cell similarity is computed to provide better preservation of global cell-cell relationships.

UMI: Unique Molecular Identifier. UMIs are essentially a large number of randomly synthesized, unique nucleotide combinations that are attached to DNA fragments before any PCR amplification steps during sequencing library preparation. The purpose of using UMIs is to identify PCR duplicated molecules for removal.

UTR: Untranslated region of an mRNA molecule. Can be located on either the 5′ end, or the 3′ end, of the mRNA molecule.

Variant Calling: Identification of sequence difference at specific positions of an individual genome (or transcriptome) in comparison with a reference genome. Each variant usually has a corresponding Phred-scale quality score.

Whole-Genome Bisulphite Sequencing (WGBS): An application of NGS that determines DNA methylation pattern across the entire genome using bisulfite conversion. Also see *Bisulfite Conversion*.

Index

E

Q

9780367349899